NEW CELL DIFFERENTIATION RESEARCH TOPICS

NEW CELL DIFFERENTIATION RESEARCH TOPICS

HITOSHI SAITAMA
EDITOR

Nova Biomedical Books
New York

LIBRARY OF CONGRESS CATALOGING-IN-PUBLICATION DATA

New cell differentiation research topics / Hitoshi Saitama (editor).
p. ; cm.
Includes bibliographical references and index.
ISBN-13: 978-1-60021-937-5 (hardcover)
ISBN-10: 1-60021-937-3 (hardcover)
1. Cell differentiation. I. Saitama, Hitoshi.
[DNLM: 1. Cell Differentiation. QU 375 N53135 2008]
QH607.N49 2008
571.8'35--dc22

2007030961

Published by Nova Science Publishers, Inc. ✢ New York

Contents

PREFACE

This new book presents reserch developments from around the globe in the field of cellular differentiation which is a concept from developmental biology describing the process by which cells acquire a "type". The morphology of a cell may change dramatically during differentiation, but the genetic material remains the same, with few exceptions. A cell that is able to differentiate into many cell types is known as pluripotent. These cells are called stem cells in animals and meristematic cells in higher plants. A cell that is able to differentiate into all cell types is known as totipotent. In mammals, only the zygote and early embryonic cells are totipotent, while in plants, many differentiated cells can become totipotent with simple laboratory techniques.

Chapter I - Differentiation is a complex multistep process that affects many biological pathways that regulate the expression of specific genes, and cell proliferation. All adult tissues are made up of lineages of cells consisting of tissue stem cells and their progeny (transit-amplifying cells and terminally differentiated cells). Recent data suggest that cancers arise from rare self-renewing stem cells that are biologically distinct from their more numerous differentiated progeny. Growing evidence suggests that pathways regulating normal stem cell self-renewal and differentiation are also present in cancer cells and cancer stem cells (CSCs). Malignant tumors can be viewed as an abnormal organ in which a small population of tumorigenic CSCs have escaped the normal limits of self-renewal giving rise to abnormally differentiated cancer cells that contribute to tumour progression and growth. This new model has important implications for the study and treatment of cancer. Understanding the molecular circuitry which contributes to the maintenance of stem cells may provide an insight into the molecular mechanisms of cancer and thus new approaches for elimination or differentiation therapy. The unique properties of stem cells are opening the door to the development of new therapeutic approaches notably in oncology and regenerative medicine. Adult mesenchymal stem cells (MSCs) possess particularly attractive properties, as they are easily expanded in vitro and possess the potential to differentiate into multiple cell lineages. Recent investigations show both the fundamental properties of MSCs and applicable therapies derived from their use, particularly in cardiology but also in cancer and immunomodulation. Moreover, circulating endothelial progenitor cells (EPC) isolated from bone marrow or peripheral blood of cancer patients are capable of contributing to the neovascularisation of tumors participating in angiogenesis process. Classical agents used in differentiation therapy of cancer such as retinoic and butyric acid mixed as esters with hyaluronan primed the expression of cardiogenic genes and elicited a remarkable increase in cardiomyocyte yield in mouse embryonic stem cells. The authors have recently reported the synthesis and anticancer properties of several 5-fluorouracil derivatives. These compounds increased microtubule stability and HLA class I expression in rhabdomyosarcoma cells. The authors results suggest that there may be significant potential advantages in the use of these new differentiating agents for the treatment of cancer. Treating cancer cells with these compounds may provide a non-toxic therapeutic modality when the options for chemotherapy have been exhausted. This review compares cancer cells to embryonic stem cells and to adult tissue stem cells. Moreover, the authors review the current status of the CSC research and propose the targets for CSC cell-surface molecules, signal transduction pathways, the stem cell niche, the stem cell differentiation and drug resistance caused by designed and synthesized molecules.

Chapter II - The therapy of leukaemia based on the use of compounds able to reinitiate the differentiation program has been extensively studied. The use of calcitriol ($1,25D_3$) or trans-retinoic acid (ATRA) in myeloid or monocytic leukemic cell differentiation has been also clinically taken into consideration. Both are natural compounds accepted at relatively high doses by the organism, however, several other secondary effects take place because their other respective main functions different than regulating the differentiation program in cancerous cells. To avoid the secondary effects due to overloading of both, ATRA or $1,25D_3$, therapies based on the co-administration of other natural compounds such as vitamin C, vitamin E, polyphenols, carotenoids or other antioxidants acting as anti-cancer active compounds have been taking into consideration. In several cases, these natural compounds do not affect by themselves the differentiation program in leukemic cells but they are able to enhance the effect of the differentiation-active compounds such as ATRA or $1,25D_3$. The present chapter reviews the effect of the combination of these natural antioxidant compounds on the treatment of leukemic cells and will try to define the probable common mechanism of action for this plethora of substances.

Chapter III - Bone marrow-derived stem cells (BMDCs) gather considerable attention for their ability to trans-differentiate into various dedicated cell types, e.g. liver, heart and skeletal muscle, and brain. Thereby, BMDCs trans-differentiate in accordance to the specific tissue blueprint, which could be the local milieu (soluble factors, cell adhesion molecules and/or nanotubes), a single cell (cell fusion) or a combination of both. The local tissue blueprint has to guide the changes from a stem cell, by definition endowed with the ability for self-renewal, to a differentiated, dedicated cell exhibiting specific tissue functions. A similar process has to occur during tissue stem cell differentiation. Again, the local tissue milieu is responsible for providing a correct blueprint guaranteeing an accurate differentiation process.

The question arises what will be the result of BMDC trans- and tissue stem cell differentiation if stem cells will receive a faulty set of instructions, which might be provided by chronic inflamed tissue or tumor tissue. Both environments have lost many of the characteristics of the tissue of origin. Tumor tissue *per se* is heterogeneous and resembles chronic inflamed tissue. Chronic inflamed tissue itself is characterized by a chaotic interplay of tissue cells, immune competent cells, cytokines, chemokines, growth factors and reactive oxygen species.

An incomplete set of instructions concomitant with the combination of proliferative stimuli and DNA damaging agents represents an explosive mixture with a fatal outcome for (trans-) differentiating stem cells, especially for those stem cells adopting tissue function by cell fusion. Aneuploidy is frequently found in tumor cells and has been postulated to be the driving force in the pathogenesis of cancer. Fusion of stem cells and tumor cells will result in an increased aneuploidy in hybrid cells. Conjointly, the proliferative and DNA damaging environment will lead to the evolution of hybrid cells exhibiting new properties as e.g. a higher proliferation rate, the ability to metastasize and/ or to be drug resistant. In particular the latter property has been associated with so-called cancer stem cells, which apparently play a crucial role in metastasis formation and tumor relapse. Thus, it might be hypothesized that cancer stem cells might be the product of cell fusion between stem cells and tumor cells. The understanding of cell fusion and its consequences for the pathogenesis of cancer, possibly driving the evolution of cancer stem cells, is therefore crucial for the development of novel anti-cancer strategies.

Chapter IV - Studies exploring the potential neural transdifferentiation of bone marrow-derived stem cells (BMSCs) are of increasing value in stem cell research. New research demonstrates that in addition to providing a good model for studying adult stem cell plasticity, BMSCs have considerable potential for promoting neural repair. Use of these cells as a therapeutic model is facilitated by their easy accessibility for autologous transplantation. Methods for inducing neural differentiation include adding simple chemicals, cytokines, growth factors, and brain tissue extracts to cultured BMSCs, as well as co-culturing BMSCs with organotypic brain slices or neural cells. Induced marrow-derived stem cells resemble mature neurons in morphology, expression of neural marker genes, and electrophysiological properties, but whether these changes represent true neural differentiation merits further investigation. Notably, naïve BMSCs express several neural marker genes and exhibit small voltage-dependent membrane currents even before neural induction. This suggests a couple of possibilities. First, these properties may be related to the multidifferentiation potential of BMSCs and facilitate neural differentiation. Alternatively, the shared neural marker genes may not be as specific to neural cells as previously thought.

Chapter V - This review focuses on the development and technique of three dimensional (3D) cultures to study epithelial cell differentiation and tissue architecture. 3D cultures employ human basal epithelial cells, stem cells and heterotypic cultures to faithfully reproduce tissue complexity. These models have arisen from a requirement to model human cells in a tissue-like structure. They are more accessible than animal based models and have a more relevant complexity than 2D cell culture. A 3D environment is important to properly integrate chemical and biomechanical signals in the establishment of epithelial differentiation and architecture. Epithelial cell function is engineered by combining the influences of matrix, neighbouring cells and soluble growth factors.

In the last 10 years there has been a great expansion in the use of 3D model systems for cell research. These can be divided into two broad categories: 1) 3D spheres generated by cell aggregation using techniques such as rotary wall vessel culture, agar, pellet culture and liquid media and 2) 3D models which produce tissue-like architecture through culturing within gel or synthetic scaffolds. More recently 3D models are being employed to define the stem cell niche, allowing the control of self renewal and commitment to differentiation to be explored.

Chapter VI - SCF and c-*kit* signaling is essential for both human and murine mast cell development, while IL-3 is required for murine mast cell hyperplasia that occurs in response to various stimuli. Relatively stable numbers of mast cells within connective and mucosal tissues are maintained by the constitutive expression of SCF in its membrane-bound and soluble forms by stromal cells (such as fibroblasts, epithelial cells, endothelial cells, bronchial smooth muscle cells) and by the expression of c-*kit* on mast cells at all stages of differentiation. Besides SCF and IL-3, the Th2 cytokines IL-4, IL-9, IL-10 and IL-13, also play critical roles in mast cell development by synergistically promoting the actions of SCF and IL-3 on mast cell proliferation and differentiation. However, the Th2 cytokines alone are unable to support neither the proliferation nor survival of mast cells. In contrast, the Th1 cytokine IFN-γ suppresses mast cell development from various murine and human sources. Interestingly, the above generally-accepted roles of the Th1 and Th2 cytokines on mast cell development are currently being challenged by controversial findings showing that mast cell development is up-regulated by IFN-γ and down-regulated by IL-4, IL-10 and IL-13. This review summarizes the recent progress in our understanding of how the Th1 and Th2 cytokines regulate mast cell development either in a positive or negative manner, with a particular focus on IL-4-induced mast cell apoptosis.

Chapter VII - Cellular differentiation is the commitment of cells in response to internal and external signals, and is often accompanied by morphological changes. Hematopoiesis is a well-understood differentiation system that leads to the formation of the various blood cell types, which are all derived from the hematopoietic stem cell (HSC). Research about this process has exploded in the last few decades with the discovery of cell surface antigen-based isolation methods and cytokine response characterizations. However, the kinetics of these responses and the specific fates of individual cells remain unclear. This is because the data are based on the average of cell populations and/or conditions that observe cell behaviour outside of their normal environment. Therefore, new technologies must be developed and utilized, following single cells, to dissect the exact nature of this differentiation process.

It is becoming increasingly clear that the bone marrow has supporting cells that influence and control blood differentiation. Research has only begun to understand what cells types and factors are involved. The data so far have recognized that some "niche cells" allow HSC to self-renew and differentiate in a controlled manner that, in turn, leads to the mature adult hematopoietic system.

Although a few components of this complex niche are now recognized, the exact nature of the HSC-niche relationship is still unclear, as technical obstacles have yet to be overcome. Much of the data are based on fixed tissues that show the localization and behaviour of a HSC at a single point in time. This neglects the temporal component to this dynamic relationship between the HSC and its niche. Furthermore, researchers have had to measure the presence of HSC by indirect functional readouts for its progeny, thus ignoring the individual characteristics and behaviours of these cells. So far, HSC have not been directly observed as live single cells, with the individual characterization of their behaviour and differentiation within their normal niche.

To understand the temporal and spatial relationships the HSC has with the various supporting cell types, HSC must be visualized within their niche over time, under constant video surveillance. Only then will we be able to understand the HSC-microenvironment relationship. This review will discuss the need for new technologies that will allow for a greater understanding of the HSC-niche relationship that influences normal blood differentiation.

Chapter VIII - Complex human tissues harbour stem cells and/or precursor cells, which are responsible for tissues' development or tissues' repair. These cells provide new insights into dental development and are important for characterizations of cellular differentiation processes under in vitro conditions. In recent times undifferentiated cells were isolated from dental tissues like the periodontal ligament, the dental papilla or the dental follicle. These cells are realistic sources for the construction for example of an engineered periodontium, which submits the connection between implanted tooth and jaw. Therefore, dental precursor cells are interesting for novel treatments of periodontitis or dental implantation. However, the authors need more information about their differentiation mechanisms for the evaluation of dental treatments. Dental follicle precursor cells (DFPCs) are derived from an embryonic dental tissue of extracted wisdom teeth. These cells are easy accessible and ideal for testing cell differentiation mechanisms during dental development. However, there are many publications about dental precursor cells or stem cells derived from dental ectomesenchymal tissues. This review gives also a short overview about known ectomesenchymal cells of dental origin and classifies these cells.

Chapter IX - In Podarcis, undifferentiated, pre-differentiated and differentiated follicle cells can be recognised by shape, size, position and function. Undifferentiated (stromal) and pre-differentiated (small) cells are stem cells destined to proliferate and differentiate while differentiated cells (pyriforms) are nurse cells programmed to sustain oocyte growth and then degenerate at the end of previtellogenesis.

The present work clarifies whether competence to undergo apoptosis in pyriforms is acquired at the time they originate from the small cells or later during aging. For this purpose the expression of Fas, Fas-L, caspase 3 and DNase have been investigated by immunocytochemistry in small and pyriform cells. Results indicate that all these cell death factors are present exclusively in pyriforms. Their synthesis starts as soon as cells differentiate and that while aging, Fas and Fas-L are down and up regulated respectively. The authors conclude that pyriforms are committed for apoptosis since differentiation and that death occurs via mechanisms alternative to receptor or mitochondrial mediated pathways.

Chapter X - Schwann cells (SCs) contribute to nerve regeneration by providing regeneration strands, many kinds of nerve growth factors and tight junctions to stabilize cell contact with injured nerve cells, as well as gap junctions to facilitate traffic of substances between the cells. It is recognized that Schwann cells play a crucial role in the process of peripheral nerve development and regeneration, even in the construction of tissue-engineered nerve grafts. However, the difficulty is to obtain sufficient numbers of autologous SCs in a short time. Use of allogeneic cells are involved in immunological rejection. Bone marrow stromal cells (MSCs) are nonhematopoietic multipotent stem cells that differentiate into connective tissue lineages including osteoblasts, chondrocytes, and adipocytes under special experimental conditions. In particular, MSCs showed potential to be differentiated to SCs in vitro and in vivo, which enhance peripheral nerve regeneration. In this charpter, the authors will summarize the various strategies that can be used to differentiate MSCs into Schwann cells in vitro and in vivo. Many different assays and techniques may be used to define and characterize MSCs which were differentiated along SCs lineage. Functional assays that may be used to further define the differentiated MSCs and determine if they can promote peripheral nerve regeneration are described.

In: New Cell Differentiation Research Topics
Editor: Hitoshi Saitama, pp. 1-6

ISBN: 978-1-60021-937-5

Expert Commentary A

OCT-3/4 Expressing Cells as Cancer Stem Cells in Human Immature Teratoma: Cancer Differentiation Potential

Akira Hara[1,*], Naoki Oka[1], Hitomi Aoki[2], Ayako Taguchi[1], Yasuhiro Yamada[1], Masayuki Niwa[3] and Hideki Mori[1]

[1]Department of Tumor Pathology, Gifu University School of Medicine, Gifu 501-1194, Japan;

[2]Department of Tissue and Organ Development, Gifu University Graduate School of Medicine, Gifu 501-1194, Japan;

[3]Medical Science Division, United Graduate School of Drug Discovery and Medical Information Sciences, Gifu University, Gifu 501-1194, Japan.

Abstract

We would like to comment on the recent cancer stem (CS) cell theory. We have focused on OCT-3/4 expressing cells as CS cells providing neoplastic cells into tumor mass of human immature teratoma. There is increasing evidence that many types of malignant tumors are initiated and maintained by a subpopulation of tumor cells that share similar biologic properties to normal adult stem cells. These specific tumor cells, which are recently called as CS cells, exhibit the ability to self-renew as well as give rise to various type of neoplastic cells. The concept of CS cell has developed in leukemia. And it has expanded to include solid tumors such as brain or breast tumors. OCT-3/4 is a key regulator of self-renewal in embryonic stem (ES) cells. Although OCT-3/4 has been introduced as a molecular marker for germ cell tumors, little is known about its expression in somatic cancers. The existence of a small subpopulation of self-renewing CS cells that are capable of expressing the pluripotent marker, OCT-3/4 are suggested in

* Correspondence concerning this article should be addressed to: Dr. Akira Hara, Department of Tumor Pathology, Gifu University Graduate School of Medicine, 1-1 Yanagido, Gifu 501-1194, Japan. E-mail: ahara@gifu-u.ac.jp; TEL +81 58 230 6225; FAX +81 58 230 6226.

some human cancer, but still controversial. Experimentally, ES cells expressing OCT-3/4 are used for the cell replacement therapy and tissue engineering. However, in vivo transplanted ES cells frequently produce immature teratoma or teratocarcinoma. This means that ES cells may have a potential to be a CS cells. Thus analysis of the tumor cell expressing OCT-3/4 in human immature teratoma is a good clue for understanding CS cell itself. An immunohistochemical study for human immature teratoma by using OCT-3/4 antibody and other cell differentiation-specific markers was helpful for confirmation of the relationship between CS cells and neoplastic differentiation.

Recent reports suggest that many types of cancer contain their own stem cells, called as cancer stem (CS) cells in a same way as normal stem cell or progenitor cell development although the concept of the CS cells is still controversial. CS cells are characterized by their self-renewing capacity and differentiation ability. The concept of CS cells has developed in leukemia. Human acute myeloid leukemia is organized as a hierarchy that originates from a primitive hematopoietic cell [1]. And it has expanded to include solid tumors such as brain [2] or breast [3] tumors. Then CS cells have resulted in the identification in several human malignancies [4-8]. These studies conclusively showed that the cancer or leukaemic clone is organized as a hierarchy in which only a rare subset of cancer cells possess the ability to initiate new tumour growth and recapitulate the original tumour heterogeneity [6]. The CS cell model has great implications for understanding of oncogenesis and treatment for cancer, however, their characteristics are still unknown.

OCT-3/4 is a key regulator of self-renewal in embryonic stem (ES) cells. Although OCT-3/4 has been introduced as a molecular marker for germ cell tumors, little is known about its expression in somatic cancers. The existence of a small subpopulation of self-renewing CS cells that are capable of expressing the pluripotent marker, OCT-3/4 are suggested in some human cancer [9,10]. The OCT-3/4 positive cells in cancers might represent the CS cells. These rare fractions of CS cells are considered responsible for maintaining the tumor mass leading to the possibility of a heterogeneous and polyclonal cell population. A hypothesis will be proposed that CS cells expressing OCT-3/4, as they have the capacity to self-renew, are possible target cells for cancer therapies [11].

Human germ cell tumors (GCTs) are heterogeneous, and comprise neoplasms found at different anatomical locations. They are very interesting not only for aspects of their tumor biology, but also from the point of view of developmental biology. GCTs show significant similarities to early germ cell development, most likely related to their cell of origin [12]. GCTs include immature teratoma, yolk sac tumor, embryonal carcinoma, choriocarcinoma and germinoma. Among them, immature teratoma particularly comprises heterogeneous cell population commonly associated with maturely differentiated tissue components.

ES cells, which are derived from the inner cell mass of blastocyst, express OCT-3/4 in the nuclei and represent a valuable source for cell replacement therapy and tissue engineering since their characteristic features include an unlimited self-renewing capacity and a multilineage differentiation potential [13,14]. Although ES cells may be attractive candidates for the cell-replacement therapy of various degenerative diseases in the future, the major limitation of its application to this therapy is teratoma formation. For example, ES cell-derived insulin-producing cells transplanted into diabetic mice produced immature teratomas after transient recovery of hyperglycemia [15]. ES cells may be useful for neural tissue regeneration in the adult mammalian retina [16]. However, in vivo transplanted ES cells, which is seeded as undifferentiated condition, frequently produce immature teratoma or teratocarcinoma [17]. This means that ES cells may have a potential to be a self-renewing CS cells as well as a source for cell-replacement therapy.

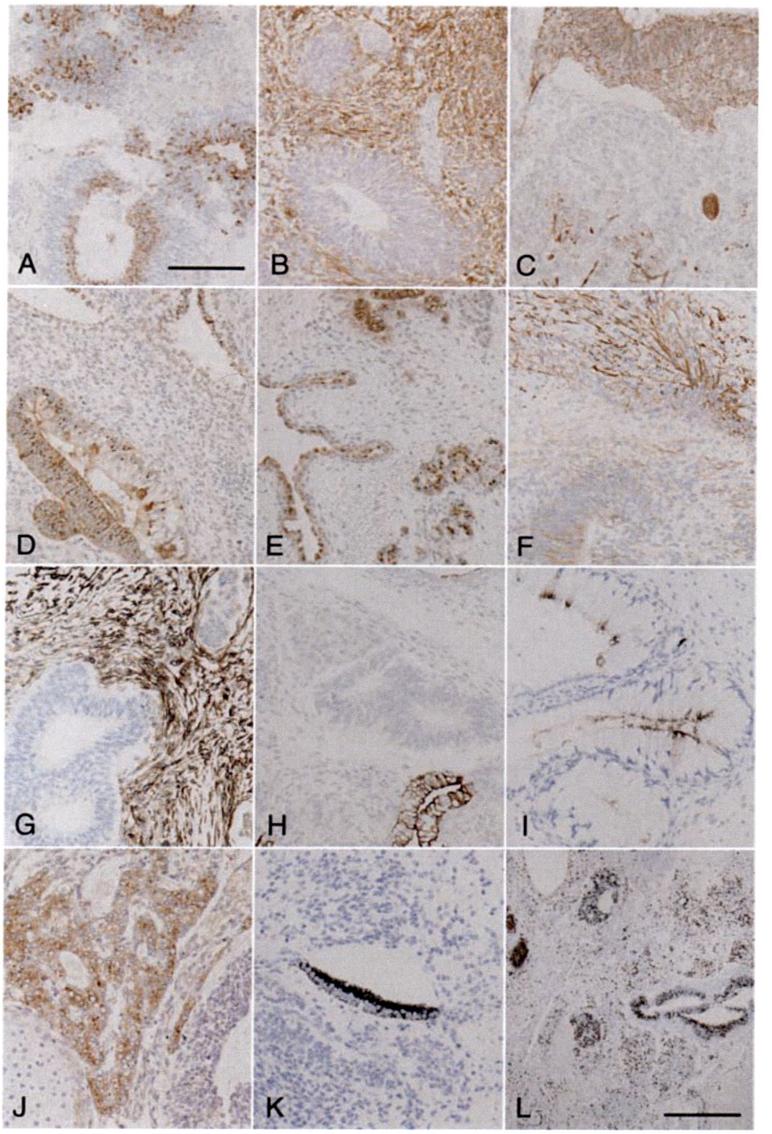

Figure 1. A-J: Representative microphotograph of immunohistochemistry for detection of various differentiatin in human immature teratoma. K: Hematoxylin staining showing intrinsic melanin pigment. L: Ki-67 immunohistochemistry demonstrating the proliferating potential of teratoma cells. A, synaptophysin; B, β-tubulin; C, neural cell adhesion molecule (NCAM); D, protein kinase A; E, syntaxin; F, Glial Fibrillary Acidic Protein (GFAP); G, Neutral endopeptidase (CD10); H, Cytokeratin 7; I, Carcinoembryonic antigen (CEA); J, α-Fetoprotein (AFP); K, melanin pigment; L, Ki-67. A-K are same magnification and L is another. Scale bar in A, 100μm; scale bar in L, 500μm.

In the present study, the existence of a small subpopulation of OCT-3/4 expressing cells in surgically resected human immature teratoma and experimentally induced mouse teratoma fron ES cells are shown by immunohistochemical analysis. Furthermore, other cell differentiation-specific markers for synaptophysin, β-tubulin, neural cell adhesion molecule (NCAM), protein kinase A, syntaxin, glial fibrillary acidic protein (GFAP), neutral endopeptidase (CD10), cytokeratin 7, carcinoembryonic antigen (CEA), α-fetoprotein (AFP) and melanin pigment were demonstrated in human immature teratoma to confirm the relationship between CS cells and neoplastic differentiation. The proliferative activity by Ki-67 labeling index within the compartment of immature teratoma was also assessed and compared. The results of immunohistochemical study for surgically resected human immature teratoma of these specific markers are summarized in Figures 1A-L.

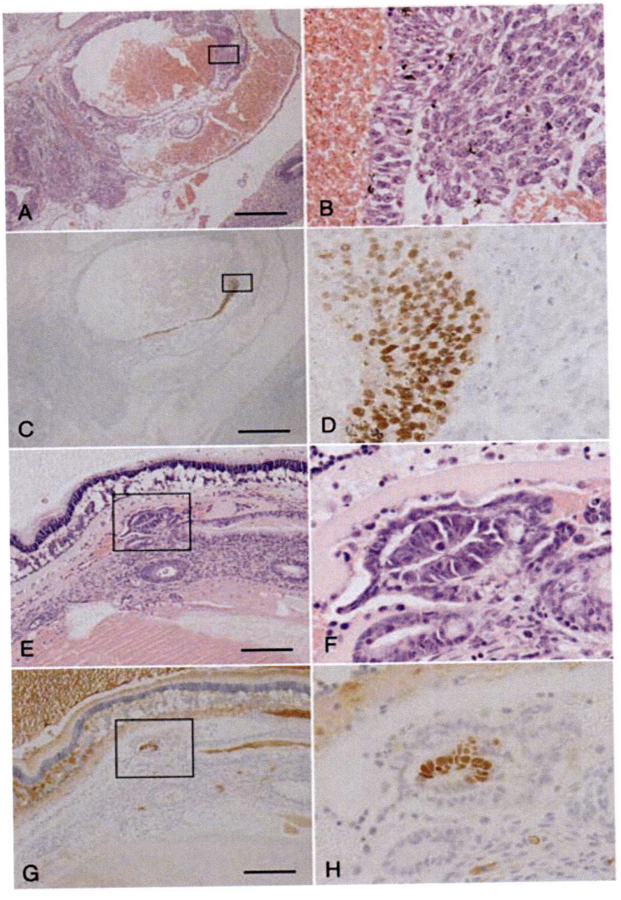

Figure 2. Representative microphotograph of human immature teratoma (A, B, C, D) and transplanted mouse ES cells (E, F, G, H) in mouse retina. Conventional hematoxylin and eosin (HE) staining (A, B, E, F), and immunohistochemistry for OCT-3/4 (C, D, G, H) are presented. Rectangles in panels A, C, E and G represent the area of high power field in B, D, F and H, respectively. OCT-3/4 immunostaining are positive in nuclei of undifferentiated cells in both human (C, D) and mouse teratoma (G, H). OCT-3/4 positive cells are clearly visible, however, the cells are not differentiated by the morphology from other OCT-3/4 negative cells in HE staining in A, B, E and F. Scale bars in A and C, 300μm and in E and G, 120μm.

OCT-3/4 immunostaining are positive in nuclei of undifferentiated cells in both human immature teratoma (Figures 2C, D) and mouse immature teratoma arising from ES cells (Figures 2G, H). OCT-3/4 positive cells are clearly visible, however, the cells are not distinguished from other OCT-3/4 negative undifferentiated cells on the morphological analysis by HE staining (Figures 1A, B and E, F, respectively). The rate of OCT-3/4 positive cells to total teratoma cells is very low and the OCT-3/4 positive cells are not shown as massively grown compartment. They exist like small islands among OCT-3/4 negative teratoma. In these circumstances, OCT-3/4 positive cells in the immature teratomas seem to exist as resorces producing a teratoma mass (Figure 3). Thus analysis of the tumor cells expressing OCT-3/4 in human immature teratoma is a good clue for understanding CS cells itself. Furthermore, OCT-3/4 positive cells in human immature teratoma and possibly other malignant tumors are a promising target of cancer treatment.

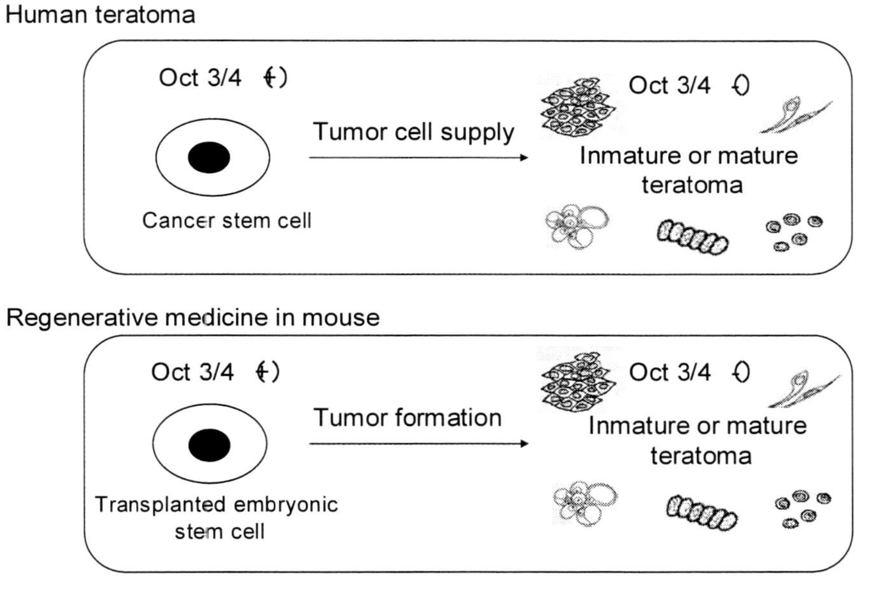

Figure 3. A schematic illustration of the concept of OCT-3/4 expressing cells as cancer stem cells in human immature teratoma in contrast with OCT-3/4 expressing ES cells used in regenerative medicine in mouse model.

REFERENCES

[1] Bonnet, D. and Dick, J.E. (1997) Human acute myeloid leukemia is organized as a hierarchy that originates from a primitive hematopoietic cell. *Nat Med*, *3*, 730-7.

[2] Singh, S.K., Hawkins, C., Clarke, I.D., Squire, J.A., Bayani, J., Hide, T., Henkelman, R.M., Cusimano, M.D. and Dirks, P.B. (2004) Identification of human brain tumour initiating cells. *Nature*, *432*, 396-401.

[3] Al-Hajj, M., Wicha, M.S., Benito-Hernandez, A., Morrison, S.J. and Clarke, M.F. (2003) Prospective identification of tumorigenic breast cancer cells. *Proc Natl Acad Sci U S A*, *100*, 3983-8.

[4] Tan, B.T., Park, C.Y., Ailles, L.E. and Weissman, I.L. (2006) The cancer stem cell hypothesis: a work in progress. *Lab Invest*, *86*, 1203-7.

[5] Fomchenko, E.I. and Holland, E.C. (2005) Stem cells and brain cancer. *Exp Cell Res*, *306*, 323-9.

[6] Collins, A.T. and Maitland, N.J. (2006) Prostate cancer stem cells. *Eur J Cancer*, *42*, 1213-8.

[7] Zhang, P., Zuo, H., Ozaki, T., Nakagomi, N. and Kakudo, K. (2006) Cancer stem cell hypothesis in thyroid cancer. *Pathol Int*, *56*, 485-9.

[8] Haraguchi, N., Inoue, H., Tanaka, F., Mimori, K., Utsunomiya, T., Sasaki, A. and Mori, M. (2006) Cancer stem cells in human gastrointestinal cancers. *Hum Cell*, *19*, 24-9.

[9] Atlasi, Y., Mowla, S.J., Ziaee, S.A. and Bahrami, A.R. (2007) OCT-4, an embryonic stem cell marker, is highly expressed in bladder cancer. *Int J Cancer*, *120*, 1598-602.

[10] Gibbs, C.P., Kukekov, V.G., Reith, J.D., Tchigrinova, O., Suslov, O.N., Scott, E.W., Ghivizzani, S.C., Ignatova, T.N. and Steindler, D.A. (2005) Stem-like cells in bone sarcomas: implications for tumorigenesis. *Neoplasia*, *7*, 967-76.

[11] Trosko, J.E. (2006) From adult stem cells to cancer stem cells: Oct-4 Gene, cell-cell communication, and hormones during tumor promotion. *Ann N Y Acad Sci*, *1089*, 36-58.

[12] Honecker, F., Oosterhuis, J.W., Mayer, F., Hartmann, J.T., Bokemeyer, C. and Looijenga, L.H. (2004) New insights into the pathology and molecular biology of human germ cell tumors. *World J Urol*, *22*, 15-24.

[13] Resnick, J.L., Bixler, L.S., Cheng, L. and Donovan, P.J. (1992) Long-term proliferation of mouse primordial germ cells in culture. *Nature*, *359*, 550-1.

[14] Thomson, J.A., Itskovitz-Eldor, J., Shapiro, S.S., Waknitz, M.A., Swiergiel, J.J., Marshall, V.S. and Jones, J.M. (1998) Embryonic stem cell lines derived from human blastocysts. *Science*, *282*, 1145-7.

[15] Fujikawa, T., Oh, S.H., Pi, L., Hatch, H.M., Shupe, T. and Petersen, B.E. (2005) Teratoma formation leads to failure of treatment for type I diabetes using embryonic stem cell-derived insulin-producing cells. *Am J Pathol*, *166*, 1781-91.

[16] Hara, A., Niwa, M., Kunisada, T., Yoshimura, N., Katayama, M., Kozawa, O. and Mori, H. (2004) Embryonic stem cells are capable of generating a neuronal network in the adult mouse retina. *Brain Res*, *999*, 216-21.

[17] Hara, A., Niwa, M., Kumada, M., Aoki, H., Kunisada, T., Oyama, T., Yamamoto, T., Kozawa, O. and Mori, H. (2006) Intraocular injection of folate antagonist methotrexate induces neuronal differentiation of embryonic stem cells transplanted in the adult mouse retina. *Brain Res*, *1085*, 33-42.

In: New Cell Differentiation Research Topics
Editor: Hitoshi Saitama, pp. 7-77
ISBN: 978-1-60021-937-5

Chapter I

THERAPEUTIC POTENTIAL OF DIFFERENTIATION IN CANCER AND NORMAL STEM CELL

Juan Antonio Marchal[1,], Houria Boulaiz[1], Macarena Peran[1,2], José Carlos Prados[1], Joaquín Campos[3], F. Jesús González[1,4], Fernando Rodríguez-Serrano[1,2], Consolación Melguizo[1], Celia Vélez[1,2], Esmeralda Carrillo[1], Fidel Hita[1,2], Raúl Ortiz[1], Antonio Martínez-Amat[1,2], Octavio Caba[1,2], Carlo Ventura[5] and Antonia Aránega[1]*

[1]Biopathology and Medicine Regenerative Institute (IBIMER), Department of Human Anatomy and Embryology, Faculty of Medicine, University of Granada, Granada E-18071, Spain;
[2]Department of Health Sciences, University of Jaén, Jaén E-23071, Spain;
[3]Department of Pharmaceutical and Organic Chemistry, Faculty of Pharmacy, University of Granada, Granada E-18071, Spain;
[4]Laboratory of Research-University hospital Virgen de la Victoria, Malaga E-29010, Spain;
[5]Laboratory of Molecular Biology and Stem Cell Engineering, National Institute of Biostructures and Biosystems, University of Bologna, Bologna, Italy.

* Correspondence concerning this article should be addressed to: Prof. Juan Antonio Marchal Corrales, MD, PhD, Biopathology and Medicine Regenerative Institute (IBIMER), Department of Human Anatomy and Embryology, Faculty of Medicine, University of Granada, Granada E-18071, Spain. E-mail: jmarchal@ugr.es.

ABSTRACT

Differentiation is a complex multistep process that affects many biological pathways that regulate the expression of specific genes, and cell proliferation. All adult tissues are made up of lineages of cells consisting of tissue stem cells and their progeny (transit-amplifying cells and terminally differentiated cells). Recent data suggest that cancers arise from rare self-renewing stem cells that are biologically distinct from their more numerous differentiated progeny. Growing evidence suggests that pathways regulating normal stem cell self-renewal and differentiation are also present in cancer cells and cancer stem cells (CSCs). Malignant tumors can be viewed as an abnormal organ in which a small population of tumorigenic CSCs have escaped the normal limits of self-renewal giving rise to abnormally differentiated cancer cells that contribute to tumour progression and growth. This new model has important implications for the study and treatment of cancer. Understanding the molecular circuitry which contributes to the maintenance of stem cells may provide an insight into the molecular mechanisms of cancer and thus new approaches for elimination or differentiation therapy. The unique properties of stem cells are opening the door to the development of new therapeutic approaches notably in oncology and regenerative medicine. Adult mesenchymal stem cells (MSCs) possess particularly attractive properties, as they are easily expanded in vitro and possess the potential to differentiate into multiple cell lineages. Recent investigations show both the fundamental properties of MSCs and applicable therapies derived from their use, particularly in cardiology but also in cancer and immunomodulation. Moreover, circulating endothelial progenitor cells (EPC) isolated from bone marrow or peripheral blood of cancer patients are capable of contributing to the neovascularisation of tumors participating in angiogenesis process. Classical agents used in differentiation therapy of cancer such as retinoic and butyric acid mixed as esters with hyaluronan primed the expression of cardiogenic genes and elicited a remarkable increase in cardiomyocyte yield in mouse embryonic stem cells. We have recently reported the synthesis and anticancer properties of several 5-fluorouracil derivatives. These compounds increased microtubule stability and HLA class I expression in rhabdomyosarcoma cells. Our results suggest that there may be significant potential advantages in the use of these new differentiating agents for the treatment of cancer. Treating cancer cells with these compounds may provide a non-toxic therapeutic modality when the options for chemotherapy have been exhausted. This review compares cancer cells to embryonic stem cells and to adult tissue stem cells. Moreover, we review the current status of the CSC research and propose the targets for CSC cell-surface molecules, signal transduction pathways, the stem cell niche, the stem cell differentiation and drug resistance caused by designed and synthesized molecules.

1. INTRODUCTION

Differentiation is a multistep process of cell specialization that begins with the installation of a genetic program, named determination, specific for a cell lineage. Development of the differentiation program includes the expression of specific genes that regulate the biological functions associated with the cellular type and that distinguish the specialized cells. In this process complex regulatory pathways act involving both positive and negative regulation control gene expression, which includes a synergistic cooperation between different signal transduction pathways (MacCorkle and Tan, 2005). Terminal differentiation is the end stage of this process where the cells irreversibly lose their proliferative capacity.

Between determination and terminal differentiation have been described several steps from many reversible states to other irreversible ones, occurring later in the differentiation process (Scott, 1997). These steps include the growth and proliferation of determined cells, the entrance into a quiescent pre-differentiation state, the expression of lineage-specific transacting factors that activate genes required for lineage-specific biological functions or the repression of growth factor responsiveness. Finally, the activation of molecular mechanisms for terminal differentiation occurs. Thus, in the differentiation process, cells typically first enter a pre-differentiation state of quiescence in the G_1 phase of the cell cycle. Next, activation of the expression of one or more lineage-specific transacting factors or a set of factors occurs, inducing lineage specific genes that are required for the function of that cell lineage. Moreover, the repression of growth factor response also occurs. Under some conditions, differentiated cells can be included to return to the cell cycle, whereas in other instances, such cells become terminally differentiated and irreversibly withdrawn from the cell cycle. Data of these steps have been defined in cultures of stem cells that differentiate into adipocytes (Otto and Lane, 2005), on myoblast differentiation (Li et al., 2005), in melanocyte development (Pla et al. 2004).

In embryonic development, differentiation begins with early determination processes on the cleavage divisions in the egg reaching intermediary levels of differentiation. Finally, after amplification divisions, each cell lineage matures to specialized cell types and tissues. In adults, these stages occur in tissues that need cell renewal. Thus, organs such as skin, liver, bone marrow and the gastrointestinal tract have populations of stem cells that can differentiate and repopulate that tissue. Normal tissue renewal is accomplished by tissue stem cells that divide to give rise to one daughter cell that remains a stem cell and another daughter cell that begins the process of determination (asynchronous division). In most organs, the normal replacement of terminally differentiated cells is accomplished by proliferation of progenitor cells or transit-amplifying cells (Sell, 2004).

So, stem cells represent natural units of embryonic development and tissue regeneration. The role of stem cells has long been known in reproductive organs and various tissues including the haematopoietic system and skin. During the last decade, stem cells have also been identified in other organs, both during development and in post-natal life. The homeostatic system replaces senescent differentiated cells and tissue loss following injury. This hierarchical system involves several stages of cells that have decreasing reproductive capacity and simultaneous increasing commitment to differentiation. The most primordial cell, the stem cell, has the capacity to reproduce for the life of the organ (Miller et al., 2005). Stem cell is a wide concept that share several characteristics as are long-term replicative potential and self-renewal, maintenance of undifferentiated properties and multi-lineage differentiation ability (Martinez-Climent, 2006). In general, after stem cell division, one daughters cell remains as stem cell for self-renewal and the other becomes a progenitor cell that proliferate and maturate.

Cancers express a spectrum of aberrantly differentiated cells ranging from those that appear well-differentiated to those that appear undifferentiated, and these phenotypes are commonly evident in the same tumor. This suggests that the transformation process can induce defects throughout the multistep process of differentiation. Moreover, growing evidence indicates that pathways regulating normal stem cell self-renewal and differentiation are also present in cancer cells and cancer stem cells (CSC). Malignant tumors can be viewed as an abnormal organ in which a small population of tumorigenic cancer stem cells have escaped the normal limits of self-renewal giving rise to abnormally differentiated cancer cells that contribute to tumor progression and growth (Massard et al., 2006).

The palliative and curative advances achieved in the fight against cancer with the use of different chemotherapeutical agents have been very important; nevertheless, as the mechanism of action of these drugs depends on the cytodestruction of the neoplastic cells, their beneficial effects are normally accompanied by a notable morbidity, cytotoxicity and multidrug resistance (Marchal et al., 2006). Moreover, cancer stem cells like normal stem cells are more resistant to conventional chemotherapy than other more differentiated cancer cells. CSC can acquire resistance to chemotherapy via a range of mechanisms, including the mutation or overexpression of the drug target, drug inactivation, or drug elimination from the cell. So, one potential reason for failure is that current therapies fail to kill cancer stem cells. Therapy that kills non-tumorigenic cancer cells can shrink tumors, but will not cure patients, whereas therapies that kill, or induce differentiation of cancer stem cells could better contribute to curing patients. Targeting pathways that maintain stem cell properties might be useful to transform malignancies into benign tumors and one way to handle the CSC problem is to treat cancer by inducing differentiation of cancer stem cells (Massard et al., 2006). Cellular differentiation therapy focuses on demonstrating that cancer is a reversible state in which the transformed phenotype may be suppressed by cytostatic agents and by the induction to differentiation by means of physiological and pharmacological agents. This therapy allows the cellular program to be modulated, making the cells mature, reaching the differentiating state, dying, disappearing and their replacement by normal cells. Differentiation therapy could force cancer stem cells to differentiate and lose their self-renewal property (Marchal et al, 2006).

All these evidences and the unique properties of stem cells are opening the door to the development of new therapeutic approaches notably in oncology and regenerative medicine. Diabetes, cardiovascular diseases, cancer and neurodegenerative diseases are among the illnesses that cost more lives and health-care investment in the development word. Nowadays organ transplantation is the commonly way to cope with totally tissue damage, the inconvenient of this replacement are not only immunological but also logistical. The number of patient awaiting for organs transplants increases each day as well as the difficulties to find compatible and willing donors. In addition immunosuppressant lifetime taking often have serious side effects. Taking this panorama into account cell therapy appears as an attractive and promised way to improve not only time live expectative but also live quality. The aim in cell therapy is the replacement of damage cells with healthy ones from the own body of the patient. The use of autologous cells for treating diseases represents a mainstay in regenerative medicine.

Which cells are the best candidates to be used? There are no a sole answer for this question, in fact there are many different roads open, some researchers bet for embryonic stem cells (ESC), others focus their efforts studying adult stem cell (ASC) therapeutic potentiality; in addition, somatic adult cell transdifferentiation and nuclear reprogramming are other methods that are being tested to gain the same goal of cell therapy and regenerative medicine.

Stem cells have been found in early embryo, the foetus, the placenta, the umbilical cord and in many tissues of the body. Embryonic stem cells derived from the inner cell mass of the blastocyst, were first isolated in 1998 by Thomson et al. ESC are totipotent, with the capacity of generate all cell types of the body (Odorico *et al.*, 2001), for these reason ESC were believed to be the panacea in cell therapy, in theory they could be induce to become whatever tissue of the body and even to be able to form a complete organ. Nevertheless, there are many controversial issues surrounding the use of ESC, not only from an ethical, legal and politic point of view (de Wert and Mummery, 2003) but also methodological and even medical, preliminary trials carried out using animal models ended within tumor induction (Wu et al., 2007).

Consequently, tissue-specific stem cells, which do not involve the employ of embryos, have lately gained interest by the scientific community. Tissue-specific stem cells (often referred to as adult or somatic stem cells, ASC) are undifferentiated cells found among differentiated cells present in most organs of the body. The primary role of the adult stem cell in a living organism is to maintain and repair the tissue in which they are found. The presence of stem cells in many more mature tissues that one was believed, have been well documented (reviewed in Preston et al., 2003). Thus, it not adventured to postulate that all the tissues in the body own their specific stem cells. Somatic stem cells could subsequently be isolated, cultivated and use to regenerate the specific damage tissue which they come from.

Taking all this into account, here, we overview our current understanding of the development and biology of stem cells and the latest approaches in cell therapy, summarizing the different strategies to obtain differentiated cells types that have been proved. We discuss the possibilities of cell differentiation as therapeutics tools in cancer and regenerative medicine.

2. Normal Stem Cells Versus Cancer Stem Cells

Normal stem cells (SCs) are defined by an extensive capacity for self-renewal and by their ability to undergo a broad range of differentiation events to clonally reconstitutive all of the functional elements within the tissue. The concept of the cancer stem cell arose from observation of similarities between the self-renewal mechanism of stem cells and those of cancer cells. In fact, both normal stem cells and tumorigenic cells have extensive proliferative potential and the ability to differentiate into other cells. Because most tumors have a clonal origin, tumorigenic cancer cells must give rise to phenotypically diverse progeny, including cancer cells with limited or no proliferative potential. This suggests that tumorigenic cancer cells undergo processes that are analogous to the self-renewal and differentiation of normal stem cells (Reya et al., 2001).

2.1. Normal Stem Cells

SCs are special kinds of cells that distinguish from other kinds of cells in the body. All SCs have some general properties: they divide infrequently, although they are capable of dividing and renewing themselves for long periods; they are unspecialized; and they can give rise to specialized cell types (such as cells of the heart, brain, bone, etc.), and frequently reside in a specialized physical locale termed a niche. Their proliferative capacity combined with the ability to become specialized makes SC unique (http://stemcells.nih.gov/info/basics/, 2001). SC function involves self-replication (maintenance of the stem cell population), as well as production of offspring to maintain all lineages within the tissue during times of normal homeostasis as well as injury.

Classically characteristic features of SCs are: (1) SCs comprise a small subpopulation of a given tissue, (2) SCs are ultrastructurally unspecialized, with a large nuclear-to-cytoplasmic ratio and few organelles, (3) SCs can be pluripotent, (4) SCs are slow-cycling, but may be induced to proliferate more rapidly in response to certain stimuli, (5) SCs have a proliferative reserve that exceeds an individual's lifetime, (6) because SCs cycle slowly, and represent only a small percentage of a cellular population, an intermediate group of more rapidly proliferating transient amplifying cells exists, that form clonal expansions resulting in the final, differentiated cell population and (7) the microenvironment of a SC plays a critical role in its homeostasis and in the differentiation of its progeny (Miller et al., 1993; 2005).

There are four kinds of SCs depending on their tissue of origin: embryonal, germinal, somatic or adult stem cells (Arey, 1974) and umbilical cord blood stem cells.

a) Embryonal stem cells (ESC) are derived from the first five or six divisions of the fertilized egg. The progeny of ESC are the precursors for all of the cells of the adult organs. An ESC is derived from a group of cells called ICM, which is part of the early (4- to 5-day) embryo called the blastocyst. The blastocyst is the stage of embryonic development prior to implantation in the uterine wall. The ICM cells are no longer totipotent, but have the ability to generate any cells type of the body (pluripotency). Once removed from the blastocyst, the cells of the ICM can be cultured into ESC. Cultured under appropriate conditions, the ICM derived cells can continue to proliferate and replicate themselves indefinitely and still maintain the developmental potential to form any cell type of the body (Rossant, 2007). These properties reflect their origin from cells of the early embryo at a stage during which the cellular machinery is geared toward the rapid expansion and diversification of cell types. Human ESC *in vitro* are pluripotent; they can produce cell types derived from all three embryonic germ layers. All of the human ESC lines generated to date are pluripotent (Thomson and Odorico, 2000).

b) Germinal stem cells in the adult produce eggs and sperm and are responsible for reproduction. An embryonic germ cell is derived from fetal tissue. Specifically, they are isolated from the primordial germ cells of the gonadal ridge of the 5- to 10-week fetus. Later in development, the gonadal ridge develops into the testes or ovaries and the primordial germ cells give rise to eggs or sperm. Embryonic germ cells in culture, like cultured embryonic stem cells, can give rise to cells of all three embryonic germ layers, indicating that embryonic germ cells may represent another source of pluripotent stem cells.

c) Somatic stem cells are also known as foetal/adult stem cells (ASCs) or tissue-derived stem cells. Somatic (adult) stem or progenitor cells are considered more limited in their potential, and they produce cells that differentiate into mature functioning cells and that are responsible for normal tissue renewal. Their primary functions are to maintain the steady state functioning of a cell—called homeostasis—and, with limitations, to replace cells that die because of injury or disease. An ASC is an undifferentiated cell that occurs in a differentiated tissue, renews itself, and becomes specialized to yield all of the specialized cell types of the tissue from which it originated. These cells are capable of making identical copies of themselves for the lifetime of the organism. ASCs are present in most tissues and usually divide to generate progenitor or precursor cells, which then differentiate or develop into "mature" cell types that have characteristic shapes and specialized functions. Stem cells exist in most adult; source of ASC include bone marrow, blood, the cornea and the retina of the eye, brain, skeletal muscle, dental pulp, liver, skin, the lining of the gastrointestinal tract, and pancreas being (Serakinci and Keith, 2006; http://stemcells.nih.gov/info/basics/, 2001; 2006; Sell, 2004).

d) Human umbilical cord blood (HUCB) is known as a rich source of hematopoietic stem cells (HSC) (Mayani and Lansdorp 1998). There do not appear to be any qualitative differences between the stem cells obtained from umbilical cord blood and those obtained from bone marrow or peripheral blood. Besides the HSC, a variety of different stem cell types have been identified within HUCB. MSC, endothelial stem cells (ESC) and a not yet fully characterized population of multipotent cells can be found in the cord blood. The unrestricted somatic stem cell (USSC) has recently emerged as the most promising candidate for such multipotent stem cells (Kogler et al., 2004). Another candidate for multipotent HUCB stem cells is a lineage-restricted cell subset that bears the markers of embryonic stem cells. These so-called “cord-blood-derived embryonic-like stem cells” (CBEs) can expand hematopoietic progenitors and differentiate into hepatocytes and neuroglial progenitors (McGuckin et al., 2004; 2005). Although these multipotent stem cells are very rare and may differentiate into a limited number of cell types, they may represent a promising tool for cell therapy (Koblas et al., 2005; Murohara et al., 2000).

The first entity of life, the fertilized egg, has the ability to generate an entire organism. This capacity, defined as totipotency, is retained by early progeny of the zygote up to the eight-cell stage of the morula. The zygote and its embryonic stem cell progeny are totipotent and give rise to branches of cells that form diverse differentiated organs. During this process, the progeny of the ESC lose potential and acquire differentiated properties, in a process named determination (Sell, 2004; Wobus and Boheler, 2005). In both embryonic and adult stem cells, differentiation is controlled by epigenetic mechanisms, and the plasticity of differentiation in these cells is associated with transcription accessibility for genes expressed in different normal tissues. Abnormalities in genetic and/or epigenetic controls can lead to development of cancer, which is maintained by self-renewing cancer stem cells (Lotem, 2006).

2.1.1. Normal Stem Cells, Precursor Cells and Progenitor Cells

A SC is an undifferentiated, clonogenic cell that originates from embryonic or adult stem cell sources and is able to asymmetric division, producing two daughter cells, one like the mother cell (long-term self-renewal) and a second one, giving rise to multiple types of differentiated cells that represent the three primitive embryonic germ layers: ectoderm, mesoderm and endoderm (Chandross and Mezey, 2001) (Table 1).

Habitually, progenitor and precursor cells terms are used indistinctly, although theirs significances are not similar. Precursor cells are the primitive cells derived from the second daughter cells, therefore those that classically proliferate before they differentiate. Progenitor cells are cells with greater developmental potential than precursor cells (Preston et al., 2003; Snykers et al., 2007).

2.1.2. Symmetric and Asymmetric Division

Stem cell function involves self-replication (maintenance of the SC population), as well as production of offspring to maintain all lineages within the tissue during times of normal homeostasis as well as injury. The different types of SCs not only have different potentials, but they also proliferate differently. Regulatory aspects of SC conduct may be intrinsic, extrinsic or, most likely, both. They include decisions to replicate and in what way. Cells can divide symmetrically, producing two identical SCs, whereby each daughter cell retains the properties of the parental cells, or asymmetrically, whereby one daughter cell retains the properties of the parental stem cell, whereas the other daughter cell begins the process of determination (Miller et al., 2005). It is possible that the molecular distinction between symmetric and asymmetric divisions lies in differential segregation of cell membrane proteins between the daughter cells.

ESCs, typically, divide symmetrically, resulting in a logarithmic expansion of cells during early embryonic growth. Then, as the germ layers of the early embryo form and the process of determination begins, the cells proliferate asymmetrically. The tissue determined stem (progenitor cells) divide asymmetrically as one daughter cell remains to continue the process of cell renewal and the other daughter cell starts the process of differentiation (transit amplifying cells). Transit amplification is one way of reducing the number of times a stem cell must divide to an extent that depends on haw many divisions intervene before differentiation (Gadner, 2007).

ASCs are primitive cells that undergo asymmetric division, thereby giving rise to one clonogenic, self-renewing cell and one cell able to undergo multipotent differentiation (transit amplifying cells). The determined transit amplifying cells retain the ability to divide for several differentiation stages and are the major contributors to normal tissue renewal. These cells are known as progenitor cells in the bone marrow, and as transit-amplifying cells in the skin and other tissues (Sell, 2004). Tissue progenitor (ASCs), or transit amplifying cells, provide an expanded population of a proliferating tissue that differentiate into more mature cells with specific functions and eventually cease to proliferate (Serakinci and Keith, 2006).

Alterations of this process of division by epigenetic alterations may lead in uncontrolled growth, formation of self-renewing malignant stem cells and eventually cancer (Snykers et al., 2007).

Table 1. Embryonic germ layers from which differentiated tissues develop

Embryonic germ layer	Differentiated tissues
Ectoderm	Skin
	Neural tissue (neuroectoderm)
	Adrenal medulla
	Pituitary gland
	Connective tissue of the head and face
	Eyes, ears
Endoderm	Thyroid, parathyroid glands
	Larynx, trachea, lung
	Thymus
	Urinary bladder, vagina, urethra
	Gastrointestinal organs, as liver and pancreas
	Lining of the GI tract
	Lining of the respiratory tract
Mesoderm	Bone marrow (blood)
	Adrenal cortex
	Heart and blood vessels
	Lymphatic tissue
	Cardiac, skeletal and smooth muscle
	Connective tissues (including bone, cartilage)
	Urogenital system

2.1.3. Plasticity

The differentiation potential or plasticity is the ability of a stem or progenitor cell to produce progeny that can express various mature phenotypes (Sell, 2004). Similarly, it is the ability of an ASC from one tissue to generate the specialized cell type(s) of another tissue. The maintenance of many tissues and organs is achieved by tissue-specific SCs. In general, SCs divide very rarely, but in the presence of an appropriate stimulus from, for example, an increasing demand for cells, they proliferate and differentiate. However, SCs possess varying degrees of potential which include the totipotency of the zygote, the pluripotency of ESCs, the multipotentiality of stem cells fetal or adult tissue, such as adult MSCs and the unipotentiality of a specific cell type, such as epidermal stem cells.

There is no formally accepted name for this phenomenon in the scientific literature. It is variously referred to as "plasticity" (Krause et al., 2001), "unorthodox differentiation" (Bianco and Cossu, 1999) or "transdifferentiation" (Anderson et al., 2000). Recent years have seen a surge of scientific research re-examining ASC plasticity. The term totipotent should be reserved for the SCs that can give rise to all of the differentiated tissues of the body as well as the placenta and membranes, as ESCs and germinal stem cells. Most scientists use the term pluripotent to describe SCs that can give rise to cells derived from all three embryonic germ layers—mesoderm, endoderm, and ectoderm. Multipotent and pluripotent are practically synonymous, and indicate the capacity of a SC to form many different cell types. During develop, the totipotent embryonal stem cells give rise to multi/pluripotent progenitor cells of the germ layers. These, in turn, give rise to oligopotent progenitor cells of the developing organs. Most adult tissues have multipotential SCs; cells capable of producing a limited range of differentiated cell lineages appropriate to their location, e.g. small intestinal stem cells are cuadripotent (paneth, goblet, absorptive columnar and enteroendocrine); central nervous system SCs are tripotent generating neurones, oligodendrocytes and astrocytes (Sanai et al., 2004); SCs of the heart can give rise to cardiomyocytes, endothelial cells and smooth muscle (Beltrami et al., 2003); the bile ductular progenitor cells of the liver appear to give rise to two cell types (bipotent), duct cells and hepatocytes.

All tissues have SCs, but these SCs do not all have identical potentials for producing different cell types. Many studies suggest that, given the right environment, some ASCs are capable of being "genetically reprogrammed" to generate specialized cells that are characteristic of different tissues. The existence of multipotent stem cells in skin, neuronal tissue and bone marrow, giving rise to other cell types than their tissue of origin (Preston et al. 2003). Many evidences support the idea that certain ASCs can engraft alternative locations, particularly when the recipient organ is damaged, and transdifferentiate into cell types with functions appropriate to their new location (Alison et al., 2004; 2006). Under specific experimental conditions, ASCs from bone marrow generated cells that resemble neurons and other cell types that are commonly found in the brain. Hematopoietic stem cells may be able to adopt certain nonhematopoietic phenotypes, such as endothelial, neural, or skeletal muscle phenotypes, without entirely losing their initial hematopoietic identity (Udani, 2006). This is not transdifferentiation, really the more correct term would be “pleopotency” or unpredicted plasticity; some cells may have more potential than previously appreciated (less than totipotent, but more than multi/pluripotency) (Sell, 2004). The finding of SC plasticity carries significant implications for potential cell therapy.

Different pathways have been proposed to explain differentiation process in adult tissues relating with the plasticity:

a) Tissue determined adult stem cells (classical model)
b) Pleo/pluripotency adult stem cells
c) Dedifferentiation
d) Transdifferentiation
e) Fusion (explanation in Figure 1).

2.2. Cancer Stem Cells

Actually, the cancer stem cell (CSC) hypothesis has two separate but related components. The first, concerns the cellular origin of tumors, including the question of whether tumors arise from tissue SCs. The concept that cancer might arise from a rare population of cells with stem cell properties was proposed about 150 years ago (Cohnheim, 1875; Sell, 2004), suggesting itself later that normal tissue stem cell may be the cell of origin of cancer and that cancer represented a maturation arrest of SCs. This concept is endorsed for several reason, SCs by their long-lived nature are subject to the accumulation of multiple mutations that are required for carcinogenesis, also shares with the cancer cells properties as the capacity for self-renewal, ability to differentiate and activation of antiapoptotic pathways. The second component of this hypothesis is that tumors are driven by cellular component that display SC properties. It seems that several SC markers may be shared by CSCs in multiple tumor types, including CD44, alpha-6 integrin, beta-1 integrin and CD133 (Wicha et al., 2006).

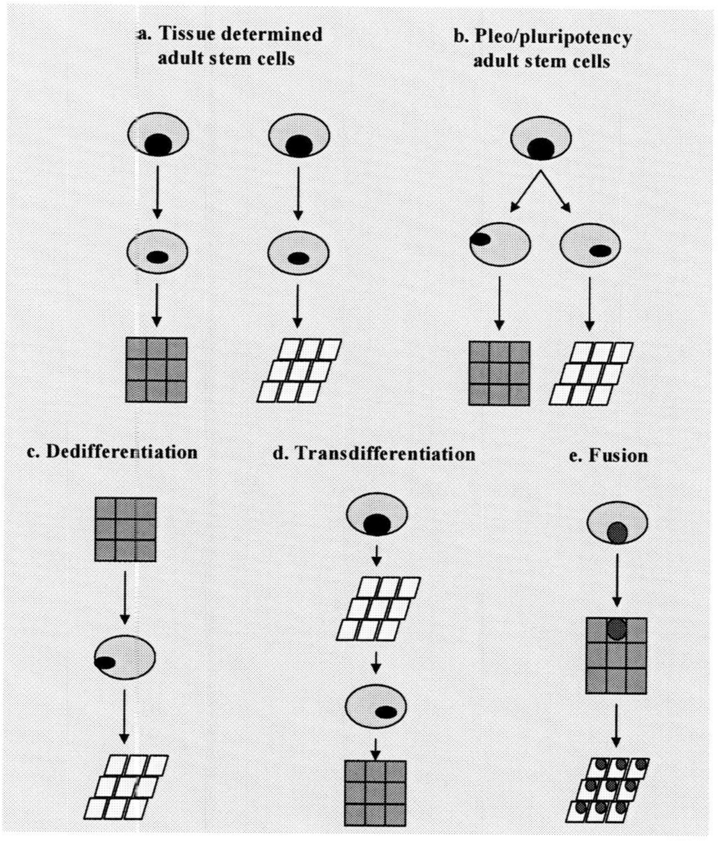

Figure 1. Probable pathways of tissues formation in adult: a) Progenitor cells become determined for a specific tissue type. b) The tissue SC retains the potential to give rise to progeny that can differentiate into different cell lineages. c) Mature tissue cells undergo backwards differentiation and give rise to progeny that can differentiate into a different cell type. d) Progenitor cells that have become determined for one tissue type change their determination and give rise to a different cell type. e) SCs from one tissue may fuse with cells of another tissue to form "hybrid" cells that may show new differentiation characteristics.

2.2.1. Origin of Cancer Stem Cells

CSCs were first described in patients with acute myeloid leukaemia (AML) and later in breast and central nervous system (Singh et al., 2004), which can be separated on the basis of cell surface marker antigen expression by flow cytometry, represented in only a small percentage of the total of tumor cells.

The cellular origin of CSCs has not been clearly determined. CSCs are derived either from transformed normal SCs or from more differentiated progenitor cells that have acquired the ability to self-renew as a result of oncogenic mutations. In the first case, it was suggested that maturation arrest of SC differentiation is a common pathway for the cellular origin of teratocarcinomas (Sell and Pierce, 1994) and substantial evidence links childhood CSCs and deregulation of normal development (Dyer and Bremner, 2005). It is believed that due to the longevity and specific self-renewing properties of SCs, they have a greater propensity to accumulate carcinogenic mutations compared to short-lived, more differentiated cells and thus would be the ideal target of the carcinogenic process (Gudjonsson and Magnusson, 2005).

Different studies realized in several organs have shown that CSCs appear to have a phenotype very similar to that of normal SC (Singh et al., 2003). However, when CSCs and normal SCs from a given organ are compared, they are differences in: 1) surface markers, CSCs in acute myeloid leukaemia (AML) present as marker CD90-, CD117- and CD123+ that are not share by normal hematopoietic stem cell (HSCs) (Guzman and Jordan, 2004), 2) expression of tumor suppressor genes, interferon factor-1 and death-associated protein kinase-1, only appears in CSCs from AML patients (Guzman et al., 2001b), 3) the anti-apoptotic factor MCl-1, and a constitutively active survival factor NF-κβ (Guzman and Jordan, 2004; Guzman et al., 2001a), none of which are found in normal HSCs (Miller et al., 2005).

The Hedgehog, Wnt and Notch pathways that often promote cancer cell proliferation also promote normal stem cell self-renewal (Molofsky et al., 2004; Taipale and Beachy, 2001). Conversely, tumor suppressors that inhibit cancer cell proliferation—such as p53, p16INK4a and p19ARF—also inhibit SC self-renewal (Molofsky, 2005; Lowe and Sherr, 2003). A recent study (Yilmaz et al., 2006) supports the existence of genes or pathways that have different effects on normal and CSCs. Thus, the tumor suppressor PTEN, that inhibits proliferation and survival is commonly deleted or otherwise inactivated in diverse cancers, including haematopoietic malignancies (Cheong et al., 2003). PTEN deletion causes the generation of leukaemic stem cells with ability to transfer disease on transplantation into irradiated mice, whereas that this deletion leads to the depletion of normal HSCs, that show less self-renewal potential, provided a rare distinction between the mechanisms that regulate the maintenance of normal compared with SCs.

The Wnt pathway is implicated in the pathogenesis of a number of cancers, usually by activating mutation that result in a ligand-independent state of constitutive activity (Taipale and Beachy, 2001). Overexpression of the Wnt pathway stimulates stem cell proliferation in a number of organ systems. In Drosophila, the wingless gene (autologous to Wnt in mammals) is necessary for maintaining somatic ovarian stem cells. Downregulation of the wingless gene results in a rapid stem cell loss, while constitutive signaling causes overproliferation and abnormal differentiation of somatic follicle cells (Song and Xie, 2003).

The hedgehog (hb) pathway also is involved in SC proliferation, survival and growth and constitutive overexpression of hb is observed in a number of cancers. The hb pathway regulates SC turnover in several organ systems. In the Drosophila ovary, it acts as a somatic SC factor; and in its absence, Drosophila somatic SC cannot proliferate, while excess hb system signaling stimulates the production of somatic SCs (Zhang and Kalderon, 2001). Constitutive overexpression of the hb pathway is observed in a number of cancers. Increased hb system signaling appears to be necessary present in a subset of small cell lung cancers as well as in many digestive tract tumors, including those of the esophagus, stomach, billiard tract and pancreas (Watkins et al., 2003).

2.2.2. Therapeutic Implications of Cancer Stem Cells

The cancer stem cell hypothesis implies a novel point of view on therapy, in which these minority tumor-initiating/maintaining cells as a critical target. An important implication of the theory that cancers are maintained by CSCs is that genes playing fundamental roles in normal stem cells might be essential in CSCs as well. If this is correct, SC genes might represent a novel target for cancer therapy (Valk-Lingbeek et al., 2004).

By virtue of their fundamental importance in organogenesis, normal SCs have involved mechanisms that promote their survival and resistance to apoptosis. There are several molecular mechanisms that may account for the resistance to apoptosis of CSCs. These include: a) cell cycle kinetics, many CSCs are nor cycling and are not in G0 and thus are resistant to cell cycle-specific chemotherapy agents (Venezia et al., 2004); b) DNA replication and repair mechanism, SCs may be resistant to DNA-damaging agents by virtue of being able to undergo asynchronous DNA synthesis in addition to displaying enhanced DNA repair (Park and Gerson, 2005); c) during asynchronous DNA synthesis, the parental immortal DNA strand always segregates with the SCs and not the differentiating progeny. This prevents the SC compartment from accumulating mutations associated with replication or from being affected by DNA-damaging agents (Rambhatla et al., 2005); d) antiapoptotic proteins, SCs express higher levels of antiapoptotic proteins, such as members of the Bcl-2 family and inhibitors apoptosis, than do differentiated cells (Wang et al., 2003); e) transporter proteins, SCs express high levels of transporter proteins, such as ABCG2 (BCRP), as well as P-glycoprotein (Wicha et al., 2006).

Just a normal SC may be resistant to the induction of apoptosis by cytotoxic agents and radiation therapy, CSCs may display increased resistance to these agents compared with more differentiated cells that comprise the bulk of tumors. Thus, has been shown that leukaemic and myeloma SCs are more resistant to chemotherapy and other many therapies used than the rest of cells that are into the tumor (Matsui et al., 2004).

2.2.3. Telomerase Activity in Normal and Cancer Stem Cells

Telomerase is a complex of a reverse transcriptase protein encoded by the TERT (telomerase reverse transcriptase) gene and a template RNA TERC (telomerase RNA component). Telomerase can add telomeric repeats onto the chromosome ends, and prevents the replication-dependent loss of telomere and cellular senescence in highly proliferative cells of the germinal and in the majority of cancer (Blasco, 2005). Thus, telomerase activity and telomere maintenance are associated with the immortality of cancer cells and ESCs (Hiyama and Hiyama, 2007). In human adult SCs including hematopoietic ad non-hematopoietic SCs such as neural, kidney, adrenal cortex and MSCs they are low levels of telomerase activity. In HSCs, telomere shortening is considered to be a risk factor that contributes to the development of chromosomal instability and malignant transformation. CSCs to maintain their characteristic replicative ability should have telomere-lengthening mechanism. In fact they might have acquired immortality by mutational events in telomere-lengthening mechanisms typically the activation of telomerase (Ohyashiki et al., 2002), considering that telomere shortening and cellular senescence are inevitable in these original cells, except for some MSCs. In human and murine MSCs, increased telomerase activity may be responsible, at least in part, for the malignant transformation (Serakinci et al., 2004; Miura et al., 2006).

2.2.4. Asymmetric Division of Stem Cells and Cancer

The ASCs normally divide asymmetrically, whereby one daughter cell retain the properties of the parental stem cell and the other daughter cell begins differentiation (Sherley, 2002). Recently, has been shown that the loss of polarity and impairment of asymmetric division in SCs leads to tumorigenesis (Morrison and Kimble, 2006). In Drosophila neuroblasts containing mutation in genes that control cell division, such as Raps, Mira, Numb, Brat o Pros, developed aggressive tumors that could be retransplanted into new host (Caussinus and Gonzalez, 2005). Asymmetric division was disrupted in mutant SCs, whereby both daughter cells grow and behave like neuroblasts leading to the formation of tumors. This transformation was associated with genome instability and centrosome alterations. Together, these data indicate that invertebrate cells rapidly develop cancer upon mutation of the asymmetric division machinery, by adopting an aberrant mode of asymmetric division (Morrison and Kimble, 2006). This hypothesis may be enhanced because has been shown that adenomatous polyposis coli gene, is required for the asymmetric division of Drosophila spermatogonial SCs but also is a tumor suppressor gene in human colon cancer. Also alterations in the Numb gene may leads to cancer by alterations in asymmetric cell division (Yamashita et al., 2005; Martinez-Climents, 2006).

3. Molecular Control of Embryonic Stem Cell Renewal

Three groups of stem cells are known to exist, namely embryonic, germinal and somatic stem cells (Sell, 2004). Embryonic stem cells (ESC) are derived from the inner cell mass of the blastocyst-stage embryos (Martin, 1981), and are the ancestors of all cells in the body. ESC possess the unique property of being able to retain their capacity for self-renewal - can be maintained and expanded as pure populations of undifferentiated cells for extended periods of time, possibly indefinitely, in culture, although, they can generate tumor cell lines- and potential to form cells of all three embryonic germ layers: endoderm, mesoderm and ectoderm. They are pluripotent, possessing the capacity to generate every cell type in the body (Keller, 2005). This proliferative and pluripotent capacity offer the prospective that banks of ESC could generate specific cell phenotypes, to a consistent quality, for use in clinical treatments for a range of human diseases (Shufaro, 2004).

3.1. Genetic Control of Pluripotency Mediated by Oct-4, Sox-2 and Nanog

Signaling pathways eventually lead to the nucleus of ESC transcriptional induction or repression of genes that are responsible for implementing stem-cell pluripotency. Recent efforts using chromatin immunoprecipitation combined with genome-wide methodologies

(ChIP-Chip and ChIP-PET studies) have begun to identify large groups of genes in mouse and human ESC that are associated with three transcription factors and they play a central role in the regulation of pluripotency and self-renewal (Boyer et al., 2005; Loh et al., 2006). These factors include the POU (Pit/Oct/Unc) domain-containing protein Oct-4 (Rosner et al., 1990), Sox-2 (Avilion et al., 2003) and the homeoprotein Nanog (Mitsui et al., 2003). All three transcription factors are expressed in high levels in pluripotent cells and are considered markers of primitive stem cells (Smith, 2001). They regulate the expression of other genes during development and are found at high levels in the pluripotent cells of the inner cell mass (ICM). Its downregulation correlates with the loss of pluripotency and self-renewal, and the beginning of subsequent differentiation steps. The roles of Nanog, Oct-4 and Sox-2 have not been fully elucidated. They are important in embryonic development and recent studies suggest that function in distinct pathways might converge to regulate certain common genomic targets. It is likely that the interplay among these factors is critical for early cell fate decisions: self-renewal or differentiation.

3.1.1. The Transcription Factor Oct-4

Oct-4, also known as Pou5f1 and Oct-3, was the first gene to be discovered whose function was crucial in maintaining ESC pluripotency. Oct-4, encodes a class V POU transcription factor initially identified in embryonic cancer cells (Rosner et al., 1990). This homeodomain POU possesses a specific region or amino-acid sequence in a protein that is associated with specific sequences such as the octamer motif (5'-ATGCAAAT-3') on DNA. The transcription factor Oct-4 is expressed in all pluripotent cells of the mammal and is downregulated upon formation of extraembryonic and somatic lineages. The Oct-4 levels that display ESC are going to determine the three possible fates: self-renewal (normal expression), trophectoderm (down-regulation) or extraembryonic endoderm and mesoderm (overexpression) (Niwa et al., 2000). Also, Oct-4 has also been established as a marker for human pluripotent ESC. Microarrays and quantitative reverse-transcription (Q-RT)-PCR analyses indicate that it is downregulated in the differentiated versus the undifferentiated state of stem cells (Brandenberger et al., 2004).

3.1.2. The Transcription Factor Sox-2

The Sox-2 (high mobility group (HMG)-box) transcription factor is implicated in the regulation of transcription and chromatin architecture (Boiani et al., 2005) and plays an important role in the maintenance of pluripotency and lineage specification. Sox-2 is expressed in the ICM, early primitive ectoderm, anterior primitive ectoderm, germ cells and multipotent extraembryonic ectoderm cells. Its expression is not restricted to pluripotent cells, because Sox-2 is also found in early neural lineages (Avilion et al., 2003). Sox-2 expression is controlled by the interaction with Oct-4 and Nanog, suggesting a positive feedback mechanism that could be related to the maintenance of ESC self-renewal (Loh et al., 2006).

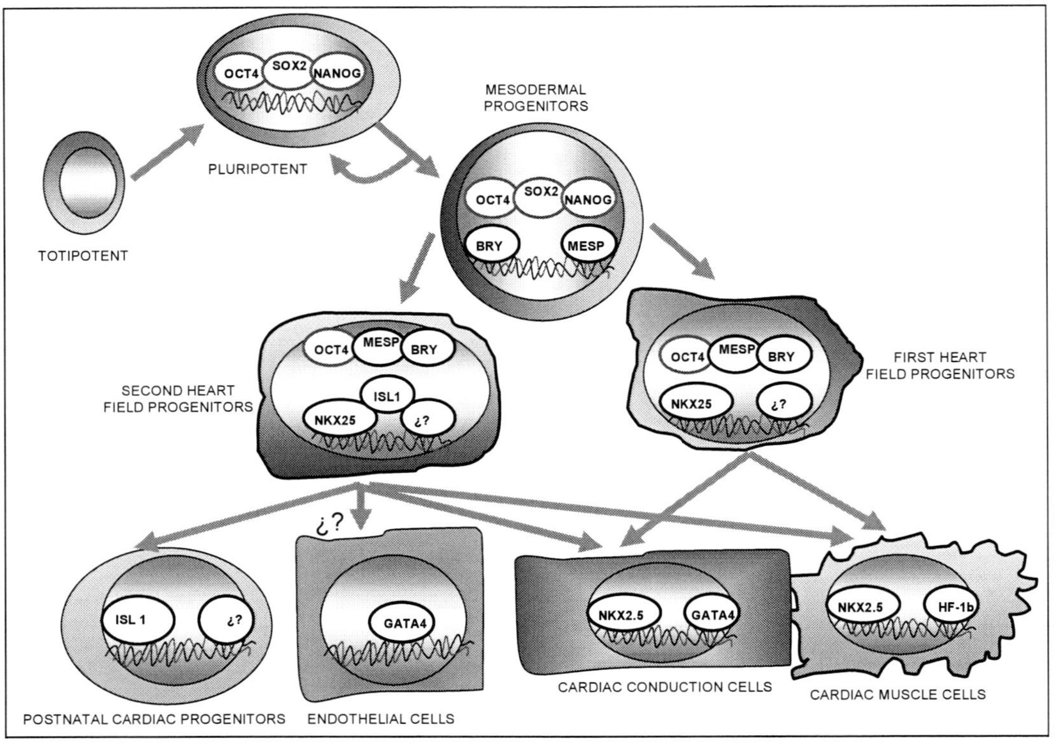

Figure 2. Importance of Oct-4, Sox-2 and Nanog transcription factors in self-renewal of ESCs and their interactions with other transcription factors during heart development.

3.1.3. The Transcription Factor Nanog

It has been the last transcription factor implied in ESC self-renewal discovered. Nanog is a novel gene that encoding a homeodomain in a protein that consists of about 60 aminoacid residues and recognizes and binds to specific AT-rich DNA sequences. It contains the transcription factor that confers to the leukemia inhibitory factor (LIF) independent ability for cell renewal and pluripotency on mouse ESC (Chambers et al., 2003). Similarly to Oct-4 and Sox-2, Nanog expression is high in ESC and is downregulated when cells differentiate. The lack of expression in ES cells leads to a significant down-regulation of Oct-4 and loss of ESC-surface antigens of pluripotency and differentiation into trophectoderm lineage (Hyslop et al., 2005). Its overexpression maintains ESC in the undifferentiated state over several passages in a feeder-free system, without the requirement of conditioned medium, and induces differentiation into primitive endoderm and mesoderm (Darr et al., 2006). Recently, it has been shown that transcription of Nanog is regulated by the Oct-4 and Sox-2 binding to the Nanog promoter (Rodda et al., 2005) (Figure 2).

3.2. Epigenetic Control

The field of epigenetic is concerned with influences on gene expression that occur independently of DNA sequence per se. Though each cell has identical DNA content, the way in which it is packaged with chromosomal proteins (i.e. chromatin) differs greatly from cell to cell. The basic subunit of all chromatin is the nucleosome, which consists of a histone octamer containing a pair of each of the standard histones H2A, H2B, H3, and H4 and 146 base pairs of DNA.

Chromatin can be broadly divided into two fractions: euchromatin, which is permissive for transcription, and heterochromatin, which is repressive. Chromatin acts not only as a support for the double helix but also as a kind of gene "gatekeeper." Chromatin structure can be modified in three different but related ways: first, nucleosomes may be repositioned by ATP-dependent protein complexes; second, histone variants may replace core histones; and third, histone tails may be covalently modified (Felsenfeld and Groudine, 2003). It is becoming increasingly clear that histone modifications and chromatin remodeling are critical for gene expression, reason why chromatin reorganization is essential for the establishment of new heritable gene expression programs that accompany lineage specification. By examining the chromatin in mouse ESC across the genome, scientists discovered an unusual pair of overlapping molecular tags in the chromatin structure, which together comprise what they called a "bivalent domain" (Azuara et al., 2006). These domains reside in the sections of chromatin that control the most evolutionarily conserved portions of DNA, particularly the key regulatory genes for embryonic development (Bernstein et al., 2006). They consist of repressive histone H3K27me3 and activating histone H3K4me3 modifications at a large set of developmentally important genes that are silent in ESC but activated upon differentiation.

The additional observation that Oct-4, Sox-2 and Nanog occupied a significant subset of genes that harbor bivalent domains supports a link between the repression of developmental regulators and stem cell pluripotency. They are proteins with enzymatic activities that can induce both global and local changes in chromatin structure through bivalent domain, for example the Polycomb group (PcG). These proteins are transcriptional repressors of homeotic gene expression during embryogenesis (Ringrose et al., 2004). PcG proteins comprise at least two distinct repressor complexes: PRC1 and PRC2–PRC3. It has been demonstrated that Polycomb proteins are required for early developmental gene expression patterning (Pasini et al., 2004), the establishment of pluripotent ESC lines (O'Carroll et al., 2001), and for ASC maintenance. In human and mouse ESC Polycomb repressor proteins (e.g. Phc1, Rnf2, Suz12 and Eed) act binding approximately half of the genes that encode transcription factors important for later stages of differentiation. These domains often coincided with binding sites at which Oct-4, Sox-2 or Nanog has previously been reported and also contained H3K27me3 in their promoter regions, a repressive histone modification catalyzed by PRC2. This resulted in a global upregulation of the expression of many bivalent genes (Boyer et al., 2005) (Figure 3).

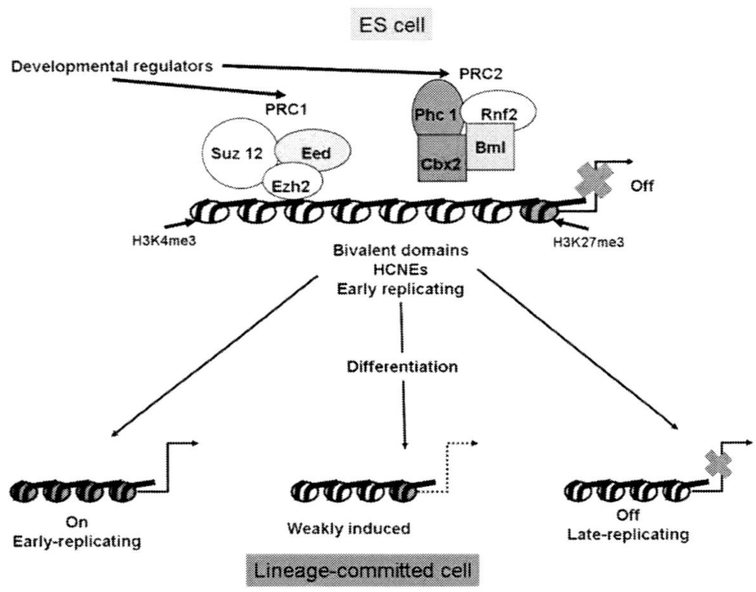

Figure 3. Epigenetic characteristics of pluripotent and lineage committed cells. PcG proteins have recently been shown to reversibly silence developmental regulators in ES cells, a process that might be necessary for the propagation of an undifferentiated state. These regulators, which are early replicating, contain highly conserved non-coding elements (HCNEs), which are rich in bivalent domains that consist of both H3K27me3 and H3K4me3 modifications. These domains might provide an epigenetic indexing system to mark genes for expression at later developmental stages. During differentiation of ESCs, the bivalent marks resolve because early-replicating genes that are expressed in the lineage-committed cells maintain or acquire activating H3K4me3 marks and late-replicating genes that are turned off in these cells possess repressive H3K27me3 modifications. Notably, genes that are weakly induced still possess bivalent domains.

3.3. Extrinsic Factors and Signaling Requirements of ESC Pluripotence

3.3.1. Leukemia Inhibitory Factor (LIF)

LIF is a member of the interleukin-6 (IL-6) family of cytokines and is known by bind to its transmembrane receptor, LIFR, which heterodimerizes with the signal-transducing receptor gp130 (glycoprotein-130). In vitro, LIF is essential for maintaining the undifferentiated state of mouse ESC. Interestingly, LIF is only able to sustain ESC in the presence of serum, suggesting that additional factors are required. LIF signaling plays a major role in maintaining the self-renewal of mouse ESCs, and as such, is routinely used in mouse ESC culture. This pluripotency of mouse ESCs depends on the intracellular signaling events that follow, including phosphorylation by the Janus family of tyrosine kinases (JAK), which leads to activation of the signal transducer protein STAT3. The STAT3 dimers then translocate to the nucleus where they bind to sites on the DNA controlling the transcription of genes important in ESC self-renewal. Recruitment and activation of STAT3 on engagement of the LIFR is necessary for self-renewal of ESC and the interference in this process inhibits ESC self-renewal (Humphrey et al., 2004). The binding of LIF to gp130 also stimulates the Ras/ mitogen-activated protein kinase (MAPK) pathway. This pathway resulting in the activation of ERK (Kolch et al., 2000) and this can phosphorylate cytoplasmic proteins and be translocated to the nucleus where they can modulate the activities of transcriptional regulators. Activation of this pathway on stimulation of the LIFR, is a pro-differentiation signal rather than contributing to self-renewal (Burdon et al., 1999). Therefore the efficiency of ESC self-renewal is due to the overall balance of the conflicting signals of STAT3 activation and ERKs signaling emanating from the LIFR on activation (Burdon et al., 1999).

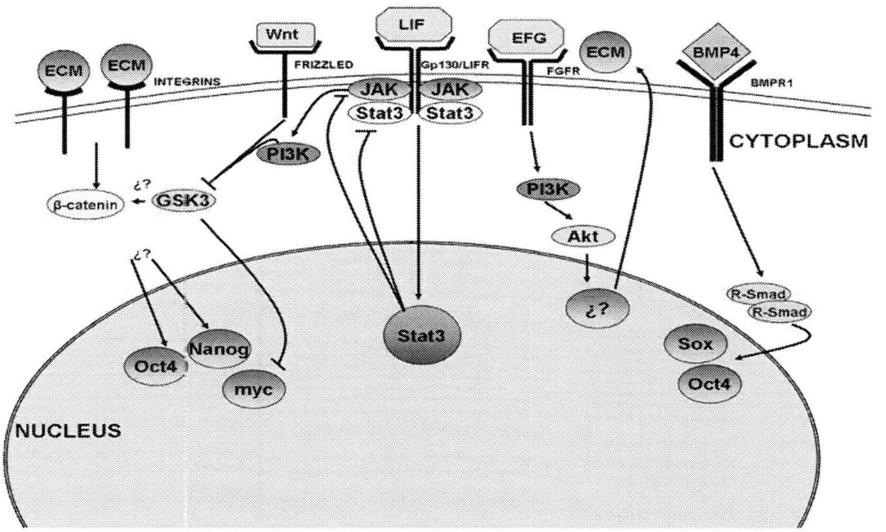

Figure 4. Relationship between different pathways that mediate self-renewal of ESCs. Several pathways are implicated in ESCs self-renewal, although mechanisms of control are relatively unknown. For example, in the figure BMP signaling induces differentiation into other cell types, but this ability is blocked by LIF/Stat3 signaling. The role of WNT signaling is unconfirmed, whether it is promoting self-renewal or differentiation and whether the effect is mediated by β-catenin or by other downstream effects of GSK-3 inhibition. Cell-surface receptors initiate signals that are conveyed to the nucleus and affect key pluripotency transcription factors such as Oct-4, Sox-2 and Nanog, and self-renewal transcription factors such as signal transducer and activator of transcription-3 (STAT3). At the nuclear level, STAT3, Oct-4 and Nanog, cause changes in gene expression that result in or counteract phenotypic traits of ESCs.

3.3.2. Transforming Growth Factor Beta-bone Morphogenic

Unlike LIF signaling, transforming growth factor beta (TGFβ) family signaling is considered to be a key of maintaining the ESC self-renewing. They are components present in serum and help LIF to self-renewing ES cells in vitro (Ying et al., 2003). In the presence of LIF, BMP4 (a member of TGFβ family) contributes to the LIF cascade, enhancing the self-renewal and pluripotency of ESC. Interestingly, it has been reported that BMP4 inhibits both MAPKs p38 and ERK (Qi et al., 2004), which further supports the growing evidence indicating that the balance between ERK and STAT3 can determine the cell fate in mouse

ESCs (Burdon et al., 1999). BMP4 has the possibility of autocrine signaling, where Smads are activated by phosphorylation, form heterodimers with Smad 4 and are translocated to the nucleus where they inhibit or activate target genes. In absence the LIF activates members of the Id (inhibitor of differentiation) gene family, which maintain the undifferentiated state, and in presence of LIF inhibits the Id proteins. In conclusion, the balance between LIF and BMP4 are jointly responsible for maintaining the undifferentiated state ESC in mouse and probably in humans (Vallier et al., 2005) (Figure 4).

3.3.3. Wnt Proteins

The Wnts comprise a large family of protein ligands that affect diverse processes including embryonic induction, the generation of cell polarity, and cell fate processes. In addition to influencing developmental processes, recent studies point to a key role for Wnt signaling during adult homeostasis in the maintenance of stem cell pluripotency. Canonical WNT signaling involves the binding of Wnt ligands to the Frizzled receptors on the cell membrane, and the activation of the pathway leads to inhibition of glycogen-synthase kinase-3 (GSK3), preventing ubiquitin mediated degradation of β-catenin, with subsequent nuclear accumulation of beta-catenin and the expression of target genes. Sato et al (2004) used a new and specific reversible inhibitor of GSK3 to demonstrate that activation of the canonical WNT pathway maintains the undifferentiated phenotype in both mouse and human ESC, and sustains expression of the pluripotent-state specific transcription factors Oct-4, REX1 and Nanog in the absence of supplemented LIF. WNT was found to stimulate human ES proliferation, but also differentiation, and beta-catenin mediated transcriptional activity was minimal in the undifferentiated state (Dravid et al., 2005), although the relationship between WNT signaling and GSK-3 activity in the regulation of human ESC was inconclusive.

In conclusion, it appears that pluripotency is achieved through the combination of properly sequenced processes that control chromatin accessibility, chromatin modifications, and activation and repression of specific genes. This is complicated by the requirement for the finely tuned regulation of the relative levels of expression.

4. Normal and Cancer Stem Cell Niches

The term niche in ecology means the habitat where an organism can live and reproduce. In the stem cell yield, niche is not a simple location for those cells, the concept should include both anatomical and functional dimensions, and in addition, it has to be seen as a dynamic structure with capacity of response against external stimuli. SCs located at their respective niche can self-renovate without overproduction of cells, and participate in tissue generation, maintenance and repair. Cell, matrix and three dimensional spaces constitute the ultrastructure of a SC niche. Extracellular matrix contains elements that modulate the stem cell pool and the presence of paracrine factors from niche cells and non-protein components of the local microenvironment are important regulators of SC function (Scadden, 2006).

The comparison of different SC niches shows that there are common features, structure and functions from Drosophila to the mammal niche (Li and Xie, 2005). So, niche is composed of cells in a specific location that lead to SC maintenance with specific interactions between SCs and niche cells or extracellular matrix, and factors that determine SC balance between proliferation and differentiation as Wnts, Notch and fibroblast growth factors (Li and Neaves, 2006).

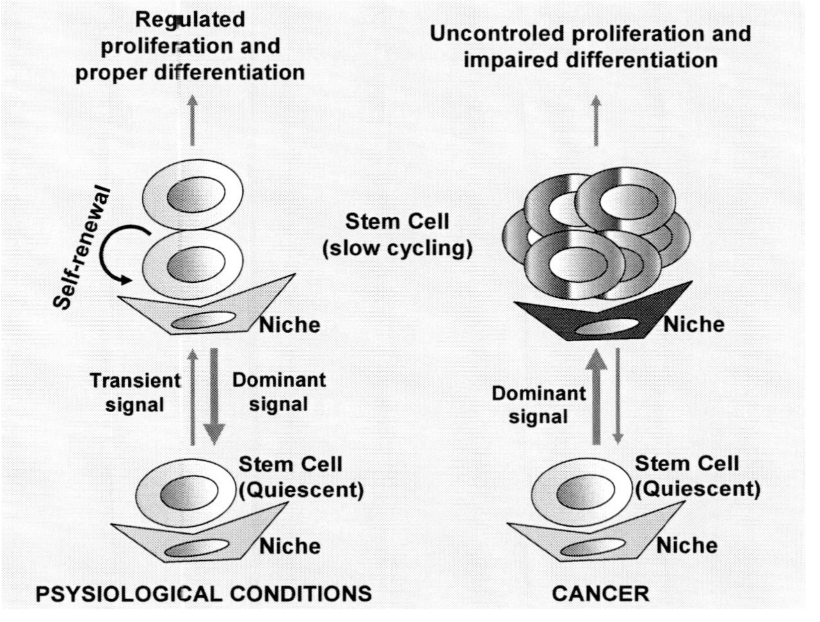

Figure 5. Comparison of normal and cancerous niches. Under physiological conditions niche cells predominantly inhibits both proliferation and differentiation. However, a transient proliferating signal supports tissue regeneration. On the other hand, CSCs are located in an environment with dominant signals toward cell proliferation and growth.

Normal somatic and cancer stem cells not only share the capacities of self-renewal and multi-lineage differentiation, but are also similarly surrounded by a specialized cell microenvironment, termed the SC niche (Martinez-Climent et al., 2006; Li and Neaves, 2006). Normally, at least in the hematopoietic, intestinal, and hair follicle system, the niche maintains stem cells primarily in a quiescent state by providing signals that inhibit cell proliferation and growth as evidenced by the ability of SCs to retain bromodeoxyuridine labeling for a relatively long period of time. Therefore, SC proliferation depends on dynamic niche signaling. Only upon receipt of a stimulation signal does the SC become activated to divide and proliferate. Maintaining a balance between the proliferation signal and antiproliferation signal is the key to homeostatic regulation of SCs, allowing to these to undergo self-renewal while supporting ongoing tissue regeneration (He et al., 2005) (Figure 5). Any genetic mutation that leads SCs to become independent of growth signals, or to resistant antigrowth signals, will cause an uncontrolled proliferation and possible tumorigenesis of them. Thus, Chepko et al. (2005) reported that deregulation in the mammary gland SC niche leads to abnormal expression of TFFα resulting in the development of breast cancer.

The similarities between SC definition and cancer cells led to the CSC theory about cancer development, which is based on the idea that the beginning of tumors instead of be derived from random tissue cells, is located in the transformation of progenitors or SCs that, with an increased proliferation rate and differentiation blockage, led to cancer expansion (Song and Miele, 2007). This definition imply that normal SCs are the origin of CSC generation, however, in some locations as in adult brain, compromised and differentiated cells as astroglia and oligodendroglia can convert toward SCs, expanding the potential substrate of oncogenic mutations (Nicolis, 2007). In fact, vacant niches can be ectopically occupied, that can result in proliferation of the ectopic cell and so, more mature descendent cells, including non-proliferating somatic cells, and can revert to a SC phenotype if they engage correctly. This theory could explain how the niche may contribute to disease, particularly neoplastic disease (Scadden, 2007).

In 1990s, it was supposed that at least 6 mutations were needed to transform a normal cell into a cancer cell, related to autonomous growth signal production, insensitivity to antigrowth signals, anti-apoptotic status, unlimited replication ability, angiogenesis and metastasis. The fact that SCs have a great life span, the accumulation of mutations is more probable, and perhaps only mutations about self-sufficiency in growth and insensitivity to antigrowth signals from the niche are enough to originate a tumor (Li and Neaves, 2006).

Metastases is a complex process that implicate specific cellular and molecular mechanisms that appear as the result of a selective advantage for cells to metastasize after genetic mutations achieved in initial or advanced phase of tumorigenesis (Li et al., 2007b). It is very important to clarify factors that control the tropism of a single cancer type. It has been described that bone marrow mesenchymal cells home to tumor-specific pre-metastatic sites to provide a permissive niche for incoming tumor cells, mainly collaborating to stromal formation, especially angiogenesis (Xiong-Zhi et al., 2007). Interestingly, Kaplan et al. (2005) have demonstrated this relation between niche formation and metastasis in mice. They found that bone marrow-derived haematopoietic progenitor cells that express vascular endothelial growth factor receptor 1 (VEGFR1) migrate and constitute a tumor-specific pre-metastatic niche and cellular clusters before the arrival of tumor cells. The elimination of VEGFR-positive cells from bone marrow or using antibodies against the receptor could avoid the formation of pre-metastatic niches and tumor metastases (Kaplan et al., 2005).

Next, some important niches will be described to show the important collaboration between stem and stromal cells for the tissue homeostasis maintenance in brain, bone marrow, digestive tract, mammary gland, cornea and skin.

In adult mammalian brain there are multipotent neural SCs situated in the subventricular zone of the lateral ventricle (astroglia) and in the dentate gyrus that through a pathway mediated by proliferating precursors that produce neurons and oligodendrocytes. Investigators have hypothesized that neurogenic niche is composed by endothelial cells and some special astrocytes, that with specific signals conduct stem cells toward differentiation. For example, bone morphogenic protein (BMP) is necessary for the glial development, but the presence of BMP antagonist lead to neuronal differentiation (Ming and Song, 2005). The importance of niche has been manifested when glial progenitors isolated from adult spinal cord and transplanted into the adult subependymal zone were able to differentiate into neurogenic fate, which demonstrates the importance of the local niche influence (Ninkovic and Gotz, 2007). It has been proposed that brain cancer stem cells contains with a specific niche in the microvasculature, where endothelial cells interacts with them and supply some factors that maintain the self-renewing and undifferentiated state. Thus, in orthotopic brain tumor xenografts, rising the number of endothelial cells or blood vessels, incremented the numbers of cancer cells and accelerated the development of tumors, and antiangiogenic therapies that reduce tumor blood vessels, arrested tumor growth (Calabrese C, 2007; Yang ZJ, 2007).

Undifferentiated HSCs reside in the endosteal region of bone marrow where osteoblasts, fibroblasts, reticular cells, and adipocytes constitute the microenvironment that supports SC quiescence, self-renewal, differentiation and trans-marrow migration. The endothelium of bone marrow sinusoids develops an important function in the constitution of a scaffold for SCs that permit that migration. This vascular niche is present in other locations as liver and spleen, what explain the viable presence of HSCs in places lacking osteoblast support, but without replicative capacity. A clear evidence of the relevancy of osteoblasts for SCs maintenance has been demonstrated in studies where osteoblast ablation led to a clear reduction of SC number in the bone marrow and a rapid expansion of extramedullary hematopoiesis. It seems that osteoblasts secrete cytokines and growth factors, such as angiopoietin-1 and SC factors that promote stem-osteoblast intercellular tight contacts that maintain "stemness". On the other hand, glycoprotein osteopontin expression by the endosteal surface contributes to SC homing and trans-marrow migration (Kaplan et al., 2007). This factor can interact with different molecules present in the SC surface as CD44, βI integrins, antigen VLA-4 and VLA-5. The blockage of that contact increases their expression, while osteopontin absence with osteoblast stimulation leads to a spectacular SC increment (Scadden, 2007).

Several of those growth factors produced by osteoblasts have to be activated by parathyroid hormone, for that reason there is currently a clinical trial focused in the evaluation of the hormone in patients without autologous HSC mobilization, with the goal to improve immune reconstitution after cord blood transplantation (Ballen, 2007). The pharmacologic use of the hormone increases the HSC mobilization, protect them against exposure to chemotherapy and improve their expansion in transplant host in mouse (Adams et al., 2007). In experimental models it has been found that the adhesiveness of HSCs from aged animals is lower than young ones, and that may be due to an elevated activity of the small RhoGTPase Cdc42, what is translated to a less efficient interaction with stroma cells of the niche. This fact could mean a reduced self-renewal potential and an altered differentiation capacity related to the maturation of SCs (Geiger et al., 2007).

On the other hand, the maintenance of mesenchymal cells isolated from blood or bone marrow in vitro is difficult, and they get terminal differentiation easily. For that purpose Sun et al. (2007) have suggested a SC niche model for those cells based on an osteoclast centralized structure, that can lead expansion and proliferation with undifferentiated properties, without lost of ulterior differentiation ability, similar to the scheme found in the hematopoietic natural niche (Sun et al., 2007). The importance of the accurate in vitro culture of stromal mesenchymal cells is patent in studies that for example have shown the mesenchymal cell contribution to HSC division by means of the interaction with β(1)-integrins, that increase the asymmetrical divisions of the SCs and accelerate division kinetics (Gottschling, 2007).

In the small intestine there are special structures developed to increase the surface of contact with intestinal content called villi. In the points of insertion flow glands or crypts of Lieberkühn are tubular structures with a continuous epithelium between villi and crypts (Sancho et al., 2003). The maintenance of epithelial cells is based on the existence of multipotent SCs located in the base of the crypts that can originate all the cells present in that epithelium as are enterocytes, globet cells, Paneth cells and enteroendocrine cells. After division, SCs produce progenitor cells that proliferate while migrate toward villi and when they reach the base of villus they suffer cell cycle arrest and start the differentiation process, that at the tip of the villus achieve the maximal level. Then, after apoptosis induction, the terminal differentiated cells are desquamated. The base of the crypt represents the appropriate niche for the intestinal SCs that may be constituted by mesenchymal cells of the underlaying lamina propria in a basement membrane of extracellular matrix (Clatworthy and Subramanian, 2001). It has been hypothesized a quiescent intestinal SC niche that inhibits migration and increases cellular adhesion for the control of SC overproduction. Thus, enteroendocrine cells at the base of the crypt have been signed as critical elements for the development of that quiescent niche. They produce somatostatin, a potent peptide with anti-proliferative activity mediated by cyclin-dependent kinase inhibitors as p21, Cip1, p27 and Kip1, and in addition, somatostatin increases cellular adhesion to basement membrane components and inhibits cell migration (Radford and Lobachevsky, 2006). The intestinal epithelial cell turnover is very fast which joined to exposure of dietary toxins make digestive tract a location with very high cancer prevalence. SC mutations may develop a selective advantage that permits them to generate aberrant crypt foci by longitudinal division of crypts, through a process that is called niche succession (Leedham et al., 2005).

Recently it has been described a possible niche of multipotent quiescent SCs in the ducts of adult human mammary gland, that after division produce proliferative progenitors that are distributed along the lobules (Villadsen et al., 2007). Those ASCs would be responsible of mammary gland homeostasis during mammary development, pregnancy, lactation and involution, and in addition, could explain how some breast tumors preserve some stem-like characteristics. Moreover the presence of estrogen receptor-alpha negative CSCs in patients would clarify how some tumors are resistant to the therapy focused in the destruction of estrogen receptor-alpha positive cells (Villadsen, 2005).

SC population in the human cornea is located in a niche in a specific region of the limbus, at the Palisades of Vogt, which is pigmented for the existence of melanocytes, and is infiltrated with antigen-presenting Langerhan´s cells and suppressor T-lymphocytes. In that location, SCs interact with the cells of the stroma, through alpha 9 integrin and N-cadherin (Li, 2007a). The SCs divide asymmetrically and generate a progenitor cell that migrates centripetally from the periphery of the cornea to form a basal cell layer, and in addition they can get terminal differentiation to constitute the apical layer of cells which are desquamated during normal life. A great evidence of the presence of this niche is that surgical removal of the limbus leads to an invasion of cells from the conjunctiva, and grafting of cells from the limbus regenerates corneal like epithelium (Bentley et al., 2007).

In the skin, epidermal regeneration needs continuous epithelial cell interactions with the subjacent niche of connective tissue that permit permanent basal cell division and loss of differentiated cells in an equilibrated manner. During the migration from stratum basale toward surface, keratinocytes experience a complex differentiation process up to complete the cutaneous barrier and organogenesis (Boehnke et al., 2007). Other SCs are located in the bulge of hair follicles. They express some markers as keratin 15, 19 and Ki-67 proliferation marker among others. It has been suggested a lower activity of those SCs than SCs of other locations as intestinal epithelium. The lower bulb matrix also contains matrix cells with proliferative capacity but with lower replicative potential than upper bulb SCs, and it seems that the former are derived from the later that suffer a migration from the bulge to the bulb. The SC niche has been demonstrated by various authors with experiments that have shown how the SC properties are doted by the microenvironment surrounding the cells at the bulge, as an assay where the bulge SC population could be re-constituted after total cell death induced by plucking by means of progenitors from the lower region of the follicle, with a more commitment fate (Waters et al., 2007). On the other hand, in the skin there are other types of SCs including resident MSCs, blood vessel-associated haematopoietic and endothelial precursors, and migratory melanocytic SCs derived from the embryonic neural crest. It has been found that the later, has a niche in the follicle dermal papillae of hair and it is supposed that are laid down in the dermis during embryogenesis. The possibility of isolation of autologous neural precursors from skin is having great expectation due to its potential therapeutic use for cell replacement strategies, as the production of functional myelinating cells that could promote growth and regeneration of axons on injured central nervous system neurons. In addition, they could be used for the differentiation toward oligodendrocytes in spinal cord injury or multiple sclerosis (Fernandes et al., 2007).

5. Therapeutic Potential of Mesenchymal Stem Cells Differentiation

In the recent years, there is an increased experimental and clinical interest in understanding MSC biology, both because of their value as a model for studying the molecular basis of differentiation, and because of their therapeutic potential for tissue repair and immune modulation. Even with, there are many controversy and discussion of how to define MSCs since they characteristics differ among laboratories and species, and there is no specific marker or combination of markers that identify MSCs either in vivo or in vitro (Javazon et al., 2004).

It is notable that, contrary to most biological systems, human MSCs are better characterized that animal MSCs. This is probably due to the fact that MSCs are easily isolated and expanded from adult human tissue collected from healthy volunteers. Nevertheless, MSCs have been isolated from a number of other species like rat, mouse, horse, cow, pig, dog, sheep and baboon that provide a better opportunity to test their therapeutic potential in large mammal systems (Devine et al., 2001; Ringe et al., 2003; Rhodes et al., 2004; Silva et al., 2005; Moscoso et al., 2005; Bosnakovski et al., 2005; Dazzi et al., 2006). MSCs were first identified in the pioneering studies of Friedenstein and Petrakova (1966), in which they isolated bone-forming progenitor cells from rat marrow. Currently, identified sources for isolation of MSCs include the adult bone marrow (BM), compact bone, peripheral blood, adipose tissue, cord blood, amniotic fluid, and other fetal tissues (Luria et al., 1971;

Nakahara et al., 1991; Gronthos et al., 2001; Kuznetsov et al., 2001; Campagnoli et al., 2001; Charbord et al., 2002; O'Donoghue et al., 2003; in't Anker et al., 2003a,b).

Table 2. Methods of differentiation and characterization of MSCs

Differentiation to	Standard medium supplemented with	Time of treatment	Differentiation Markers
Osteogenic cells	Dexamethasone, ascorbic acid-2-phosphate, beta-glycerophosphate and calcium.	For up to three weeks	- Detection of calcium phosphate deposits by staining with Alizarin-Red S solution C or silver nitrate for von Kossa staining. - Expression of genes such as osterix, cbfa1, osteopontin, osteocalcin, bone sialoprotein
Chondrogenic cells	Dexamethasone, ascorbic acid phosphate and ITS+ supplement, (bovine insulin, transferrin, selenous acid), linoleic acid and bovine serum albumin, sodium pyruvate, proline, L-glutamine and TGF-beta1.	Two to three weeks	- Upregulation of chondrogenic markers such as collagen II, collagen XI, aggrecan, perlecan and syndecan. - Staining with Safranin-O, Toluidine blue or Alcian blue to highlight acid mucopolysaccharides, glycosaminoglycans and proteoglycans respectively.
Adipogenic cells	Dexamethasone, isobutylmethylxanthine, insulin and a PPARgamma agonist such as BRL 49653	For three weeks	- Staining with oil red O and the lipophilic dye nile red - Expression of adipsin, aP2 and PPARgamma - Enzymatic dosage of the Glycerol-3-phosphate dehydrogenase, a marker of the mature adipocyte
Myogenic cells	Demethylating agent 5-azacytidine / co-cultured with murine skeletal myocytes.	Two weeks after 5-azacytidine treatment	Expression of alpha1, alpha1B, alpha1D, beta1 and beta2 adrenergic and M1 and M2 muscarinic receptors
Neural cells	-Chemical treatments such as DMSO, N2 supplement, butylated hydroxyanisole, KCl, valproic acid and forskolin. - Growth factors such as fibroblast growth factor (FGF), neuronal growth factor (NGF) and epidermal growth factor (EGF). N2 supplement, butylated hydroxyanisole, KCl, valproic acid and forskolin	One week	Expression of nestin, Tuj1, neuron specific enolase, and GFAP.

The most well studied and accessible source of MSC is bone marrow, although even in this tissue the cells are present in a very small fraction. It has been estimated that MSCs represent 0.001–0.01% of the total population of nucleated cells in marrow (Pittenger et al., 1999). However, they can be isolated and expanded with high efficiency, and induced to differentiate to multiple lineages under defined culture conditions (Table 2). MSC form characteristic adherent, fibroblast-like colonies when grown in vitro in fetal calf serum and may be expanded further in the presence of various growth factors (Dazzi et al., 2006). The resulting cultures remain morphologically heterogeneous, containing cells ranging from narrow spindle shaped to large polygonal and, in some confluent cultures, tightly packed, slightly cubical cells. Another important property, but not defining feature, of MSC populations in vitro is their ability to form colonies after low-density plating or single-cell sorting (Javazon et al., 2004).

Phenotypically, MSCs express a number of non-specific markers, none of which individually, or in combination, has been shown to achieve high level enrichment. Currently, the most accepted profile for immunophenotyping MSC is the co-expression of CD105 (SH2) and CD73 (SH3 and SH4). More recently, the co-expression of CD166 and CD105 has been suggested to differentiate MSC from other mature mesenchymal cells (Alsalameh et al., 2004) although not always with success (Dazzi et al., 2006).

Actually, their multi-lineage differentiation capacity has been extensively characterized (Pedemonte et al., 2007). They remain able to proliferate and differentiate into tissues of mesenchymal origin, as well as to tissues of both endodermal and ectodermal origin thus allowing to speculate about their pluripotency (Wislet-Gendebien et al., 2005; Krampera et al., 2007) (Table 3). In addition, they are also an essential component of the stromal scaffold of the bone marrow providing the cellular microenvironment supporting system for HSCs. Whereas in some cases the ability of MSCs to provide newly generated tissues may be ascribed to 'reprogramming' of gene expression in MSC (Sato et al., 2007) in other situations it appears that MSCs act through differentiation independent mechanisms probably mediated by soluble factors (Togel et al., 2005). A well-characterized cell that appears to be distinct from, but perhaps related to the "classic" MSC is the MAPC, (Reyes et al., 2001; Dazzi et al., 2006). These cells are initially isolated together with MSC, but subsequently grow indefinitely in nutrient-poor medium. In specific conditioning media, MAPC can differentiate into cells which express markers of endodermal, mesodermal, and ectodermal origin. The relationship of MAPC to MSC is unknown. MAPC could be MSC progenitors or may even represent a cell population generated in vitro as an artifact which has no counterpart in vivo (Dazzi et al., 2006).

Table 3. Identity features of MSCs

Phenotype: Markers used by different research groups to characterize MSC from human BM but are not limited to: CD13/ CD29/CD31/CD44/CD54/CD63/ CD73/CD106/CD140b/CD166/CD105+/CD73+/STRO-1+/CD45.
Genotype: Co-expression of specific genes of multiple cell lineages.
Differentiation: Under appropriate in vitro conditions, they must differentiate into tissue of:
1. **Mesenchymal origin:** including bone marrow stromal cells, adipocytes, osteoblasts, tenocytes, myocytes and chondrocytes,
2. **Endodermal origin:** hepatocytes
3. **Ectodermal origin:** neural cells

MSCs can be readily obtained from patients for autologous treatment, do not generate immune response, and have a limited tendency to produce tumors. All this makes them a promising source of cells for many clinical applications in the involving field of regenerative medicine. It have been demonstrated that MSCs play a role in hematopoetic itself providing the scaffolding fraction of the bone marrow for HSCs to proliferate on (Dazzi et al., 2006). Furthermore, MSCs represent an advantageous cell type for allogenic transplantation as evidence suggests that MSCs are immune-privileged, with low MHC I and no MHC II expression, therefore reducing risks of rejection and complications for transplantation. MSCs have also been found to be immunosuppressive, through a mechanism that involves paracrine inhibition of T- and B-cell proliferation (Uccelli et al., 2006).

MSCs represent one of the few ASC populations already in use in the clinic. Several studies have demonstrated the possible use of MSCs in systemic transplantation for systemic diseases, local implantation for local tissue defects, as a vehicle for genes in gene therapy protocols or to generate transplantable tissues and organs in tissue engineering protocols (Mustapha, 2004).

Some striking examples of the therapeutic use of marrow-derived MSCs have been reported recently. These address a broad spectrum of indications, including cardiovascular repair (Wollert et al., 2004; Xu et al., 2006), treatment of lung fibrosis (Ortiz et al., 2003; Rojas et al., 2005), brain (Hellmann et al., 2006), spinal cord injury (Satake et al., 2004) and bone (Quarto et al., 2001) and cartilage repair (Wakitani et al., 2002; Sottile, 2006).

6. Therapeutic Vasculogenesis or Tumor Targeting by Endothelial Progenitor Cells

Until about 1990, the postnatal vascular endothelial was considered to be an inactive tissue. In case of trauma, the repair of damage endothelium would occur by pre-existing endothelial cells (ECs) proliferating from margins of the endothelial wound towards the center of injured area, although resident ECs proliferate infrequently. But this concept of endothelial repair changed after Asahara et al. described in 1997 the existence of a subpopulation of circulating progenitor cells, denominated endothelial progenitor cells (EPC), derived from bone marrow HSCs and that can differentiate into mature EC, which contributed to repair the damage endothelium participating in the physiopathological process of vasculogenesis and angiogenesis. The origin of circulating adult EPC is the bone marrow, while the circulating adult ECs are detached from the vessels walls after vascular trauma and enter to circulating blood.

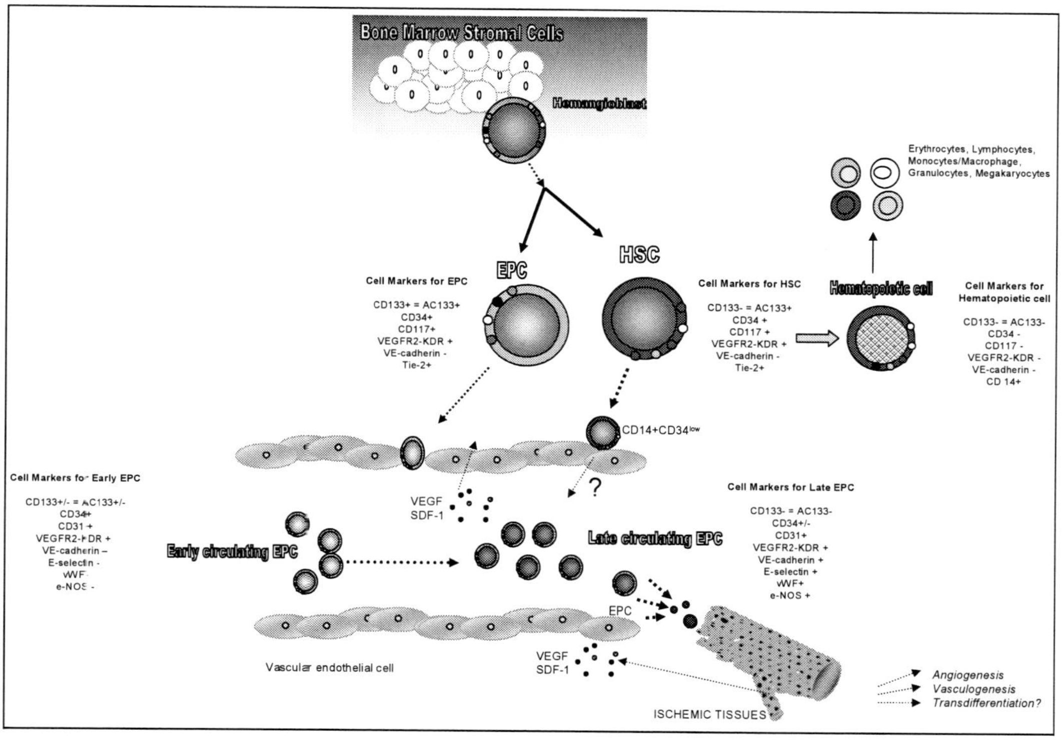

Figure 6 Endothelial progenitor cell are derived from the hemangioblast, it is mobilized into peripheral blood and are recruited by ischemic tissues to participate in vasculogenesis and angiogenesis physiopathological proccess. EPC, endothelial progenitor cells; HSC, haematopoietic stem cells; KDR, kinase insert domain receptor; e-NOS, endothelial nitric oxide synthetasa; SDF-1, stromal derived factor 1; VEGF, vascular endothelial growth factor; vWF, von Willebrand Factor.

Several angiogenic growth factors have been identified and are implicated in vasculogenesis and angiogenesis process. The physiological angiogenesis is regulated by the balance between proangiogenic (VEGF, bFGF, etc.) and antiangiogenic growth factors (endostatin, angiostatin, thrombospondin-1, etc). This balance is altered in pathological conditions such as in cancer patients, towards to a proangiogenic phenotype that promove the angiogenesis shift and the increase of tumor neovascularization (Tonini et al., 2003; Gonzalez et al, 2004). Although several cytokines and growth factors has been identificated to influence in angiogenesis process, the VEGF is the main activator of angiogenesis known. EPC mobilization from bone marrow and recruitment by ischemic tissues have been positively correlated with a increase of circulating VEGF levels, but other cytokines as stromal cell-derived factor 1 (SDF-1) and angiopoietin-1 can also be implicated (Shintani et al, 2001; Moore MAS et al, 2001) (Figure 6).

The definition of EPC is complex due to the lack of a single specific marker and different cell populations are termed EPC. The term "EPC" may define a group of cells existing in a variety of stages ranging from hemangioblasts to fully defined adult ECs (Ribatti, 2007). The hemangioblast is a common precursor for both HSCs and EPCs. CD34 is the common surface marker for cells of the hematopoietic origin, although is also is expressed on membrane of capillary ECs. With the use of CD133, an antigen specifically identifying human primitive

SCs, it is possible identified a subpopulation of stem and progenitor cells, since expression of CD133 is rapidly dowregulated when hematopoietic stem or progenitor cells differentiate into mature cells. A series of surface receptors of haematopoietic stem/ progenitor cells and ECs appear early in development of EC from EPC, such as CD34, CD133, VEGF receptor 1 (fms-like kinase, flt-1) and 2 (KDR, kinase insert domain receptor; Flk-1, fetal liver kinase receptor in mouse), vascular endothelial cadherin (VE cadherin), platelet endothelial cell adhesion molecule 1 (CD31), CD146, von Willebrand factor (vWF), Tie-1 and Tie-2 that have been used to defined the phenotype of EPCs by flow cytometry and/or immunocytochemistry techniques (Figure 6). Additionally, two different EPC subpopulations have been described, denoted as early and late EPCs. While early EPCs have a peak growth in culture after 2-3 weeks and secrete a number of angiogenic activators and inhibitors growth factors, and neuroregulatory cytokines, expressing the surface markers CD34+ CD133+/- and KDR+; the denominated as late EPCs appear in culture after 2-3 weeks and can be maintained in culture up to 12 weeks and on surface expressing CD34+/- CD133- KDR+ and typical markers of EC such as VE cadherin, vWF or E-selectin (revised by Ribatti, 2007 and Brixius et al., 2006). Other authors have described a circulating cellular subpopulation CD14+CD34^{low} (monocyte phenotype), that can be other source of circulating EPCs; these CD14+CD34^{low} cells proliferated in response to SCs factor and can differentiated in vitro into EC, but also into osteoblasts, neural cells or adipocytes (Romagnani et al., 2005). Recently, Lopez-Holgado et al. (2007) have observed in vitro the formation of EPC colonies after 7 days using EPC medium supplemented with insulin growth factor (IGF) belong to the human cells with a monocyte/macrophage phenotype.

Alternatively, EPCs have been identified by their high clonogenic and proliferative potential in adequate cell culture. In vitro, bone marrow derived or peripheral blood mononuclear cells had been isolated and can be differentiated in an endothelial cell-like phenotype when they are cultured in the adequate culture medium for EC, supplemented with growth factors as VEGF, bFGF or IGF that induce the differentiation of EPC to EC-like phenotype. The enumeration of number of EPC by the measurement of colony forming units (CFU-EC) has emerged as useful instruments in recent years (Herbrig et al., 2006; Werner and Nickening, 2006). The EPCs in vitro have been defined by double positive cells for both the *lectin Ulex europeaus agglutinin* (UEA) and the acetylated low-density lipoprotein (LDL) cholesterol.

6.1. Endotelial Progenitor Cells, Vasculogenesis and Cardiovascular Diseases

Vascular trauma and tissue ischemia induce the production and release of several cytokines, including several proangiogenic growth factors such as VEGF and bFGF that foster the mobilization from bone marrow of progenitor cells such as EPC (Gill et al., 2001). Moreover, they regulate the recruitment and homing of several circulating inflammatory cells and EPC to the site of injured vascular tissue or ischemic tissue area, promote the angiogenesis and vasculogenesis process and accelerate vascular or tissue wound healing. A decreased of the number and functionality of circulating EPC has been associated with the presence of cardiovascular diseases and are associated with several cardiovascular risk factors as circulating lipids levels, hypertension, diabetes mellitus or smoking (revised by Shantila et al., 2007). Mobilization of EPCs to peripheral has been observed in patients after acute myocardial infarction (Shintani et al., 2001) and an inadequate coronary collateral development is associated with reduced numbers of circulating EPC in patients and impaired of their chemotactic and proangiogenic activity (Lambiase et al., 2004). In animal models, the EPCs develop an endothelial phenotype in culture and they incorporated into developing collaterals of ischemia and thereby enhance neovascularization (Murohara et al, 2000). The number of circulating EPCs has been demonstrated as a useful tool to predict cardiovascular outcomes in coronary artery diseases patients (Werner et al., 2005).

Several clinical trials using granulocyte monocyte colony-stimulating factor (GM-CSF) or erythropoietin have shown the mobilization of EPCs, but different side effects have been observed in the patients after the use of these drugs (Brixius K et al, 2006).

On the other hand, in vitro models demonstrated that progenitor cells from human origin co-cultivated with rat cardiomyocytes can transdifferentiated to cells expressing cardiac genotypic and phenotypic (Rangappa et al., 2003). It is well known that isolated human EPCs, in culture in supplemented medium with adequate cytokines differentiated to ECs. Recently, it has been described that human CD133+ cells isolated by a clinical-grade immunoselection method from cord blood can differentiate in vitro into ECs and cardiomyocyte-like cells (Bonanno et al., 2007). The mechanisms by which adult EPCs acquire the cardiomyocyte phenotype are not well defined. Differentiation, transdifferentiation and fusion have been proposed as possible mechanisms. Human EPCs isolated from healthy subjects and coronary artery diseases patients also can transdifferentiated in vitro to functionally active cardiomyocytes (Badorff et al., 2003). Co-culture of circulating EPCs isolated from human blood with neonatal or adult rat cardiomyocytes to mimic the cardiac microenvironment, triggered the expression of cardiac genes and cardiac marker proteins, such as myosin heavy chain, within the EPC from human origin. Moreover, it has been observed the expression of specific cardiac proteins, when rat cardiomyocytes were fixed with paraformaldehyde, indicating the necessary presence de some proteins and/or glycoproteins of membrane for the acquirement of cardiomyogenic phenotype by human EPCs. Afterwards, Koyanagi et al. (2005) have observed in co-culture studies the participation of E-cadherin and the formation of intercellular nanotubular structures that allow in the transport of proteins and/or organelles from the cardiomyocyte to the human EPCs. This communication between human EPC and rat cardiomyocyte may contribute to the acquisition of cardiomyogenic phenotype by human EPCs isolated from healthy subjects. The process of cell fusion origin hybrid cells co-expressing cardiomyocytes and EPC markers, it has been proposed as an explication of this observation. However, others authors do not observe evidence of human EPC transdifferentiation to cardiomyocytes using co-culture methods (Gruh et al., 2006). However, recently Murosawa et al. (2005) in animal models of myocardial infarction have demonstrated the transdifferentiation "in vivo" of human EPC isolated from peripheral blood of healthy subject to cardiomyocytes. Moreover, these authors observed that human EPC injected in the tail of rat, homing to ischemic myocardial area of infarcted tissue, and transdifferentiated to cardiomyocytes, endothelial or smooth muscle cells depending of the microenvironment in the area where EPC are localized.

Although clinical studies using EPC have demonstrated moderate improvement of cardiac function (reviewed by Shantsila et al., 2007), clinical applications employing EPCs may be future possibilities for the treatment of cardiovascular diseases due to their growth and differentiation potential, and their ability for vasculogenesis and angiogenesis.

6.2. Endothelial Progenitor Cells and Tumor Angiogenesis

Tumor angiogenesis allows the formation of new vessels that are essential for tumor growth and metastasis, and the prognostic value of the measurement of angiogenesis activity in the tumor or the circulating proangiogenic growth factors levels it has been described in cancer patients (Giatromanolaki et al., 2006). Although it is known that the angiogenesis process in the tumor is regulated by the balance between activators and inhibitors angiogenic factors, several authors have shown that VEGF-dependent mobilization of bone marrow derived EPCs can contributed to tumor neovascularization and impairing the recruitment of circulating EPCs to neovasculature it is possible to block tumor angiogenesis and growth (Ribatti, 2004). The incorporation of circulating EPCs into tumor vasculature has been described in human lung cancer (Hilbe et al., 2004) and an increase in the circulating EPCs number has been detected in cancer patients (Dome et al., 2006). In the recent years, potential therapeutic benefits of antiangiogenesis therapy have lead to the development of several new anticancer drugs that are permitted their commercial use such as bevacizumab, the first anti VEGF humanized monoclonal antibody commercialized, or are in various stages of clinical investigation. In order to determine the diagnostic and/or prognostic value of the measurement of angiogenic activity in cancer patients and to assess the antiangiogenic activity of angiogenesis inhibitors drugs, it is necessary the development of new methods and surrogate markers (Gonzalez et al, 2007). Bertolini et al. (2006) have proposed that the quantification of circulating EPCs by flow cytometry multiparametric procedures has potential as a surrogate marker for monitoring drug activity of antiangiogenic agents. It remains to be determined whether the determination of the number of circulating EPCs has any prognostic and/or predictive value or if it is a good surrogate marker to evaluate of response to antiangiogenic therapy in cancer patients.

7. Reprograming Adult Cell Differentiation as Therapeutic Tool

One important characteristic of the tissue-specific SCs is in fact that they are considered to be pluripotent, thus able to generate mature cell types of one of more lineages, but not capable to reconstitute a whole organism. This opens a new perspective for the clinical use of these cells. Not only to regenerate the tissues where they originate but to regenerate other tissue types. First the ASCs should be able to differentiate into a different tissue specific SC. It has been proposed that they do so by transdifferentiation (Holden and Vogel, 2002). Although these studies support the idea of tissue-specific stem cells plasticity controversial results have also emerged (Wurmser and Gage, 2002). There are some authors that postulate cell fusion event as a more likely mechanism than transdifferentiation to explain SC differentiation plasticity (Terada et al., 2002). But no evidence from *in vivo* experiment that cell fusion occurs has been proved. Moreover, recent clinical observations have shown promising data that backup ASC transdifferentiation (Krause et al., 2001; Körbling et al., 2003; Luk et al., 2005).

The therapeutic potential of tissue specific-SC is clear (Vats et al., 2005), bone-marrow SCs are proved to be the most versatile among ASCs (Pomerantz and Blau, 2004), they can become neurons (Weimann et al., 2003); hepatocytes (Oh et al., 2007); skeletal muscle cells (Dezawa et al., 2005); cardiomyocytes (Jackson et al., 2001) and ECs (Reyes et al., 2002). In addition, there are many others examples, as neuronal SCs that can convert into skeletal and cardiac muscle cells (Crarke et al., 2000); liver SCs, called oval cells, which can turn into pancreatic islet cells (Yang et al., 2002); and adipose tissue SCs, which can give rise to cartilage, bone, neurons and muscle cells (reviewed in Satija et al., 2007).

Although the great differentiation potential of ASCs, the difficulties of their therapeutically use should not be underestimated. ASCs are usually difficult to obtain from living donors. The number of those cells in the tissues seems to be very low, and may decrease with age and in pathological conditions. In addition, the isolation and the culture of ASCs have proved to be a very difficult task. Isolated ASCs immediately start growing into their designated tissue type, which limits their replication (Boyer et al., 2005). The struggles of ASCs manipulation are trying to be overtaking by various biotechnology companies (Collas and Hakelien, 2003) and word wide research institutions and many advances have been made so far using bone marrow; muscle; liver; skin; heart and fat ASCs.

Transdifferentiation can be defined as the conversion of an already adult differentiated cell into another type of differentiated cell (Tosh and Slack, 2002; Slack and Tosh, 2001); while metaplasia enclose the switch between one cell type to another (including conversion between tissue-specific stem cells). There is some controversial using these two terms; here we use transdifferentiation as a subset of metaplasia (Figure 7).

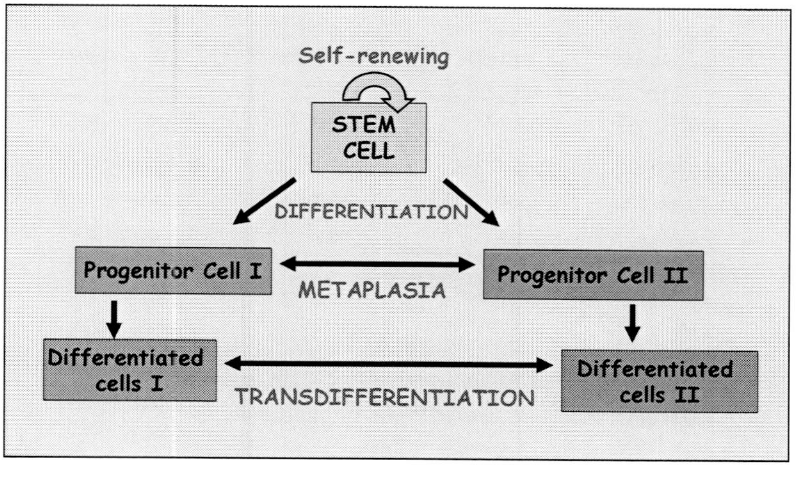

Figure 7. SCs are pluripotent with the ability of self-renewing; they can differentiate by asymmetrical division into progenitor cells. The switch between two different progenitor cells is termed as metaplasia. Already differentiated cells change to another differentiated cells by transdifferentiation.

No matter which definition is applied, the study and understanding of both, transdifferentiation and metaplasia are of vital importance in the field of the regenerative medicine. Adult differentiated cells are always available, are easily to grow in culture and can be obtained from the own patient. Thus, if we could convert them to the cell type needed to cure a disease this will means to overtake the problems of specific tissue SC availability. The cue now is to determine, how cells can change their phenotype and function; which are the master switch genes involve; and how can we reprogram adult cells differentiation.

The switch between adult differentiated cells must involve the down-regulation of one cell type genes and the up-regulation of the other cell type genes, implying that is some stage there must be a co-expression of both cell types' genes. This could end to a stage in which cells present a hybrid phenotype. To be of therapeutical use, transdifferentiated cells should be completely characterized once they have acquired a stable phenotype and a functional determination. In addition, these cells should be genetically labeled to ensure their traceability in the organism (Tosh and Slack, 2002).

A good documented example of transdifferentiation occurs between pancreatic and hepatic adult cells and *vice versa* (Burke et al., 2006; reviewed in Burke et al., 2007). The ability to interconvert between liver and pancreas reflects the close developmental relationship between these two tissues and supports the theory that transdifferentiation is driven by one or two of the so called master-switch genes that distinguish the two tissues during development (Slack and Tosh, 2001). The identification of those master switch genes will contribute to direct SC differentiation towards a hepatic or pancreatic phenotype.

Others examples of transdifferentiation are the conversion of osteoblasts into adipocytes. This has been achieved manipulating culture conditions and also inhibiting the gap junctions via 18-α-glycyrrhetinic acid (Schiller et al., 2001). The transdifferentiation of endothelial cells to cardiomyocytes has as well been proved (Condorelli et al., 2001). The plasticity of endothelial cells might create possibilities for autologous cell replacement. In addition myoblast can turn into adipocytes regulating some paracrine and autocrine factors that control cell growth and cell fate (Ross et al., 2000).

Finally, we will discuss the use of reprogramming somatic cells into SCs, which could then be redirected to become specific cells types. Somatic cell nuclear transplantation (SNCT) is a process were adult somatic cells undergo reprogramming after fusion with a mature oocyte. A functional embryo is produced which can indeed develop to term in live animals. ESCs have also been derivate from this created blastocyst (Wakayama et al., 2001). The aim of nuclear transplantation is direct the blastocyst to development into specific tissues that could be used fro therapeutic proposes. But nuclear transfer potential applications are limited by low efficiency and by the possibility of occurrence in later development of gene expression aberrancies (Tian et al., 2007).

Here we have focus our attention on the progress being made in cell therapy as a branch of regenerative medicine. The used of reprogrammed cells for cell replacement is a succulent challenge for biomedical researchers and might contribute greatly to achieve whole tissue reconstruction. Despites all the progress being made so far, little is still known about the initial steps and mechanisms leading to another differentiation pathway, about the hierarchy of molecular changes involved in switching to another program, and about the stability of these alterations in the progeny cells. Interaction with the microenvironment probably determines the long-term outcome of the re-programmed cells. Tissue integrity is reached by cell to cell adhesion and junctional complexes, transdifferentiated cells have to be able to maintain their new reprogrammed identity and to produce specific signaling to interact with neighbor cells. Studying and understanding the molecular signaling and pathways that govern organ development is essential to understand the routes that should follow reprogrammed cells to become tissue and be a reality as potential therapy in regenerative medicine. Summarizing, cells identity can be redirected by in vitro manipulation but more in vivo testing is still required before cell therapy will be a reality.

8. Multidrug Resistance and Cancer Stem Cell

CSCs have biological unknown characteristics but it has been showed that it is generally present some mechanism to avoid the death of the cell. There are several molecular mechanisms for the resistance to death of cancer stem cells such as CSCs in G_0 which are resistant to cell cycle–specific chemotherapy agents (Venezia et al., 2004); cancer stem cells resistant to DNA-damaging agents by enhanced DNA repair (Park and Gerson, 2005); SCs which express higher levels of antiapoptotic proteins such as members of the Bcl-2 family and inhibitors of apoptosis (Wang et al., 2003) or SCs expressing high levels of transporter proteins, such as P-glycoprotein (P-gp). In relation to this last mechanism, it is clear that understanding the central role played by multiple drug resistance (MDR) transporters in the protection and self-renewal of normal and CSCs may allow us to identify differences that can be exploited therapeutically. The most recent advances in cancer etiology and treatment suggest that genes related with the development of MDR in neoplastic cells and their expression generating some cellular transporters are also an essential feature that enables tumor stem cells to circumvent therapy.

8.1. Multiple Drug Resistance: an old Problem in Cancer Chemotherapy

Nowadays, it is taking place an intense research in relation to the transporter proteins expressed in the CSCs. But the knowledge of the importance of the resistance drugs in the cancer treatment is a very old and it keeps on being a major obstacle for the successful treatment of cancer. Numerous factors affect efficacy of chemotherapeutic drugs. However, the correlation between cytotoxic therapy and the development of MDR remains a problem in the tumors chemotherapy because sequential selection of cell subpopulations could result in the preferential out-growth of resistant cells. In fact, even when some benefit has occurred during cancer treatment, initial response to therapy is often superseded by the acquisition of drug resistance (Tsuruo, 2003). This ability of a small population of malignant cells to survive exposure to lethal drug doses and to develop resistance to those agents has determined the very nature of the practice of oncology. Despite the successes of chemotherapy, it is frequently obtained only temporary remissions from the disease as the same therapy that is effective in obtaining the remission has often proved ineffective at the time of relapse. The combination of drugs used in treatment programs, the duration of protocols and the dose and frequency of administration of antineoplastic agents have been determined, in part, by strategies designed to reduce the possibility of emergence of a resistant population of cells. One basic distinction identifies whether a neoplastic cell is resistant to chemotherapy at the time of its malignant transformation or whether it becomes resistant only after exposure to therapeutic cytotoxic agents (acquired resistance). Malignancies that exhibit de novo resistance are, by definition, refractory to chemotherapy and carry the worst prognoses. However, and as it will be seen afterwards, the observations related to the presence of resistance mechanism in CSCs are modifying the general idea about the inherent and acquired resistance in cancer (Dean et al., 2005).

8.1.1. Some Genes are Related to the Drug Resistance Mechanism in Cancer

The understanding of mechanism of drug resistance has increased in a fast way over recent years. Numerous genes have been related to the development of drug resistance. Many of these mechanisms may co-exist in vitro in human tumors; some are a feature of cell lineage while others appear de novo during disease progression.

Classically, MDR phenomenon was associated of a 170 kD protein, P-glycoprotein (P-gp), the product of the mdr1 gene (Callaghan et al., 2006). P-gp functions as an energy dependent transport protein which pumps out cytotoxic agents. The expression of this protein leads tumor cells to show cross-resistance to a wide variety of other structurally unrelated drugs including vinca alkaloids, anthracyclines, epipodophyllotoxins, colchicine, actinomycin D, taxol and taxotere (Ambudkar et al., 2003). Besides its resistance function, mdr1 gene has been implicated in the differentiation of tumor cells and connected with modulation of MHC molecule expression (Prados et al., 2006). This protein is a member of a family which has a physiological function in cellular detoxification and secretion mechanisms. Different classes of P-gp cDNAs have been identified in human cells which were generated by alternative splicing patterns. Differential expression of various P-gp genes or different transcriptional modifications of their RNA products result in varying levels of drug resistance (Loo and Clarke, 2005). Subsequently, a strong correlation between a high level of mdr1 gene expression -the gene encoding P-gp- and resistance has been observed even without amplification of the gene. Moreover, an increased expression of mdr1 gene has been observed in samples from patients both at levels of mRNA and the protein product (Prados et al., 1996, 1998). P-gp expression is apparently regulated on many different levels such transcriptional and translational regulation, protein stability and amplification. This last is an important mechanism of genetic change leading to overproduction of proteins encoded by these genes. In fact, in cells selected for resistance to cytotoxic drugs, mdr1 amplification is often amplified (Nielsen and Skovsgaard, 1992). However, drug resistance can arise through mechanisms other than amplification. Studies in cell lines resistant to different drugs which showed mdr1 amplification, demonstrated that levels of mdr1 gene expression were higher than those that would have been expected from the levels of amplification. For example, in human ovarian carcinoma, the finding of MDR cell lines in which P-gp was increased in the absence of higher copy numbers or increased mRNA concentrations suggests increased protein stability or translation (De Vries et al., 1993).

Besides P-gp, others members of the ATP-binding cassette (ABC) family of have been identified and involved with the multidrug resistant phenotype. The multidrug resistance-associated protein (MRP1) and breast cancer resistance protein (BCRP, MXR, ABCG2) are another members of the same superfamily. Both mediate multidrug resistance by virtue of energy-dependent efflux (Gottesman et al., 2002). MRPs transport anionic drugs, such as methotrexate and neutral drugs conjugated to glutathione, glucuronate or sulfate. The MRP1 gene was found to be overexpressed in several other P-gp negative cancer cell lines (Barrand et al., 1994) that were also resistant to multiple drugs. In out of the 10 known members of the MRP family, only MRP1 may be useful clinically (Borst et al., 2000). BCRP is more biologically complex than P-gp and MRP-1 because it is an ABC half-transporter that requires multimerization for function. This particular drug-resistance protein appears to be associated with resistance to anthracene-type drugs, such as doxorubicin and mitoxantrone (Abbott, 2006). In 1993, Scheper et al., described LRP in non-P-gp drug-resistant cell lines. Since its original description, LRP has been identified as the major vault proteins. The structure and localization of the vaults have led to the speculation that they can mediate the bidirectional transport of a variety of substrates between the nucleus and the cytoplasm. Today we know that LRP expression is associated with a redistribution of drugs from the nucleus to the cytoplasm without changes in total drug intracellular concentrations (Steiner et al., 2006).

Other MDR mechanisms have been studied in last years including altered target enzyme, decreased drug activation, increased drug degradation due to altered expression of drug-metabolizing enzymes, drug inactivation due to conjugation with increased glutathione, subcellular redistribution, drug interaction, enhanced DNA repair and failure to apoptosis as a result of mutated cell cycle proteins such as p53 (Melguizo et al., 2006). For example, topoisomerases, enzymes necessary for DNA replication, catalyze changes in the secondary and tertiary structures of DNA. Most attention has focused on topoisomerase II, an enzyme that is the target of antineoplastic drugs that act as DNA-intercalating agents. Cell lines resistant to different classes of antineoplastic agents in the absence of any apparent P-gp gene expression have abnormal topoisomerase II activity. This new «atypical MDR» revealed that drug influx and steady-state intracellular drug concentrations were no different than in the parental drug-sensitive cells, in contrast to cells with the «classic» MDR phenotype, which invariably show changes in drug transport (Lage et al., 2000). On the other hand, increased amounts of anionic glutathione S-transferase (GST) mRNA have been found in tumors such as colon cancer which exhibits de novo resistance to anticancer agents and showed that GST expression is a marker for inherent drug resistance (Beaumont et al., 1998). Finally, cytochrome P-450 is a constitutive and inducible enzyme responsible for the catalytic hydroxylation of a multitude of compounds with diverse chemical structures and mode of action linked only by their lipophilic nature. The xenobiotic-metabolizing P450 enzymes have a central role in the oxidative metabolism of a wide variety of endogenous and exogenous compounds including environmental carcinogens, plant toxins, and a wide range of anticancer drugs. This last point has been related with it intervention in MDR in cancer (Oyama et al., 2004). In conclusion, studies on mechanisms of cancer drug resistance have yielded important information about how to circumvent this resistance to improve cancer chemotherapy and have implications for pharmacokinetics of many commonly used drugs.

8.2. Stem Cells, Cancer Stem Cells and Transporter Proteins

In last years, the research in relation to the potentiality and plasticity of SCs that mediate tissue repair and maintenance has acquired an enormous importance. It is generally accepted that normal stem cells show properties that provide for a long lifespan such as relative quiescence, an active DNA-repair capacity, a resistance to apoptosis and resistance to drugs and toxins through the expression of several ATP-binding cassette (ABC) transporters. In this context, the analyses of the MDR transporter activity, has provided the principal means to recognize and purify such tissue SCs (Giangreco et al., 2004). This transporter activity has been detected in the form of a small cells termed side population (SP) in murine bone marrow cells (Goodell et al., 1996). Studies with specific substrates for the MDR transporters ABCG2 and ABCB1 have demonstrated a constitutive expression of these proteins in SP which was responsible for exclusion. These results have been confirmed by mdr1 transfection in normal murine marrow. Analysis of these transfected cells showed a 100% increase in SP cells phenotype (Bunting et al., 2000). In addition, the generation of the ABCG2 knockout abrogated the SP (Zhou et al., 2002). It is not clear the relevance of the MDR transporter in adult tissue SCs. However, it is possible that these proteins represent a mechanism to survive toxic insults that are able to eliminate progenitor cells.

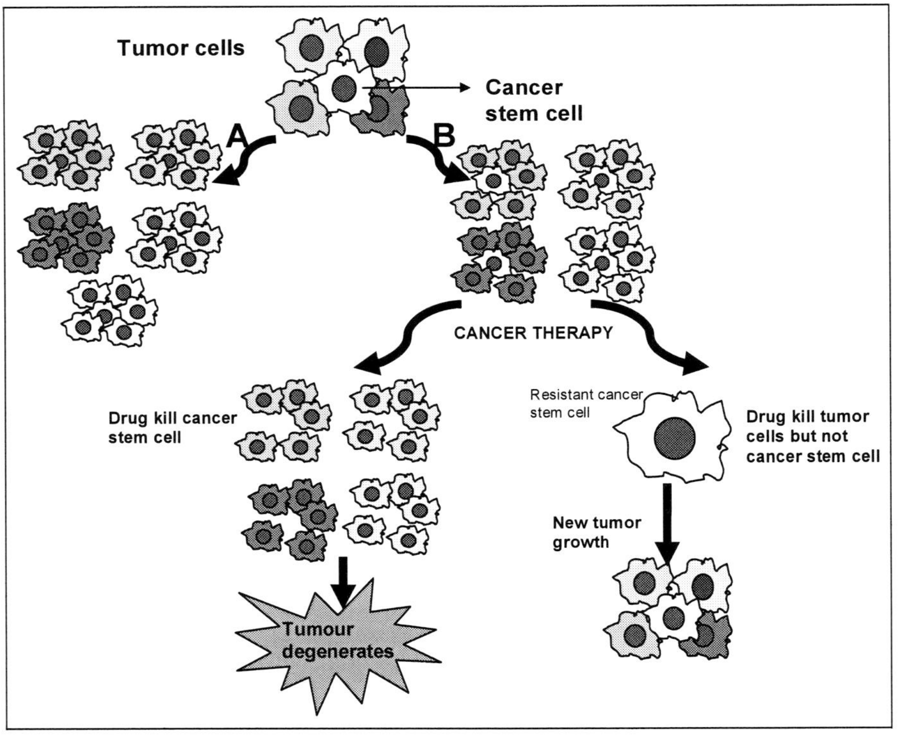

Figure 8. A simplified model of tumor growth theories. In the traditional neoplasm development all tissue tumoral cells can form new tumors (A). In the CSC theory only this cell is able to generate new tumors (B). After exposition to antineoplastic agents specifically target CSCs, tumor degenerates, but when cancer stem cells show drug resistance, after a remission due to the drug-sensitive tumoral cells elimination, resting tumor SCs are tumor growth back responsible.

Interestingly, CSCs shares many properties of the normal SCs. One of this is the presence of the resistance mechanisms. It seems that one of the defining characteristics of adult tissue SCs may be their constitutive resistance to environmental toxins, including most chemotherapeutic agents. The importance of the MDR mechanism in the normal tissues and neoplastic cells and the drug resistance originated in the SC, provides an explanation for the successes and failures of the cytotoxic antineoplastic therapy.

In this CSC hypothesis, SCs have constitutive MDR activity. When they entry into the cell cycle, MDR is transiently downregulated allowing that the cell acquire a DNA alteration. In this asymmetrical division two cells are originated; one is a MDR-protected resting SC and the other a drug-sensitive amplifying progenitor cell with high proliferative capacity. The continuous accumulation of a genetic instability by the damaged SC result in the apparition of a neoplasia which is generated by the mitotically active, drug-sensitive amplifying progenitor cells while the CSC remains protected and rarely enters in the cell cycle. When we developed a cytotoxic treatment against tumor cells, we can eliminate the drug-sensitive tumor cells but not resting tumor SCs (Figure 8). Continuous exposures to drugs lead to the overexpression of MDR transporters which generate a tumor drug resistance phenotype. So, the clinically relevant target of therapy is a resting cell with drug resistance that is not dependent on therapy-induced gene duplication or translocation (Tan et al., 2006).

On the other hand, we can explain the observation that some cancer develop an acquired resistance during the cytotoxic therapy while in others appear resistant from the start to the treatment. In both, the resting CSCs have innate drug resistance. The observation of an acquired drug resistance may be related to patient prognosis but it is not the cause for cancer recurrence. In fact, the major barrier to therapy is the quiescent tumor SC with constitutive MDR. Innate and acquired drug resistance, cross-resistance and the failure of MDR reversal strategies may be now reinterpreted by CSC hypothesis. Firstly, drug resistance is an innate characteristic of the resting tumor SC. The CSC expresses constitutive MDR activity, which is independent of drug exposure and it is downregulated in more differentiated tumor progeny. It has been proposed that selective pressure imposed by chemotherapy leads to both mutation and secondary genetic changes, including MDR upregulation. However, unless these changes occur in the self-renewing tumor SC compartment, the limited proliferative capacity of the bulky tumor ensures that they are self-limiting (Leonard et al., 2003). Secondly, the discovery of the molecular mechanism of cross-resistance allowed developing some agents with the objective to block them (verapamil or cyclosporine). However, the first in vivo experiences using these compounds have not showed to have efficacy. For example, in ovarian cancer patients treated with verapamil there was no evidence of enhanced response to adriamycin (Fojo et al., 1987). In a similar way, in a murine model, verapamil failed to increase the sensitivity of resistant P388 lymphoma cells to adriamycin (Rustum et al., 1986). This limitation for potentiating antineoplastic agents has also been observed in the second- and third-generation MDR inhibitors such as PSC 833, GF120918, VX-710 (Biricodar), and LY335979 and may be explained in part by multiple and redundant cellular mechanisms of resistance. So, MDR reversal agents can be used to increase the plasma concentration of a variety of antineoplastic agents but not to increase their therapeutic index (Tan et al., 2000).

8.3. Therapeutic and Biological Implication of the Cancer Stem Cells MDR

Most traditional cancer treatments have been developed and assayed based on their ability to kill most of the tumor population. These treatments, however, do not take into account the CSCs which have been shown in several tumor types to be more resistant to standard chemotherapeutic agents. This model explains why standard chemotherapy may result in tumor response, but most tumors recur, likely because the CSC survives and regenerates the neoplasm. Studies in CD34+ CD38- leukaemic SCs showed that they were significantly less sensitive to daunorubicin or cytarabine than the bulk population of leukaemic blast cells (Costello et al., 2000). A recent report examining mechanisms of radioresistance in gliomas showed that the fraction of CD133+ CSCs within gliomas is enriched after radiation, because these CSCs have enhanced capacity for DNA repair (Bao et al., 2006). In addition, Matsui et al. (2004) have shown that myeloma CSCs are more resistant to chemotherapy used to the classical treatment of myeloma. A prediction of the model shown it is that treatments targeted specifically to the CSC population will be required to result in an effective cure of cancer.

But, knowing the phenotypic and functional similarities between normal SCs and CSCs, can we target CSCs without harming normal SCs? Two recent reports have suggested that selective targeting of CSCs may be possible. The investigation of the role of the tumor suppressor gene PTEN in leukemia showed that a conditional deletion of PTEN in adult hematopoietic cells in mice lead to expansion of the leukaemic population of CSCs and depletion of exclusion normal HSCs (Yilmaz et al., 2006). Conversely, treatment of the leukaemic mice with the drug rapamycin, which functions to reverse the effects of PTEN deletion, blocked development of CSCs (and subsequent tumors) in these mice, but had no deleterious effect on the normal SC population. In a separate report, targeting of the CSC surface molecule CD44 using a monoclonal antibody resulted in eradication of human acute myeloid leukaemic SCs using a xenograft model in immunocompromised mice while sparing normal SCs (Jin et al., 2006).

Thus, every time is getting more relevance the hypothesis that cancer development and propagation is based in the presence of tumorigenic CSCs. Although not all the signaling pathways in the development of these cells are known, it is possible that the presence of drug resistance mechanism is of great importance in the obtained results to standard forms of therapy. The development of more effective cancer therapies will likely be dependent on targeting this important cell population.

9. Cancer Stem Cells and Differentiation Therapy

Malignant potential can be controlled forcing CSCs to differentiate and to cease proliferation. So, one way to handle the CSC problem is to treat cancer by inducing differentiation of cancer stem cells. In the following sections we present a series of natural and pharmacological agents, enhancers of differentiation used in clinical trials or recently proved in experimental assays.

9.1. Retinoic Acid and Derivatives

Retinoids are defined as biological isosteres of all-*trans* retinoic acid (ATRA, 1, Figure 9) that have been implicated in the regulation of diverse cellular functions, including survival and differentiation in a wide variety of cells. They play a critical role in embryonic morphogenesis, cellular growth, and apoptosis (Dimberg et al., 2002). Clinical trials have proved the efficacy of several natural and synthetic retinoids in the differentiation therapy of leukaemias (Egyed et al., 2001). Several natural and synthetic retinoids have proved to be clinically useful, such as ATRA (1, Figure 9) and etretinate (2, Figure 9) for the treatment of acute promyelocytic leukaemia (APL) (Castaigne et al., 1990) and psoriasis, respectively (Kagechika, 2002). High concentrations of retinoids were able to induce phenotypic changes (differentiation) and late apoptosis in pancreatic cancer cells in vitro (El-Metwally et al., 2005).

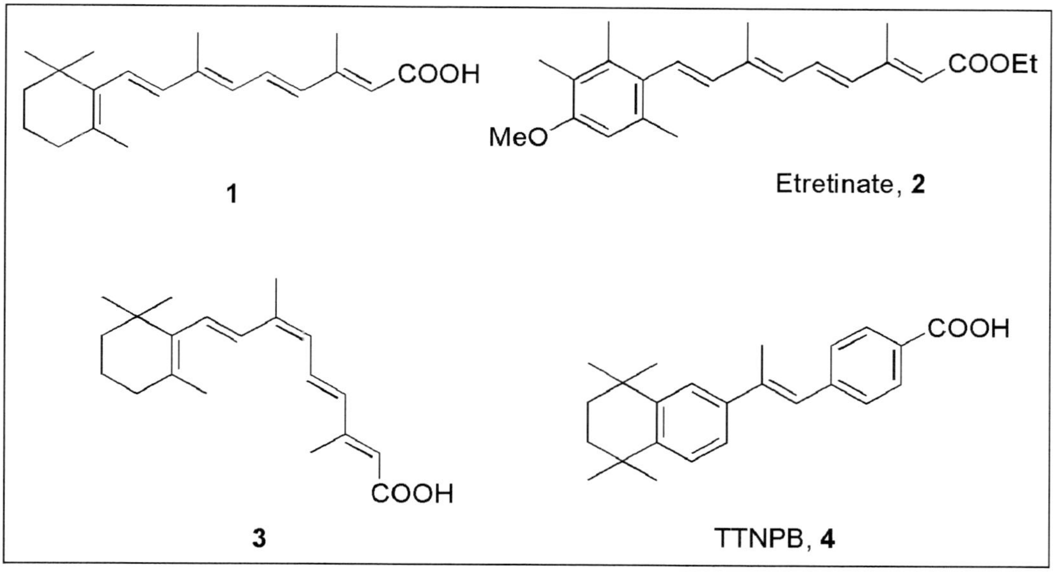

Figures 9. Chemical structures.

The receptor for retinoic acid consists of a super family of transcription factors (nuclear receptors) in which the two main families are the retinoic acid receptors (RARs) and the retinoid X receptors (RXRs). There are three distinct receptor subtypes (RARα, -β, and –γ), which possess considerable homology in their ligand-binding domains. The RXRs bind *cis* retinoic acid (9-*cis*-RA, 3, Figure 9), while the RARs bind both 9-*cis*-RA and ATRA (Chambon, 1996). The diverse range of retinoid actions, both desirable and undesirable, reflects the existence of multiple retinoid receptor subtypes. Recently, it has been reviewed the structural evolution and biological characteristics of RAR and RXR class- and subtype-selective synthetic retinoids (Dawson and Zhang, 2002).

Recent research show that the severe toxicity of retinoids is due to the non-specific activation of nuclear receptor subtypes. The synthesis of novel subtype-specific retinoids might have specific biological activities through activating only subsets of retinoid-responsive genes. In this sense, synthetic RARα agonists have proved most effective (Gianni et al., 2000) and often much more active than ATRA. The unstable polyene structure of retinoic acid analogs made them difficult to use as biological tools, and the hydrophobic hydrocarbon skeleton of retinoids such as etretinate 2 or TTNPB 4 (Figure 9) is disadvantageous in retinoid therapy, leading to long-lasting hypervitaminosis A and teratogenicity (Kagechika, 2002). Thus, a stable benzanilide-4-carboxylic acid skeleton with a polar amide linkage was designed (Kagechika, 2002). Compound Am68 (5, Figure 10) was as active as retinoic acid in the above-mentioned assay.

Another type of retinoidal benzanilide derivatives is the *N*-benzoylaminobenzoic acid, such as Am580 (7, Figure 10), which is an isomer of Am80 (6, Figure 10) with opposite aromatic substitution on the amide bond. Am580 (7, Figure 10) or Am80 (6, Figure 10) activate RARα which is necessary and sufficient to determine granulocytic maturation. Moreover, they have some binding affinity to RARβ, but none to RARγ and RXRs. Unlike ATRA, Am580 (7, Figure 10) do not induce their own catabolism and maintain elevated plasma levels following chronic administration (Gianni et al., 1996). In leukemia patients, Am80 (6) induces the clinical remission of ATRA-resistant relapses (Tobita et al., 1997).

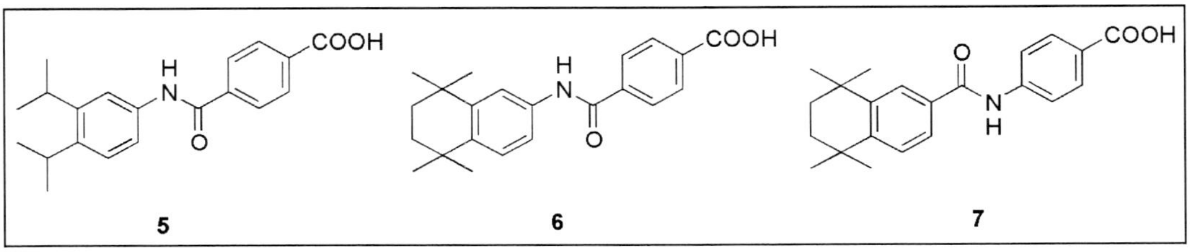

Figures 10. Chemical structures.

9.2. Vitamin D_3 and Analogs

Vitamin D_3 (8, Figure 11) is a natural hormone that is intimately involved in Ca^{2+} homeostasis. Vitamin D_3 and its active metabolite, 1α,25-$(OH)_2D_3$ (calcitriol, 9, Figure 11), generally determine growth arrest, the expression of various differentiation markers in breast and prostate carcinoma and play a role in apoptosis (Lazzaro et al., 2000). Moreover, it regulates hematopietic cell functions such as the myeloid cell line differentiation along the monocyte/macrophage pathway (Bhatia et al., 1996). At physiological concentrations, calcitriol (9, Figure 11) promoted induction of all the monocyte/macrophage immunophenotypic and morphological differentiation markers by the vitamin D_3 genomic signaling pathways (Grande et al., 2002).

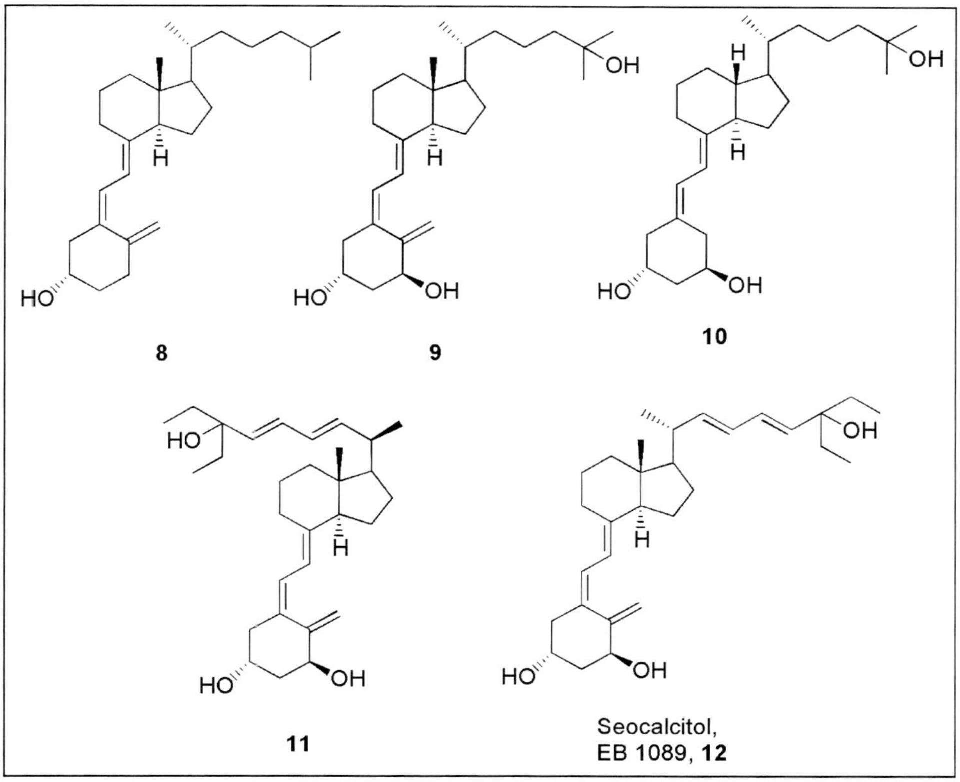

Figures 11. Chemical structures.

Recently, attempts have been made to exploit the ability of Vitamin D_3 to inhibit the growth of neoplastic cells by synthesizing more potent analogs. For example, 19-norvitamin D_3 (10, Figure 11), and 1α,25-dihydroxy-20-epivitamin D_3 (11, Figure 11) have been found to be more potent than the parent compound inducing maturation in HL-60 cells (Asou et al., 1998). In addition, attempts have been made to circumvent the hypercalcemic effects of vitamin D_3 by developing analogs free of this effect. Seocalcitol (EB 1089, 12, Figure 11) is one of such synthetic analogs (Hansen et al., 2000). Despite being 50-200 times more potent than calcitriol (9, Figure 11), with respect to regulation of cell growth and differentiation *in vitro* as well as *in vivo*, 12 displays a reduced calcemic activity *in vivo* compared to that of 9.

9.3. Histone Deacetylase Inhibitors

In cancer, altered activity of HAT or HDAC had been showed. Aberrant gene expression that results in functional inactivation of histone acetyl transferase (HAT) activity or overexpression of histone deacetylases (HDACs) can mediate tumor cell proliferation (Kouzarides, 1999).

HDAC inhibitors (HDACIs) have a great potential as new drugs initially identified by their ability to reverse the malignant phenotype of transformed cells. They have been shown to modulate transcription (Marks et al., 2000), inhibit the cell cycle, activate differentiation and induce apoptosis in a wide range of tumor-derived cell lines and to block angiogenesis and stimulate the immune system *in vivo* (Johnstone 2002). Several classes of compounds have been identified as HDACIs (Marks et al., 2001): short-chain fatty acids (e.g. butyrates), hydroxamic acids [e.g., trichostatin A (TSA, 13, Figure 12), SAHA (14, Figure 12) or oxamflatin (15, Figure 12], benzamides [e.g. CI 994 (16), and MS-275 (17)(Figure 12)], and cyclic peptides [e.g. trapoxin (18), and apicidin (19) (Figure 12)].

Recently, a series of novel selective and differentiating HDACIs have been designed (Haggarty et al., 2003). Tubacin (20, Figure 13), strongly increases α-tubulin acetylation in mammalian cells, mediating the localization of microtubule-associated proteins (Haggarty et al., 2003). The results suggest that inhibitors like tubacin might have therapeutic applications as antimetastatic and antiangiogenic agents (Monneret, 2005). Another molecule, namely histacin (21, Figure 13), is a selective inducer of histone acetylation without affecting the level of α-tubulin acetylation (Haggarty et al., 2003). All these data indicate that there was a strong correlation between HDAC inhibitory potency, induction of differentiation and antiproliferative activity.

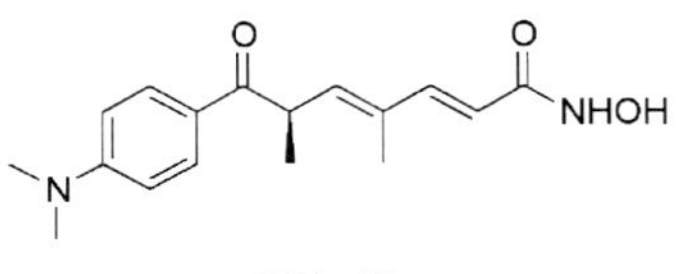

TSA, **13**

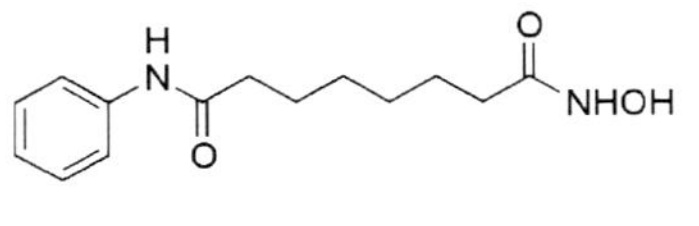

SAHA, **14**

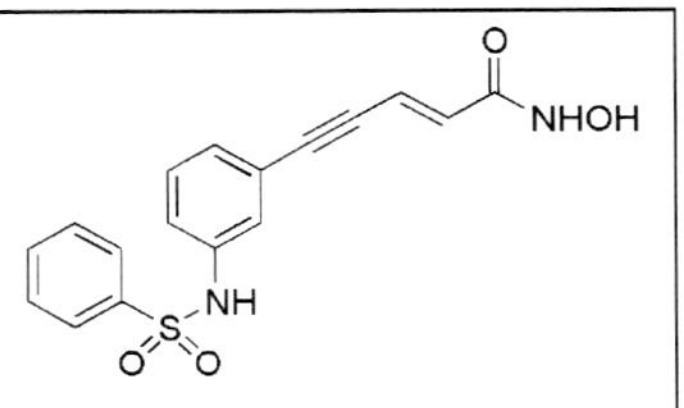

Oxamflatin, **15**

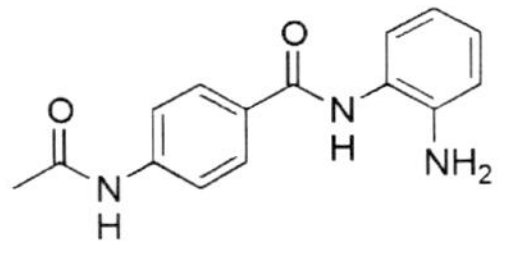

CI-994, **16**

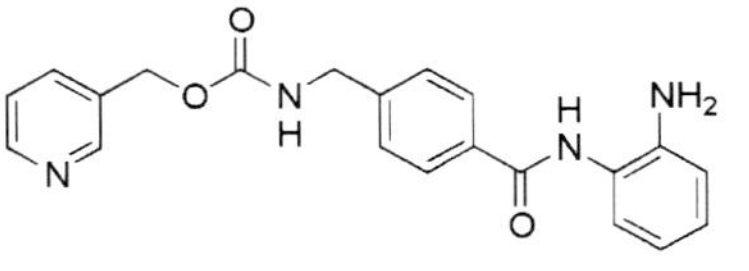

MS-275, **17**

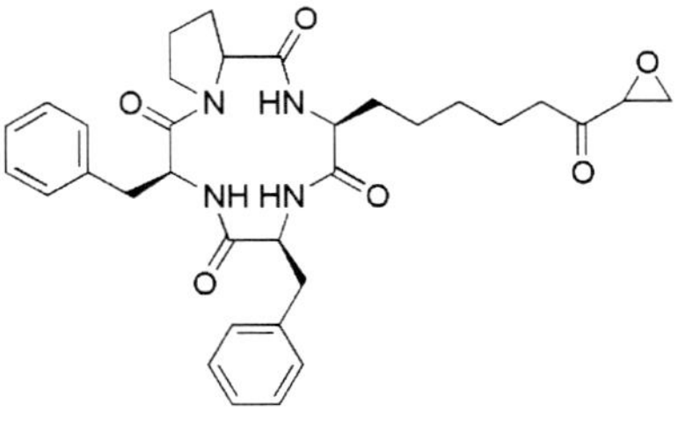

Trapoxin, **18**

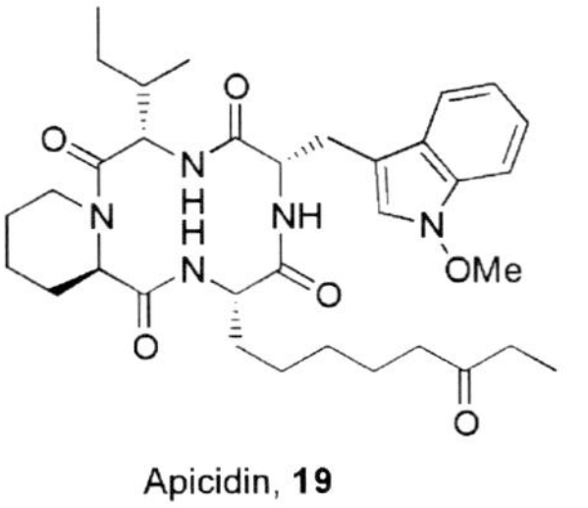

Apicidin, **19**

Figures 12. Chemical structures.

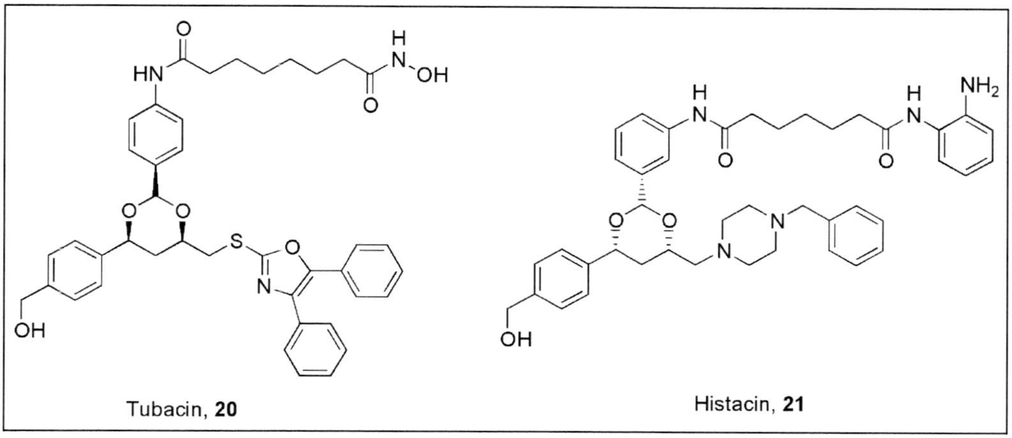

Figures 13. Chemical structures.

9.4. Hypomethylating Agents

The methylation of genes in the regions of CpG islands results in transcriptional repression and impaired induction of genes required for cellular maturation. Moreover, DNA methylation is observed in many cancer cells, and these changes occur parallel with the hyperactivation of DNA methyltransferases (DNMTs). The diverse roles of DNA methylation in cancer have encouraged the search for therapeutic agents which inhibit DNA methylation. There are two ways to inhibit DNA methylation; one is to block methylation by incorporating nucleotide analogs into the DNA and the other is to perturb DNMTs (Attwood et al., 2002).

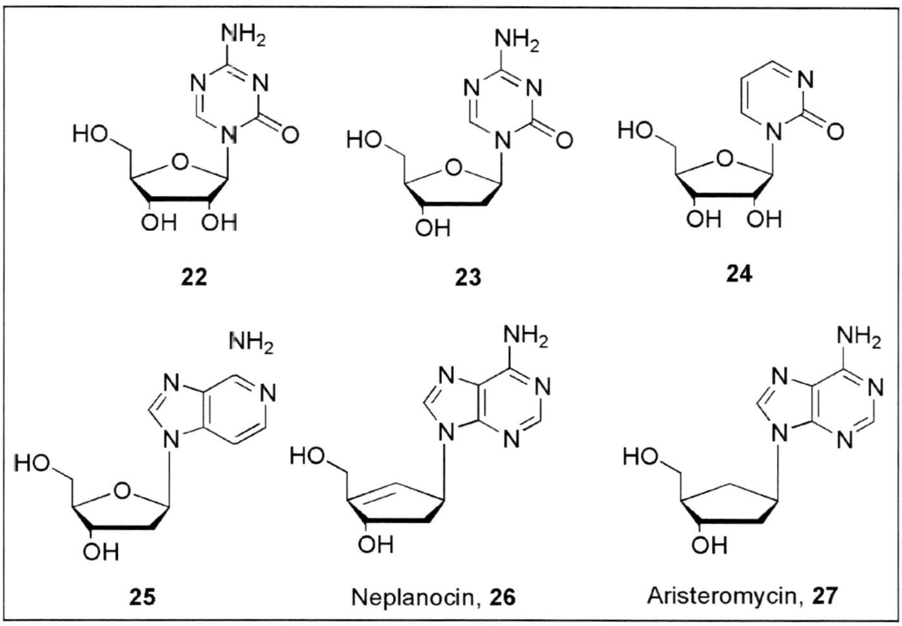

Figures 14. Chemical structures.

The nucleoside cytidine analogs 5-azacytidine (5-aza-CD, 22, Figure 14) and 5-aza-2`-deoxycytidine (5-aza-dCD, 23, Figure 14) modified in the 5 position, were the first synthesized as cancer chemotherapeutic agents in the United States in the 1970's, and have been useful in the treatment of the leukaemia and solid tumors (Wijermans et al., 2000). 5-Aza-CD triggers *in vitro* differentiation of leukemic cell lines (HL-60, K-562 and MEL) or in patients with myelodysplastic syndromes (Silverman et al., 1993). 5-aza-dCD, a potent inhibitor of DNA methylation, was able to induce *in vitro* differentiation of myleloid leukemic cell lines and the formation of functional fat muscle and cartilage from mouse ESCs in cultures. Tsuji-Talayama et al. (2004) have demonstrated that 5-azaC could reverse the differentiation of ESCs morphologically, phenotypically, and genetically, indicating the correlation of both differentiation reversing activity and demethylating activity of 5-azaC (Tsuji-Takayama et al., 2004). Recently, the new pyrimidinone ribonucleoside zebularine (24, Figure 14) has been developed with an increased chemical stability that allows oral delivery in clinical studies (Cheng et al., 2003).

Similarly, hypomethylating compounds inhibitors of S-adenosylhomocysteine (SAH) hydrolase such as 3-deazaadenosine (25), neplanocin A (26) or aristeromycin (27) (Figure 14) have been reported to induce differentiation of various cell lines (Niitsu and Honma, 1999).

9.5. Cyclin-dependent Kinase Inhibitors

Checkpoints at the G_1/S and G_2/M phases serve to maintain the proper sequence of cell cycle events and allow the cell to respond to proliferation signals or decide between the alternate pathways leading to cytokinesis, differentiation, quiescence, and cell death. The loss of proper checkpoints control in cancer cells contributes to tumorigenesis. It is reasonable to assume that small molecule inhibitors that target these pathways would be effective modulators of cell proliferation.

The therapeutic outcome might be considered when developing a strategy for cyclin-dependent kinase (CDK) modulation. Frequently, diminished CDK activity accompanies loss of proliferative potential, exit from the cell cycle, and entry of the cell into a differentiated state G_0. Thus, an additional consequence of CDK inhibition could be the induction of a differentiated cellular phenotype (Sausville et al., 2000). Examples include the positive effects of CDK5 and CDK9 and negative effects of CDK4 and CDK2 in the MyoD-mediated differentiation of muscle cells; the involvement of CDKs in the differentiation and the cell cycle exit of model systems by selective inhibitors of CDK2 and CDK6; suppression of HIV-1 transcription by flavopiridol or roscovitine is accompanied by inhibition of podocyte proliferation and by re-expression of podocyte differentiation markers (Nelson et al., 2001). Hence, small molecule CDK inhibitors can be used to alter the balance between cell growth and differentiation in cancer cells.

The first group of CDK inhibitors normally includes natural products or first-generation of compounds developed after the screening of compound libraries. Examples of these non-selective inhibitors are flavopiridol (28), butyrolactone I (29), staurosporine (30), indirubin

(31), indirubin-3'-monoxime (32), *N*-substituted adenine derivatives like purvalanols A (33) and B (34), the synthetic paullones (35-37), and hymenialdisine (38) (Figure 15).

The semisynthetic flavonoid, flavopiridol (28) with potent activity in the NCI's *in vitro* drug screen (mean IC_{50} value of 60 nM), displayed potent *in vitro* inhibition of all CDKs tested while arresting cells at the G_2/M and the G_1/S transitions and decreasing the rate of progression through the S phase (Toogood, 2002). Other studies demonstrated the capacity of flavopiridol to induce apoptosis, promote cellular differentiation, and inhibit angiogenesis. Importantly, flavopiridol treatment of MCF-7 human breast carcinoma cell line reduced cyclin D protein levels as a consequence of downregulation of cyclin D1 mRNA (Carlson et al., 1999).

The paullones (35-37, Figure 15) represent a series of benzazepines with submicromolar potencies for CDK1/cyclin B and CDk2/cyclin which inhibit the cell cycle progression of serum-stimulated cells through G_1 into S phase.

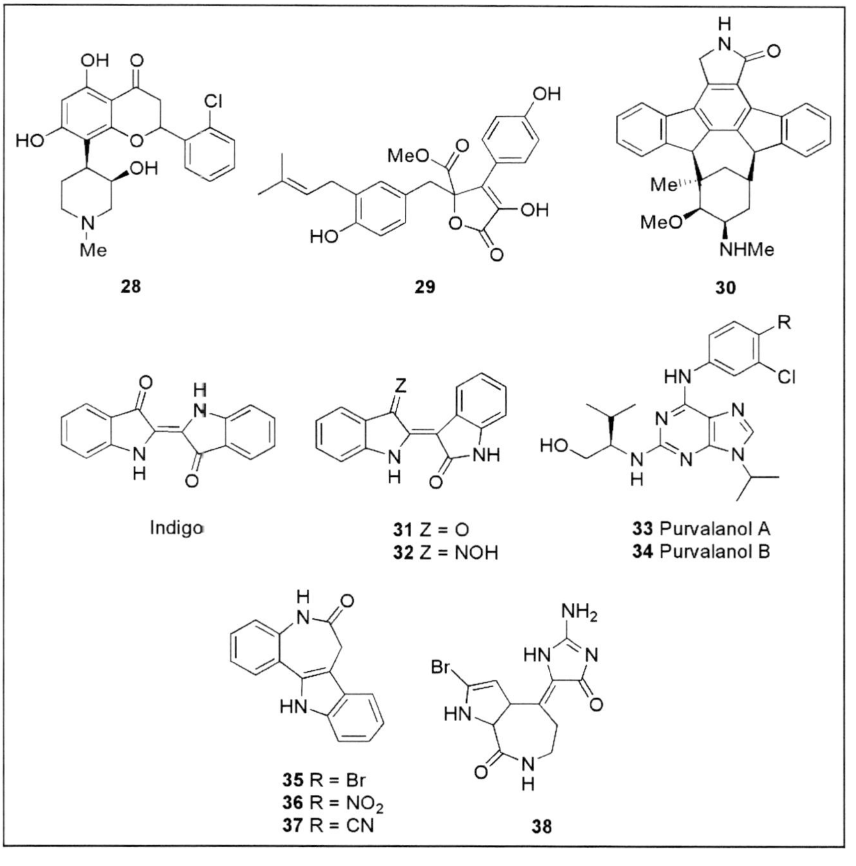

Figures 15. Chemical structures.

Among the earliest CDK inhibitors described are a class of purines, some of the most potent members of this series have been prepared using combinatorial approaches by the Schultz group (Chang et al., 1999). After intensive screening of purine-related structures, olomoucine (39, Figure 16) and roscovitine (40, Figure 16, IC_{50} = 0.7 μM against cyclin

B/CDK1) were identified. The binding mode of 2,6,9-trisubstituted purines to the ATP-binding site of the human CDK2 has proved to be useful in the development of selective and potent inhibitors of the human CDK2/cyclin A complex (Gray et al., 1998).

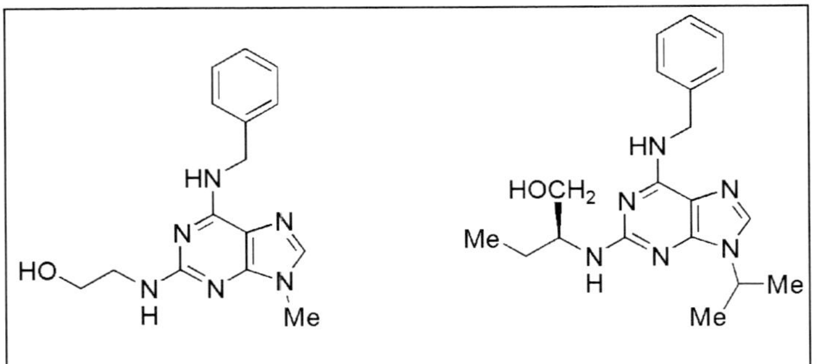

Figures 16. Chemical structures.

9.6. Antimetabolites and Cytotoxic Drugs

It is well established that different anticancer drugs may exert their therapeutic effects by combining differentiating and cytotoxic actions in several types of cancer (Marchal et al., 1997; Camarasa et al., 2005). Moreover, classical chemotherapy possesses undesirable effects such as citotoxicity, with affectation of healthy cells, and the development of drug resistance, with overexpression of active drug-efflux pumps (Gottesman et al., 2002).

We have previously shown that actinomycin D (41, Figure 17), an antibiotic used as a choice drug in the treatment of rhabdomyosarcomas (RMS) and some types of leukaemia, inhibited the proliferation and induced the phenotypic and biochemical differentiation in the RD rhabdomyosarcoma cell line (Marchal et al., 1997). In patients, the increased induced-cytodifferentiation correlates with a favorable outcome for RMS (Smith et al., 2002).

Purine and pyrimidine analogs inhibit the growth of tumor cells, interfere in the synthesis and metabolism of nucleic acids acting on key enzymes for the synthesis of DNA (Gómez-Vidal et al., 2004; Marchal et al., 1997). Some of these analogues induce cell differentiation. For example, ara-C (42, Figure 17) is effective in differentiation therapy for patients with acute myeloid leukemia or myelodysplastic syndrome, with some, albeit limited, success (Honma, 2003).

5-Fluorouracil (5-FU, 43, Figure 17) had played an important role in standard chemotherapy protocols for a range of solid tumors, including breast and colorectal cancers. The antitumor activity of 5-FU is dependent on the ability of the drug to bind and inactivate the enzyme thymidylate synthase, causing genetic miscoding, DNA damage and the inhibition of rRNA maturation (Parker and Cheng, 1990). Moreover, in human erythroleukemia K562 cells, 5-FU induced cell differentiation associated with a biphasic change of p210bcr/c-Abl, and the *abl*-associated kinase activities mediated at the translational level (Yang and Chang, 1995).

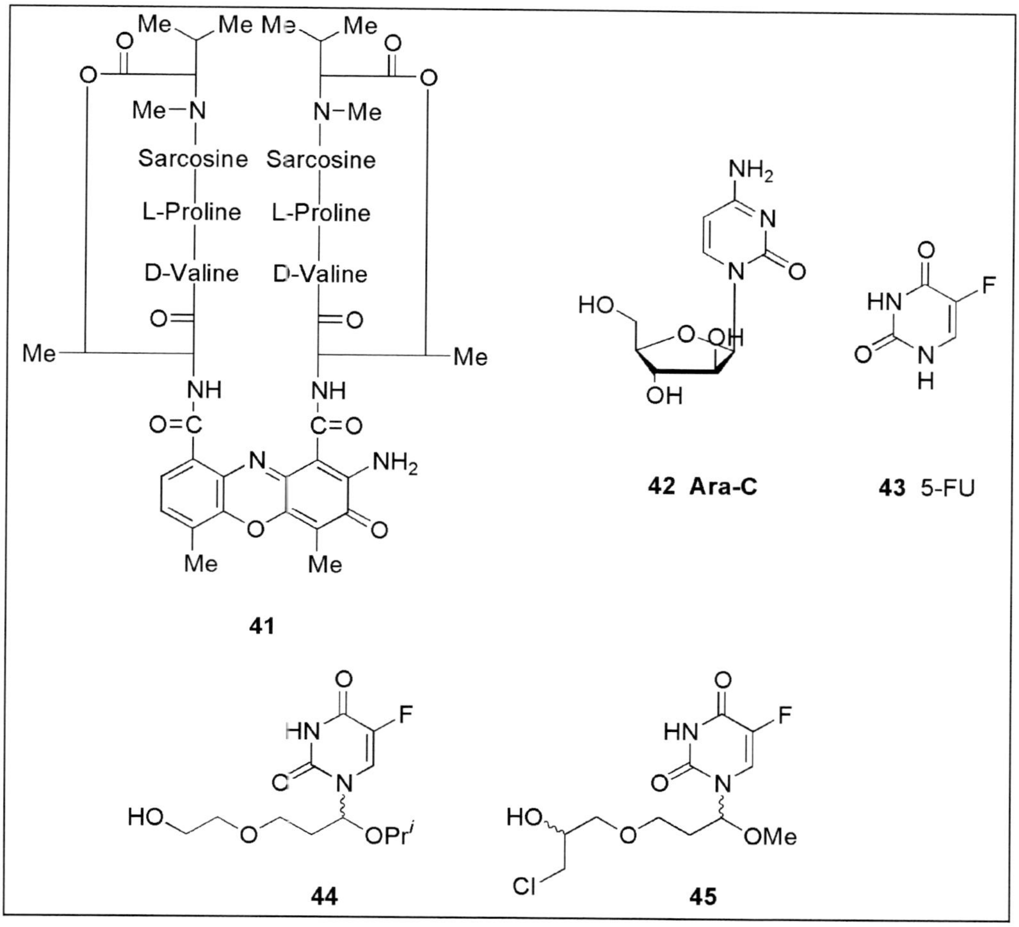

Figures 17. Chemical structures.

Finally, we have developed a series of 5-FU prodrugs such as pyrimidine acyclonucleoside-like compounds (44 and 45, Figure 17) that possess antitumor activity and present low toxicity *in vivo* (Marchal et al., 2004). We used specific markers of normal muscular differentiation to decide if these compounds led to the re-entry of the tumoral cells into the normal pathway of development. These 5-FU derivatives induced myogenic differentiation in the rhabdomyosarcoma human cells (Marchal et al., 1999; Marchal et al., 2000) not inducing the multidrug resistance phenomenon, indicating that this type of tumor - the most frequent soft tissue malignancy in paediatrics patients- may be amenable to treatment by differentiation therapy. Camarasa et al. (2005) have recently reported that differentiating agents in leukaemic cells may affect the susceptibility of tumor cells to cytotoxic drugs modulating the expression of genes involved in drug resistance and that they inhibit the expression of drug efflux mechanisms of resistance. Moreover, we have showed recently that the 5-FU derivatives increased microtubule stability and HLA class I expression in rhabdomyosarcoma cells. These data seem to be related to both the antitumor immune response and the re-entry into the normal myogenesis program (Marchal et al., 2007).

10. Cancer Differentiating Agents in the Stem Cell-based Therapy

In this regard, SCs may provide innovative methods for drug delivery of biologics, immunotherapy, and tissue regenerative or replacement engineering (Gage et al., 1998; Dove et al., 2002). The translation of this discipline to medical practice has tremendous potential, but in many applications technological issues need to be overcome. In order to solve and improve the difficulties appearing in these approaches, some new techniques are developing to help and overcome the results obtained with the SC-based therapy. One of these techniques consist of using SCs pre-treated by cancer differentiating agents such us retinoic derivatives, hypomethylating agents, antimetabolites and cytotoxic drugs, etc. Successful results have been obtained in some clinical fields, especially in cardiomyopathies, diabetes mellitus type I and blood diseases.

10.1. Retinoic acid and Derivatives

10.1.1. Retinoic Acids in Cardiac Cell Therapy

Loss of cardiomyocytes due to myocardial infarction or hereditary cardiomyopathies may represent causative factors in the progression toward heart failure. Cardiomyocyte loss is associated with hibernation of underperfused heart cells. In mice, human MSCs of bone marrow (BMhMSCs) have been found to enhance functional recovery in infarcted hearts (Orlic et al, 2001a, 2001b), suggesting that delivery of these cells may contribute to the regeneration of myocardial tissue.

Recently, we have shown that a mixed ester of hyaluronan with butyric and retinoic acids (HBR) (a well known agent used in the differentiation-based therapy) remarkably increased the yield of cardiomyocytes derived from mouse ES cells (Ventura et al., 2004), and these HBR pre-treated SCs obtain clear benefits compared with the non pre-treated ones. According to *in vitro* designs, it has been shown that HBR increased the expression of cardiac lineage promoting genes without affecting MyoD mRNA. This observation seems to exclude a generalized activation of repressed genes. The ability of HBR to enhance cardiogenesis *in vitro* is also inferred by the observation that the mixed ester enhanced the amount of cells expressing MHC, and α-sarcomeric actinin, and by the finding that Cx43 positive cells only developed from HBR-treated human MSCs (Ventura et al., 2007).

Interestingly, exposure to HBR remarkably enhanced the transcription of genes tightly involved in vasculogenesis and angiogenesis and led to the development of vWF expressing endothelial cells (Ventura et al., 2007). It is also known that human MSC-like profiles in both amnion and chorion cells of foetal membranes (FMhMSCs) secrete large amounts of VEGF, and HGF, and that HBR consistently increased the secretion of each trophic mediator, and besides VEGF, HGF has not only angiogenic but also cardioprotective effects, including antiapoptotic, mitogenic and antifibrotic activities (Nakamura et al., 1989; Nakamura, 2000), showing that *in vitro* responses to HBR were considerably more accentuated in FMhMSCs than in BMhMSCs or human dental pulp MSCs (DPhMSCs).

In vivo experiments were made to assess whether FMhMSCs may be used to repair infarcted rat hearts and whether their rescuing potential may be enhanced by cell pretreatment with HBR. The transplantation of HBR pretreated FMhMSCs led to increased left ventricle to body weight ratio and diastolic thickness, as well as further reduction in infarct size, as compared with untreated cells. The injection of HBR-treated cells was followed by a significant increase in density of capillaries positive for anti-human vWF, compared to the capillary density observed in the tissue receiving untreated cells (Ventura et al., 2007).

Finally, in the hearts injected with HBR-exposed cells, the number of FMhMSCs positively stained with a human-specific anti-vWF antibody remarkably exceeded the number of vWF-positive cells detected in samples from the untreated group, and a consistent organization of vWF positive FMhMSCs into capillary structures was only observed in hearts transplanted with HBR-treated cells (Ventura et al., 2007), concluding that HBR-treated cells may also contribute to neovascularization and heart rescue through their ability to generate capillary-like structures.

Therefore, there is the possibility to use a novel glycoconjugate to enhance both vascular differentiation and the secretion of angiogenic, antiapoptotic, and mitogenic factors, from FMhMSCs. These pretreated cells do not induce allogenic or xenogenic lymphocyte proliferation and actively suppress lymphocyte responsiveness. Moreover, FMhMSC transplantation in neonatal swine and rats resulted in human microchimerism in various organs and tissues (Bailo et al., 2004), suggesting that FMhMSCs may be administered as allogenic grafts, without the need of bone marrow harvests on recipient subjects (Ventura et al., 2007).

10.1.2. All-trans-retinoic Acids in Diabetes Mellitus Type I

Diabetes mellitus affects 4%–5% of the world's population and is the most common metabolic disorder in humans. The number of people with diabetes is predicted to exceed 350 million by 2010. Type I diabetes mellitus results from the autoimmune destruction of the ß cells in pancreatic islets. Until now, pancreatic islet cell transplantation is the only effective approach to cure type I diabetes instead of insulin injection (Serup et al., 2001). However, this therapy is not widely used because of the severe shortage of transplantable donor islets.

One of the most interesting approaches consists of the generation of functional ß cells from ESCs. It has been suggested that some factors are able to promote definitive endoderm differentiation. Activin A, a member of the transforming growth factor–beta (TGF-ß) superfamily, is critical for mesoderm and endoderm formation during gastrulation. When used at a high concentration, it primarily induces endoderm formation (Kumar et al., 2003; Hill, 2001). RA is a well-characterized signaling molecule that acts in anteroposterior patterning of neuroectoderm and mesoderm in vertebrates (Maden, 2001). Recent evidence indicates that RA is also involved in the regulation of the embryonic endoderm differentiation pattern, especially the early pancreas bud formation, and it is able to improve insulin expression in pancreatic ß cells and the INS-1 cell line (Stafford and Prince, 2002). It has been demonstrated that the combination of activin A and RA was able to induce the *Xenopus* presumptive ectoderm region of the blastula to differentiate into pancreatic insulin-positive cells (Moriya et al., 2000).

A three-step protocol based on combination of activin A, RA, and other factors that mature pancreatic ß cells has been described. It can induce ESCs to differentiate into insulin-producing cells within only 2 weeks. These insulin-positive cells express characteristic pancreatic ß-cell marker genes such as *insulinI, pdx1, glut2, hnf3ß,* and *is11*. This strategy can indeed induce ESCs to differentiate into insulin-producing cells instead of uptaking insulin, which was one of the problems found in the past experiences, and those insulin-producing cells are able to fully rescue the streptozocin (STZ)–induced diabetic mice when they are transplanted under their renal capsules (Shi et al., 2005). Therefore, this three step protocol, in which ATRA is included, provides a potential source of insulin-producing cells for transplantation therapy of type I diabetes mellitus.

10.2. Hypomethylating Agents

It has been recently reported (Yoon et al, 2004; Saito et al., 2003) that MSCs transplantation resulted in an improved cardiac function. 5-azacytidine treated MSCs transplanted into the myocardial prescar tissue connected with the adjacent host cardiomyocytes with gap junction. The MSCs-derived cardiomyocytes formed junction gaps with host cardiomyocytes *in vivo*, making it likely to reduce the incidence of arrhythmia.

Recently, several publications have reported that the adipose tissue contains a putative stem cell population, termed adipose-derived stem cells (ADSCs), able to differentiate into different cell types including adipocytes, osteoblasts, myoblasts, and chondroblasts (Aust et al., 2004). ADSCs are similar to MSCs in growth kinetics, cell senescence, multi-lineage differentiation capacity, and gene transduction efficiency (De Ugarte et al., 2003a; 2003b). On the other hand, ADSCs are differentiated into endothelial cells and cardiomyocytes *in vitro* under controlled conditions (Zang et al., 2005a; 2005b).

To investigate the potential efficacy of ADSC implantation as well as the role of 5-azacytidine in ADSC differentiation, one study was designed by transplanting both the 5-azacytidine- pretreated (9 μmol/L and cultured for 4 days before transplantation so that enough cells would differentiate *in vivo* in the scar and not *in vitro)* and untreated ADSCs into myocardial scars produced by ligation of the left anterior descending coronary artery in a rabbit model. The transplant was made into myocardial scar tissues by direct intramyocardial injection (Duan-Zhen et al., 2007), avoiding some of the disadvantages of the transcoronary injection like coronary embolism (Saito et al., 2003). Whereas both untreated ADSCs and 5-azacytidine-pretreated ADSCs were observed to be differentiated into cardiomyogenic cells in the prescar by the milieu of the host myocardium, only 5-azacytidine-pretreated ADSCs were differentiated into numerous cardiomyogenic cells in the middle of the scar.

It is also demonstrated that both untreated and 5-azacytidine-pretreated ADSCs transplanted into myocardial scar tissue contributed to an improvement of cardiac function that probably mainly resulted from a combination of factors including improvement of local myocardial perfusion (there were more microvessels in the ADSC-implanted hearts than in the control hearts) and elasticity, but as a whole, the cardiac functions of rabbits transplanted with 5-azacytidine- pretreated ADSCs were better than those with untreated ADSCs, and only the left ventricular diastolic function and transmural scar size of rabbits transplanted with 5-azacytidine-pretreated ADSCs improved significantly compared with the control. The effectiveness of 5-azacytidine pretreatment in improving cardiac function is attributed to an increased conversion of ADSCs to myogenic cells, with their differentiation into cardiomyocytes, 5-azacytidine-pretreated ADSCs might contain more contractile structures and elastic properties than untreated ADSCs (Duan-Zhen et al., 2007). Both of them were stained positively for the troponin I, but only 5-azacytidine-pretreated MSCs transplanted into the myocardial infarction scar tissue in a swine model expressed the muscle specific proteins such as actinin and troponin T, and attenuated the systolic paradoxical wall motion (Tomita et al., 1999; Tomita et al., 2002).

So, it is a fact that that transplanted ADSCs promote angiogenesis and vessel-like structure formation and form islands of cardiac tissue in myocardial scar tissue, resulting in improved cardiac function. But the transplantation of 5-azacytidine (an agent that is used in differentiation-based anticancer therapy by inhibiting DNA methylation) pre-treated ADCs into the scar of the failing heart is more efficient than untreated ADSCs in preservation of cardiac function (Duan-Zhen et al., 2007).

11. Conclusion

Understanding the molecular mechanism responsible for the multistage process of differentiation, embryonic development, maintenance of stem cells and neoplastic transformation may provide new approaches for elimination or differentiation therapy. A detailed understanding of these processes will require continued studies of the mechanisms of embryonic and adult stem cell biology and the identification of those factors and signaling components that are necessary to generate and to manipulate stem cell progeny for therapeutic applications. Thus, investigation of the mechanisms and signaling pathways that support stem cell renewal in normal and malignant tissue may provide new targets for therapies designed to complement existing approaches and reduce tumor recurrence. There is evidence across species that regeneration of tissues can occur. Both animal and human studies suggest that SCs capable of improving organ functions exist in adults. This might be part of a natural repair process. In this chapter we have analyzed characteristic features of both normal and cancer SCs and the molecular pathways for self-renewal or differentiation. Moreover, we showed that although SC niches have been shown to play an instructive and pivotal role in the regulation of stem cells, their implication in the CSC formation remains to be elucidated. The CSC hypothesis is a promising new paradigm that could potentially influence cancer management. Current cancer treatments (surgery, chemotherapy, radiotherapy) have been evaluated for their ability to kill proliferating cells. However, these therapies can fail due to the survival of cancer stem cells and their beneficial effects are normally accompanied by a notable morbidity, citotoxicity and multidrug resistance. The unique properties of stem cells are opening the door to the development of new therapeutic approaches notably in oncology and regenerative medicine. Novel therapies must therefore first aim at killing all proliferating tumor cells, and secondarily differentiating or eliminating CSCs.

12. References

Abbott, BL. ABCG2 (BCRP): a cytoprotectant in normal and malignant stem cells. *Clin Adv Hematol Oncol*, 2006, 4, 63-72.

Adams, GB; Martin, RP; Alley, IR; et al. Therapeutic targeting of a stem cell niche. *Nat Biotechnol*, 2007, 25, 238-243.

Alison, MR; Lovell, MJ; Direkze, NC; et al. Stem cell plasticity and tumour formation. *Eur J Cancer*, 2006, 42, 1247-1256.

Alison, MR; Poulsom, R; Otto, WR; et al. Recipes for adult stem cell plasticity: fusion cuisine or readymade? *J Clin Path*, 2004, 57, 113-120.

Alsalameh, S; Amin, R; Gemba, T; Lotz, M. Identification of mesenchymal progenitor cells in normal and osteoarthritic human articular cartilage. *Arthritis Rheum*, 2004, 50, 1522-1532.

Ambudkar, SV; Kimchi-Sarfaty, C; Sauna, ZE; Gottesman, MM. P-glycoprotein: from genomics to mechanism. *Oncogene*, 2003, 22, 7468-7485.

Anderson, DJ; Gage, FH; Weissman, IL. Can stem cells cross lineage boundaries? *Nat Med*, 2001, 7, 393-395.

Arey, LB. *Developmental anatomy: a textbook and laboratory manual of embryology*. 7th ed. London: WB Saunders Co.; 1974.

Asahara, T; Murohara, T; Sullivan, A; et al. Isolation of putative progenitor endotelial cell for angiogenesis. *Science*, 1997, 275, 964-967.

Asou, H; Hoike, M; Elstner, E et al. 19-nor vitamin-D analogs: a new class of potent inhibitors of proliferation and inducers of differentiation of human myeloid leukemia cell lines. *Blood*, 1998, 92, 2441-2449.

Attwood, JT; Yung, RL; Richardson, BC. DNA methylation and the regulation of gene transcription. *Cell Mol Life Sci*, 2002, 59, 241-257.

Aust, L; Devlin, B; Foster, SJ; et al. Yield of human adipose-derived adult stem cells from liposuction aspirates. *Cytotherapy*, 2004, 6, 7-14.

Avilion, AA; Nicolis, SK; Pevny, LH; et al. Multipotent cell lineages in early mouse development depend on SOX2 function. *Genes Dev*, 2003, 17, 126-140.

Azuara, V; Perry, P; Sauer, S; et al. Chromatin signatures of pluripotent cell lines. *Nat Cell Biol*, 2006, 8, 532-538.

Badorff, C; Brandes, RP; Popp, R; et al. Transdifferentiation of blood-derived human adult endothelial progenitor cells into functionally active cardiomyocytes. *Circulation*, 2003, 107, 1024-1032.

Bailo, M; Soncini, M; Vertua, E; et al. Engraftment potential of human amnion and chorion cells derived from term placenta. *Transplantation*, 2004, 78, 1439-1448.

Ballen, K. Targeting the stem cell niche: squeezing blood from bones. *Bone Marrow Transplant*, 2007, 39, 655-60.

Bao, S; Wu, Q; McLendon, RE; et al. Glioma stem cells promote radioresistance by preferential activation of the DNA damage response. *Nature*, 2006, 444, 756-60.

Barrand, MA; Heppell-Parton, AC; Wright, KA; et al. A 190-kilodalton protein overexpressed in non-P-glycoprotein-containing multidrug-resistant cells and its relationship to the MRP gene. *J Natl Cancer Inst*, 1994, 86, 110-117.

Beaumont, PO; Moore, MJ; Ahmad, K; et al. Role of glutathione S-transferases in the resistance of human colon cancer cell lines to doxorubicin. *Cancer Res*, 1998, 58, 947-995.

Beltrami, AP; Barlucchi, L; Torella, D; et al. Adult cardiac stem cells are multipotent and support myocardial regeneration. *Cell*, 2003, 114, 763-776.

Bentley, AJ; Nakamura, T; Hammiche, A; et al. Characterization of human corneal stem cells by synchrotron infrared micro-spectroscopy. *Mol Vis*, 2007, 13, 237-242.

Bernstein, BE; Mikkelsen, TS; Xie, X; et al. A bivalent chromatin structure marks key developmental genes in embryonic stem cells. *Cell*, 2006, 125, 315-326.

Bertolini, F; Shaked, Y; Mancuso, P; Kerbel, RS. The multifaceted circulating endothelial cell in cancer: towards marker and target identification. *Nat Rev Cancer*, 2006, 6, 835-845.

Bhatia M; Kirkland JB; Meckling-Gill KA. 1,25-dihydroxyvitamin D3 primes acute promyelocytic cells for TPA-induced monocytic differentiation through both PKC and tyrosine phosphorylation cascades. *Exp Cell Res*, 1996, 222, 61-69.

Bianco, P; Cossu, G. Uno, nessuno e centomila: searching for the identity of mesodermal progenitors. *Exp Cell Res*, 1999, 251, 257-263.

Blasco, MA. Telomeres and human disease: ageing, cancer and beyond. *Nat Rev Genet*, 2005, 6, 611-622.

Boehnke, K; Mirancea, N; Pavesio, A; et al. Effects of fibroblasts and microenvironment on epidermal regeneration and tissue function in long-term skin equivalents. *Eur J Cell Biol*, 2007, in press.

Boiani, M; Scholer, HR. Regulatory networks in embryo-derived pluripotent stem cells. *Nat Rev Mol Cell Biol*, 2005, 6, 872-884.

Bonanno, G; Mariotti, A; Procoli, A; et al. Human cord blood CD133+ cells immunoselected by a clinical-grade apparatus differentiate in vitro into endothelial- and cardiomyocyte-like cells. *Transfusion*, 2007, 47, 280-289.

Borst, P; Evers, R; Kool, M; Wijnholds, J. A family of drug transporters: the multidrug resistance-associated proteins. *J Natl Cancer Inst*, 2000, 92, 1295-1302.

Bosnakovski, D; Mizuno, M; Kim, G; Takagi, S; et al. Isolation and multi lineage differentiation of bovine bone marrow mesenchymal stem cells. *Cell Tissue Res*, 2005, 319, 243-253.

Boyer, LA; Lee, TI; Cole, MF; et al. Core transcriptional regulatory circuitry in human embryonic stem cells. *Cell*, 2005, 122, 947–956.

Boyer, LA; Lee, TI; Cole, MF; et al. Core transcriptional regulatory circuitry in human embryonic stem cells. *Cell*, 2005, 122, 947-956.

Brandenberger, R; Wei, H; Zhang, S; et al. Transcriptome characterization elucidates signaling networks that control human ES cell growth and differentiation. *Nature Biotechnol*, 2004, 22, 707-716.

Brixius, K; Funcke, F; Graf, C; Bloch, W. Endothelial progenitor cells: a new target for the prevention of cardiovascular diseases. *Eur J Cardiovasc Prev Rehabil*, 2006, 13, 705-710.

Bunting, KD; Zhou, S; Lu, T; Sorrentino, BP. Enforced Pglycoprotein pump function in murine bone marrow cells results in expansion of side population stem cells in vitro and repopulating cells in vivo. *Blood*, 2000, 96, 902-909.

Burdon, T; Chambers, I; Stracey, C; et al. Signalling mechanisms regulating self-renewal and differentiation of pluripotent embryonic stem cells. *Cells Tissues Organs*, 1999a, 165, 131-143.

Burdon, T; Stracey, C; Chambers, I; et al. Suppression of SHP-2 and ERK signalling promotes selfrenewal of mouse embryonic stem cells. *Dev Biol*, 1999b, 210, 30-43.

Burke, ZD; Shen, CN; Ralphs, KL; Tosh, D. Characterisation of liver function in transdifferentiated hepatocytes. *J Cell Physiol*, 2006, 206, 147-159.

Calabrese, C; Poppleton, H; Kocak, M; et al. A perivascular niche for brain tumor stem cells. *Cancer Cell*, 2007, 11, 69-82.

Callaghan, R; Ford, RC; Kerr, ID. The translocation mechanism of P-glycoprotein. *FEBS Lett*, 2006, 580, 1056-1053.

Camarasa MV; Castro-Galache MD; Carrasco-García E et al. Differentiation and drug resistance relationships in leucemia cells. *J Cell Biochem*, 2005, 94, 98-108.

Campagnoli, C; Roberts, IA; Kumar, S; et al. Identification of mesenchymal stem/progenitor cells in human first-trimester fetal blood, liver, and bone marrow. *Blood*, 2001, 98, 2396-2402.

Carlson B; Labusen T; Singh S et al. Downregulation of cyclin D1 by transcriptional repression in MCF-7 human breast carcinoma cells induced by flavopiridol. *Cancer Res*, 1999, 59, 4634-4641.

Castaigne S; Chomienne C; Daniel MT et al. All-trans retinoic acid as a differentiation therapy for acute promyelocytic leukemia. I. Clinical results. *Blood*, 1990, 76, 1704-1709.

Caussinus, E; González, C. Induction of tumor growth by alterated stem-cells asymmetric division in Drosophila melanogaster. *Nat Genet*, 2006, 57, 1125-1129.

Chambers, I; Colby, D; Robertson, M; et al. Functional expression cloning of Nanog; a pluripotency sustaining factor in embryonic stem cells. *Cell*, 2003, 113, 643-655.

Chambon PA. A decade of molecular biology of retinoic acid receptors. *FASEB J*, 1996, 10, 940-954.

Chandross, KJ; Mezey, E. *Plasticity of adult bone marrow stem cells*. Mattson, M.P. and Van Zant, G. eds. (Greenwich, CT: JAI Press); 2001.

Chang YT; Gray NS; Rosania GR et al. Synthesis and application of functionally diverse 2,6,9-trisubstituted purine libraries as CDK inhibitors. *Chem Biol*, 1999, 6, 361-375.

Charbord, P; Oostendorp, R; Pang, W; et al. Comparative study of stromal cell lines derived from embryonic, fetal, and postnatal mouse blood-forming tissues. *Exp Hematol*, 2002, 30, 1202–1210.

Cheng JC; Matsen CB; Gonzales FA et al. Inhibition of DNA methylation and reactivation of silenced genes by zebularine. *J. Natl. Cancer Inst*., 2003, 95, 399-409.

Cheong, JW; Eom, JI; Maeng, HY; et al. Phosphatase and tensin homologue phosphorylation in the C-terminal regulatory domain is frequently observed in acute myeloid leukaemia and associated with poor clinical outcome. *Br J Haematol*, 2003, 122, 454-456.

Chepko, G; Slack, R; Carbott, D. Differential alteration of stem cell and other cell populations in ducts and lobules of TGFalpha and c-myc transgenic mouse mammary epithelium. *Tissue Cell*, 2005, 37, 393-412.

Clatworthy, JP; Subramanian, V. Stem cells and the regulation of proliferation, differentiation and patterning in the intestinal epithelium: emerging insights from gene expression patterns, transgenic and gene ablation studies. *Mech Dev*, 2001, 101, 3-9.

Cohnheim, J. Congenitales, quergestreiftes muskelsarkon der nireren. *Virchows Arch*, 1875, 65, 64.

Collas, P; Hakelien AM. Teaching cells new tricks. *Trens Biotechnol*, 2003, 21, 354-361.

Condorelli, G; Borello, U; De Angelis, L; et al. Cardiomyocytes induce endothelial cells to trans-differentiate into cardiac muscle: implications for myocardium regeneration. *Proc Natl Acad Sci USA*, 2001, 98, 10733-10738.

Costello, RT; Mallet, F; Gaugler, B. Human acute myeloid leukemia CD34_/CD38_ progenitor cells have decreased sensitivity to chemotherapy and Fas-induced apoptosis, reduced immunogenicity, and impaired dendritic cell transformation capacities. *Cancer Res*, 2000, 60, 4403-4411.

Crarke, DL; Johansson, CB; Wilbertz, J; et al. Generalized potential of adult neural stem cells. *Science*, 2000, 288, 1660-1663.

Darr, H; Mayshar, Y; Benvenisty, N. Overexpression of NANOG in human ES cells enables feederfree growth while inducing primitive ectoderm features. *Development*, 2006, 133, 1193–1201.

Dawson MI; Zhang X-k. Discovery and design of retinoic acid receptor and retinoid X receptor class- and subtype-selective synthetic analogs of all-*trans*-retinoic acid and 9-*cis*-retinoic acid. *Curr Med Chem*, 2002, 9, 623-637.

Dazzi, F; Ramasamy, R; Glennie, S; et al. The role of mesenchymal stem cells in haemopoiesis. *Blood Rev*, 2006, 20, 161-171.

De Ugarte, DA; Alfonso, Z; Zuka, PA; et al. Differential expression of stem cell mobilization-associated molecules on multi-lineage cells from adipose tissue and bone marrow. *Immunol Lett*, 2003a, 89, 267-270.

De Ugarte, DA; Morizono, K; Elbarbary, A; et al. Comparison of multi-lineage cells from human adipose tissue and bone marrow. *Cells Tissues Organs*, 2003b, 174, 101-109.

De Vries, EG; Hamilton, TC; Lind, M; et al. Advanced ovarian cancer. Drug resistance, supportive care and dose intensity. *Ann Oncol*, 1993, 4, 57-62.

de Wert, G; Mummery, C. Human embryonic stem cells: research, ethics and policy. *Hum Reprod*, 2003, 18, 672-682.

Dean, M; Fojo, T; Bates, S. Tumour stem cells and drug resistance. *Nat Rev Cancer*, 2005, 5, 275-284.

Devine, SM; Bartholomew, AM; Mahmud, N; et al. Mesenchymal stem cells are capable of homing to the bone marrow of non-human primates following systemic infusion. *Exp Hematol*, 2001, 29, 244-255.

Dezawa, M; Ishikawa, H; Itokazu, Y; et al. Bone Marrow Stromal Cells Generate Muscle Cells and Repair Muscle Degeneration. *Science*, 2005, 309, 314-317.

Dimberg A; Bahram F; Karlberg I et al. Retinoic acid-induced cell cycle arrest of human myeloid cell lines is associated with sequential down-regulation of c-myc and cyclin E and posttranscriptional up-regulation of p27 (Kip1). *Blood*, 2002, 99, 2199–2206.

Dimberg, A; Bahram, F; Karlberg, I; et al. Retinoic acid- induced cell cycle arrest of human myeloid cell lines is associated with sequential down-regulation of c-myc and Cyclin E and posttranscriptional up-regulation of p27 (kip1). *Blood*, 2002, 99, 2199-2206.

Dome, B; Jozsef, T; Dobos, J; et al. Identification and clinical significance of circulating endothelial progenitor cells in human non-small cell lung cancer. *Cancer Res*, 2006, 66, 7341-7347.

Dove A. Cell-based therapies go live. *Nat Biotechnol*, 2002, 20, 339-343.

Dravid, G; Ye, Z; Hammond, H; et al. Defining the role of Wnt/beta-catenin signaling in the survival; proliferation and self-renewal of human embryonic stem cells. *Stem Cells*, 2005, 23, 1489–1501.

Duan-zhen, Z; Lu-yue, G; Hong-wei, L; et al. Transplantation of autologous adipose-derived stem cells ameliorates cardiac function in rabbits with myocardial infarction. *Chin Med J*, 2007, 20, 120, 300-307.

Dyer, MA; Bremner, R. The search for the retinoblastoma cell origin. *Nat Rev Cancer*, 2005, 5, 91-101.

Egyed, M; Mihalyfalvi, Z; Kollar, B; et al. Effect of the retinoic acid on the cytogenetic remission in the first chronic phase of chronic myeloid leukemia treated with interferon. *Orv Hetil*, 2001, 142, 2421-2425.

El-Metwally TH; Hussein MR; Pour PM et al. High concentrations of retinoids induce differentiation and late apoptosis in pancreatic cancer cells in vitro. *Cancer Biol Ther*, 2005, 4, 602-611.

Felsenfeld, G; Groudine, M. Controlling the double helix. *Nature*, 2003, 421, 448-453.

Fernandes, KJ; Toma, JG; Miller, FD. Multipotent skin-derived precursors: adult neural crest-related precursors with therapeutic potential. *Philos Trans R Soc Lond B Biol Sci*, 2007, in press.

Fojo, A; Hamilton, TC; Young, RC; Ozols, RF. Multidrug resistance in ovarian cancer. *Cancer*, 1987, 60; 2075-2080.

Foster, V; Sanz, J. Gene therapy and stem cell therapy for cardiovascular diseases today: a model for translational research. *Nature clinical practice, Cardiovasc Med*, 2007, 4, 1-8.

Friedenstein, AJ; Piatetzky-Shapiro, II; Petrakova, KV. Osteogenesis in transplants of bone marrow cells. *J Embryol Exp Morphol*, 1966, 16, 381-390.

Gadner, RL. Stem cells and regenerative medicine: principles, prospects and problems. *CR Biologies*, 2007, in press.

Gage FH. Cell therapy. *Nature*, 1998; 392, 518-524.

Geiger, H; Koehler, A; Gunzer, M. Stem cells, aging, niche, adhesion and Cdc42: a model for changes in cell-cell interactions and hematopoietic stem cell aging. *Cell Cycle*, 2007, 6, 884-887.

Giangreco, A; Shen, H; Reynolds, SD; Stripp, BR. Molecular phenotype of airway side population cells. *Am J Physiol Lung Cell Mol Physiol*, 2004, 286, 624-630.

Gianni M; Ponzanelli I; Mologni L et al. Retinoid-dependent growth inhibition, differentiation and apoptosis in acute promyelocytic leukemia cells. Expression and activation of caspases. *Cell Death Differ*, 2000, 7, 447-460.

Giatromanolaki, A; Sivridis, E; Koukourakis, MI. Angiogenesis in colorectal cancer: prognostic and therapeutic implications. *Am J Clin Oncol*, 2006, 29, 408-417.

Gill, M; Dias, K; Hattori, ML; et al. Vascular trauma induces rapid but transient mobilization of VEGFR2(+) AC133 (+) endothelial precursor cells. *Circ Res*, 2001, 88, 167-174.

Gómez-Vidal JA; Campos J; Marchal JA et al. Actual targets in cytodifferentiation cancer therapy. *Curr Topics Med Chem*, 2004, 4, 175-202.

Gonzalez, FJ; Quesada, AR; Sevilla, I; et al. Prognostic value of serum angiogenic activity in colorectal cancer patients. *J Cell Mol Med*, 2007, 11, 294-300.

Gonzalez, FJ; Rueda, A; Sevilla, I; et al. Shift in the balance between circulating thrombospondin-1 and vascular endothelial growth factor in cancer patients: relationship to platelet alpha-granule content and primary activation. *Int J Biol Markers*, 2004, 19, 221-228.

Goodell, MA; Brose, K; Paradis, G; et al. Isolation and functional properties of murine hematopoietic stem cells that are replicating in vivo. *J Exp Med*, 1996, 183, 1797-1806.

Gottesman MM; Fojo T; Bates SE. Multidrug resistance in cancer: role of ATP-dependent transporters. *Nat Rev Cancer*, 2002, 2, 48-58.

Gottesman, MM. Mechanisms of cancer drug resistance. *Annu Rev Med*, 2002, 53, 615-627.

Gottschling, S; Saffrich, R; Seckinger, A; et al. Human mesenchymal stromal cells regulate initial self-renewing divisions of hematopoietic progenitor cells by a beta1-integrin-dependent mechanism. *Stem Cells*, 2007, 25, 798-806.

Grande A; Montarani M; Tagliafico E; et al. Physiological levels of 1,alpha, 25 dihydroxyvitamin D3 induce the monocytic commitment of CD34+ hematopoietic progenitors. *Leukoc Biol*, 2002, 71, 641-651.

Gray NS; Wodicka L; Thunnisen AM et al. Exploiting chemical libraries, structure, and genomics in the search for kinase inhibitors. *Science*, 1998, 281, 533-538.

Gronthos, S; Franklin, DM; Leddy, HA; et al. Surface protein characterization of human adipose tissuederived stromal cells. *J Cell Physiol*, 2001, 189, 54-63.

Gruh, I; Beilner, J; Blomer, U; et al. No evidence of transdifferentiation of human endothelial progenitor cells into cardiomyocytes after coculture with neonatal rat cardiomyocytes. *Circulation*, 2006, 113, 1326-1334.

Gudjonsson, T; Magnusson, MK. Stem cell biology and the cellular pathways of carcinogenesis. *APMIS*, 2005, 113, 922-929.

Guzman, ML; Jordan, CT. Considerations for targeting malignant stem cells in leukemia. *Cancer Control*, 2004, 11, 97-104.

Guzman, ML; Neering, SJ; Upchurch, D; et al. Nuclear factor κβ is constitutively activated in primitive human acute myelogenous leukaemia cells. *Blood*, 2001a, 98, 2301-2307.

Guzman, ML; Upchurch, D; Grimes, B; et al. Expression of tumor suppressor genes interferon regulatory factor 1 and death-associated protein kinase in primitive acute myelogenous leukaemia cells. *Blood*, 2001b, 97, 2177-2179.

Haggarty SJ; Koeller KM; Wong JC et al. Domain-selective small-molecule inhibitor of histone deacetylase 6 (HDAC6)-mediated tubulin deacetylation. *Proc Natl Acad Sci USA*, 2003, 100, 4389-4394.

Hansen CM; Hamberg KJ; Binderup E; Binderup L. Seocalcitol (EB 1089): a vitamin D analogue of anti-cancer potential. Background, design, synthesis, pre-clinical and clinical evaluation. *Curr Pharm Design*, 2000, 6, 803-828.

He, XC; Zhang, J; Li, L. Cellular and molecular regulation of hematopoietic and intestinal stem cell behaviour. *Ann NY Acad Sci*, 2005, 1049, 28-38.

Hellmann, MA; Panet, H; Barhum, Y; et al. Increased survival and migration of engrafted mesenchymal bone marrow stem cells in 6-hydroxydopamine-lesioned rodents. *Neurosci Lett*, 2006, 395, 124-128.

Herbrig, K; Pistrosch, F; Foerster, S; Gross, P. Endothelial progenitor cells in chronic renal insufficiency. *Kidney Blood Press Res*, 2006, 29, 24-31.

Hilbe, W; Dirnhofer, S; Oberwasserlechner, F; et al. CD133 positive endothelial progenitor cells contribute to the tumour vasculature in non-small cell lung cancer. *J Clin Pathol*, 2004, 27, 965-969.

Hill, CS. TGF-ß signalling pathways in early Xenopus development. *Curr Opin Genet Dev*, 2001, 11, 533-540.

Hiyama, E; Hiyama, K. Telomere and telomerase in ítem cells. *Br J Cancer*, 2007, 96, 1020-1024.

Holden, C; Vogel, G. Stem cells Plasticity: time for a reappraisal? *Scienc*e, 2002, 296, 2126-2129.

Honma Y. Adenine analogs as potencial differentiation therapy agents for acute myeloid leukemia. *Drug Dev Res*, 2003, 59, 14-22

Humes, HD. Stem cells. The next therapeutic frontier. *Trans Am Clin Climatol Assoc*, 2005, 116, 167-184.

Humphrey, RK; Beattie, GM; Lopez, AD; et al. Maintenance of pluripotency in human embryonic stem cells is STAT3 independent. *Stem Cells*, 2004, 22, 522-530.

Hyslop, L; Stojkovic, M; Armstrong, L; et al. Downregulation of NANOG induces differentiation of human embryonic stem cells to extraembryonic lineages. *Stem Cells*, 2005, 23, 1035–1043.

In 't Anker, PS; Noort, WA; Scherjon, SA; et al. Mesenchymal stem cells in human second-trimester bone marrow, liver, lung, and spleen exhibit a similar immunophenotype but a heterogeneous multilineage differentiation potential. *Haematol,* 2003a, 88, 845-852.

In 't Anker, PS; Scherjon, SA; Kleijburg-van der Keur, C; et al. Amniotic fluid as a novel source of mesenchymal stem cells for therapeutic transplantation. *Blood*, 2003b, 102, 1548-1549.

Jackson, KA; Majka, SM; Wang, H; et al. Regeneration of ischemic cardiac muscle and vascular endothelium by adult stem cells. *Clin Invest*, 2001, 107, 1395-1402.

Javazon, EH; Beggs, KJ; Flake, AW. Mesenchymal stem cells: paradoxes of passaging. *Exp Hematol*, 2004, 32, 414-425.

Jin, L; Hope, KJ; Zhai, Q; et al. Targeting of CD44 eradicates acute myeloid leukemic stem cells. *Nature Med*, 2006, 12, 1167-1174.

Johnstone RW. Histone-deacetylase inhibitors: novel drugs for the treatment of cancer. *Nat Rev Drug Discovery*, 2002, 1, 287-299.

Kagechika H. Novel synthetic retinoids and separation of the pleiotropic retinoidal activities. *Curr Med Chem*, 2002, 9, 591-608.

Kaplan, RN; Psaila, B; Lyden, D. Niche-to-niche migration of bone-marrow-derived cells. *Trends Mol Med*, 2007, 13, 72-81.

Kaplan, RN; Riba, RD; Zacharoulis, S; et al. VEGFR1-positive haematopoietic bone marrow progenitors initiate the pre-metastatic niche. *Nature*, 2005, 438, 820-827.

Keller, G. Embryonic stem cell differentiation: emergence of a new era in biology and medicine. *Genes Dev*, 2005, 19, 1129-1155.

Koblas, T; Harman, SM; Saudek, F. The application of umbilical cord blood cells in the treatment of diabetes mellitus. *Rev Diabetic Studies*, 2005, 2, 228-234.

Kogler, G; Sensken, S; Airey, JA; et al. A new human somatic stem cell from placental cord blood with intrinsic pluripotent differentiation potential. *J Exp Med*, 2004, 200, 123-135.

Kolch, W. Meaningful relationships: the regulation of the Ras/Raf/MEK/ERK pathway by protein interactions. *Biochem J*, 2000, 351, 289-305.

Körbling, M; Estrov, Z; Champlin, R. Adult stem cells and tissue repair. *Bone Marrow Transplant*, 2003, 32, 23-24.

Kouzarides T. Histone acetylases and deacetylases in cell proliferation. *Curr Opin Genet Dev*, 1999, 9, 40-48.

Koyanagi, M; Brandes, RP; Haendeler, J; et al. Cell-to-cell connection of endothelial progenitor cells with cardiac myocytes by nantotubes: A novel mechanism for cell fate changes? *Cir Res*, 2005, 96, 1039-1041.

Krampera, M; Marconi, S; Pasini, A; et al. Induction of neural-like differentiation in human mesenchymal stem cells derived from bone marrow, fat, spleen and thymus. *Bone*, 2007, 40, 382-390.

Krause, DS; Theise, ND; Collector, MI; et al. Multi-organ, multi-lineage engraftment by a single bone marrow-derived stem cell. *Cell*, 2001, 105, 369-377.

Kumar, M; Jordan, N; Melton, D; et al. Signals from lateral plate mesoderm instruct endoderm toward a pancreatic fate. *Dev Biol*, 2003, 259, 109-122.

Kuznetsov, SA; Mankani, MH; Gronthos, S; et al. Circulating skeletal stem cells. *J Cell Biol*, 2001, 153, 1133-1140.

Lage, H; Helmbach H; Dietel, M; Schadendorf, D. Modulation of DNA topoisomerase II activity and expression in melanoma cells with acquired drug resistance. *Br J Cancer*, 2000, 82, 488-491.

Lambiase, PD; Edwards, RJ; Anthopoulos, P; et al. Circulating humoral factors and endothelial progenitor cells in patients with differing coronary collateral support. *Circulation*, 2004, 109, 2986-2992.

Lazzaro G; Agadir A; Qing W et al. Induction of differentiation by 1a-hydroxyvitamin D5 in T47D human breast cancer cells and it interaction with vitamin D receptors. *Eur J Cancer*, 2000, 24, 129-137.

Leedham, SJ; Thliveris, AT; Halberg, RB; et al. Gastrointestinal stem cells and cancer: bridging the molecular gap. *Stem Cell Rev*, 2005, 1, 233-241.

Leonard, GD; Fojo, T; Bates, SE. The role of ABC transporters in clinical practice. *Oncologist*, 2003, 8, 411-424.

Li WC; Yu WY; Quinian JM et al. The molecular basis of transdifferentiation. *J Cell Mol* Biol, 2005, 9, 569-582.

Li, F; Tiede, B; Massague, J; Kang, Y. Beyond tumorigenesis: cancer stem cells in metastasis. *Cell Res*, 2007, 17, 3-14.

Li, L; Neaves, WB. Normal stem cells and cancer stem cells: the niche matters. *Cancer Res*, 2006, 66, 4553-4557.

Li, L; Xie, T. Stem cell niche: structure and function. *Annu Rev Cell Dev Biol*, 2005, 21, 605-631.

Li, W; Hayashida, Y; Chen, YT; Tseng, SC. Niche regulation of corneal epithelial stem cells at the limbus. *Cell Res*, 2007, 17, 26-36.

Loh, YH; Wu, Q; Chew, JL; et al. The Oct4 and Nanog transcription network regulates pluripotency in mouse embryonic stem cells. *Nat Genet*, 2006, 38, 431-440.

Loo, TW; Clarke, DM. Recent progress in understanding the mechanism of P-glycoprotein-mediated drug efflux. *J Membr Biol*, 2005, 206, 173-185.

Lopez-Holgado, N; Alberca, M; Sanchez-Guijo, F; et al. Short-term endothelial progenitor cell colonies are composed of monocyte and do not acquire endothelial markers. *Cythotherapy,* 2007, 9, 14-22.

Lotem, J; Sachs, L. Epigenetics and the plasticity of differentiation in normal and cancer stem cells. *Oncogene*, 2006, 25, 7663-7672.

Lowe, SW; Sherr, CJ. Tumor suppression by Ink4a--Arf: progress and puzzles. *Curr Opin Genet Dev*, 2003, 13, 77-83.

Luk, JM; Wang P; Carol, KL; et al. Hepatic potential of bone marrow stromal cells: Development of in vitro co-culture and intra-portal transplantation models. *J Immunol Methods*, 2005, 305, 39-47.

Luria, EA; Panasyuk, AF; Friedenstein, AY. Fibroblast colony formation from monolayer cultures of blood cells. *Transfusion*, 1971, 11, 345-349.

MacCorkle RA and Tan H. Mitogen-activated protein kinases in cell-cycle control. *Cell Biochem Biophys*, 2005, 43, 451-461.

Maden, M. Role and distribution of retinoic acid during CNS development. *Int Rev Cytol*, 2001, 209, 1-77.

Marchal JA; Boulaiz H; Rodríguez-Serrano F et al. 5-fluorouracil derivatives induce differentiation mediated by tubulin and HLA class I modulation. *Med Chem*, 2007, 3, 233-239.

Marchal JA; Boulaiz H; Suarez I et al. Growth inhibition, G_1-arrest, and apoptosis in MCF-7 human breast cancer cells by novel highly lipophilic 5-fluorouracil derivatives. *Inv New Drugs*, 2004, 22, 379-389.

Marchal JA; Melguizo C; Prados J et al. Modulation of myogenic differentiation in a human rhabdomyosarcoma cell line by a new derivative of 5-fluorouracil (QF-3602). *Jpn J Cancer Res*, 2000, 91, 934-940.

Marchal JA; Prados J; Melguizo C et al. Actinomycin D treatment leads to differentiation and inhibits proliferation in rhabdomyosarcoma cells. *J Lab Clin Med*, 1997, 130, 42-50.

Marchal JA; Prados J; Melguizo C et al. GR-891: a novel 5-fluorouracil acyclonucleoside prodrug for differentiation therapy in rhabdomyosarcoma cells. *Br J Cancer*, 1999, 79, 807-813.

Marchal JA; Rodríguez-Serrano F; Campos J; et al. Differentiation: an encouraging approach to anticancer therapy. *Ital J Anat Embryol*, 2006, 111, 45-64.

Marks PA; Rifkind RA; Richon VM et al. Histone deacetylase and cancer: causes and therapies. *Nature Rev Cancer*, 2001, 1, 194-202.

Martin, G. Isolation of a pluripotent cell line from early mouse embryos cultured in medium conditioned by teratocarcinoma stem cells. *Proc Natl Acad Sci*, 1981, 78, 7635-7638.

Martinez-Climent, JA; Andreu, EJ; Prosper, F. Somatic stem cells and the origin of cancer. *Clin Transl Oncol*, 2006, 8, 647-663.

Massard C; Deutsch E; Soria JC. Tumour stem cell-targeted treatment: elimination or differentiation. *Ann Oncol*, 2006, 17, 1620-1624.

Matsui, W; Huff, CA; Wang, Q; et al. Characterization of clonogenic multiple myeloma cells. *Blood*, 2004, 103, 2332-2336.

Mayani, H; Lansdorp, PM. Biology of human umbilical cord blood-derived hematopoietic stem/progenitor cells. *Stem Cells*, 1998, 16, 153-165.

McGuckin, CP; Forraz, N; Allouard, Q; Pettengell, R. Umbilical cord blood stem cells can expand hematopoietic and neuroglial progenitors in vitro. *Exp Cell Res*, 2004, 295, 350-359.

McGuckin, CP; Forraz, N; Baradez, MO; et al. Production of stem cells with embryonic characteristics from human umbilical cord blood. *Cell Prolif*, 2005, 38, 245-255.

Melguizo, C; Carrillo, E; Prados, J; et al. Non-classic cancer mechanism of multidrug resistance. In: Marchal, JA; Gaforio, JA; Aránega, A (eds). *Trends in basic and applied oncology*. Granada: Editorial Universidad de Granada, 2006, 286-298.

Miller, SJ; Lavker, RM; Sun, TT. Interpreting epithelial cancer biology in the context of stem cells: Tumor properties and therapeutic implications. *Biochem Bioph Acta*, 2005, 1756, 25-52.

Miller, SJ; Lavker, RM; Sun, TT. Keratinocyte stem cells of cornea, skin and hair follicle: common and distinguishing features. *Sem Develop Biol*, 1993, 4, 217-240.

Ming, GL; Song, H. Adult neurogenesis in the mammalian central nervous system. *Annu Rev Neurosci*, 2005, 28, 223-50.

Mitsui, K; Tokuzawa, Y; Itoh, H; et al. The homeoprotein Nanog is required for maintenance of pluripotency in mouse epiblast and ES cells. *Cell*, 2003, 113, 631-642.

Miura, M; Miura, Y; Padilla-Nash, HM. Accumulated chromosomal inestability in murine bone maroow mesenchymal ítems cells leads to malignant transformation. *Stem Cells*, 2006, 24, 1095-1103.

Molofsky, AV; He, S; Bydon, M; et al. Bmi-1 promotes neural stem cell self-renewal and neural development but not mouse growth and survival by repressing the p16Ink4a and p19Arf senescence pathways. *Genes Dev*, 2005, 19, 1432-1437.

Molofsky, AV; Pardal, R; Morrison, SJ. Diverse mechanisms regulate stem cell self-renewal. *Curr Opin Cell Biol*, 2004, 16, 700-707.

Monneret C. Histone deacetylase inhibitors. *Eur J Med Chem*, 2005, 40, 1-13.

Moore, MAS; Hattori, K; et al. Mobilization of endothelial and hematopoietic stem and progenitor cells by adenovector-mediated elevation of serum levels of SDF-1, VEGF, and angiopoietin-1. *Ann NY Acad Sci*, 2001, 938, 36-47.

Moriya, N; Komazaki, S; Takahashi S; et al. In vitro pancreas formation from Xenopus ectoderm treated with activin and retinoic acid. *Dev Growth Differ*, 2000, 42, 593-602.

Morrison, SJ; Kimble, J. Asymmentric and symmetric stem cell division in development and cancer. *Nature*, 2006, 441, 1068-1074.

Moscoso, I; Centeno, A; Lopez, E; et al. Differentiation "in vitro" of primary and immortalized porcine mesenchymal stem cells into cardiomyocytes for cell transplantation. *Transplant Proc*, 2005, 37, 481-492.

Moustapha, K. Mesenchymal Stem Cells: Biological Characteristics and Potential Clinical Applications. *Cloning and Stem Cells*, 2004, 6, 369 -374.

Murasawa, S; Kawamoto, A; Horii, M; et al. Niche-dependent translineage commitment of endothelial progenitor cells, not cell fusion in general, into myocardial lineage cells. *Arterioscler Thromb Vasc Biol*, 2005, 25, 1388-1394.

Murohara, T; Ikeda, H; Duan, J; et al. Transplanted cord blood-derived endothelial precursor cells augment postnatal neovascularization. *J Clin Invest*, 2000, 105, 1527-1536.

Nakahara, H; Dennis, JE; Bruder, SP; et al. In vitro differentiation of bone and hypertrophic cartilage from periosteal-derived cells. *Exp Cell Res*, 1991, 195, 492-503.

Nakamura, T; Mizuno, S; Matsumoto, K; et al. Myocardial protection from ischemia/reperfusion injury by endogenous and exogenous HGF. *J Clin Invest*, 2000, 106, 1511-1519.

Nelson, PJ; Gelman, IH; Klotman, PE. Suppression of HIV-1 expression by inhibitors of cyclin-dependent kinases promotes differentiation of infectd podocytes. *J Am Soc Nephrol*, 2001, 12, 2827-2831.

Nicolis, SK. Cancer stem cells and "stemness" genes in neuro-oncology. *Neurobiol Dis*, 2007, 25, 217-29.

Nielsen, D; Skovsgaard, T. P-glycoprotein as multidrug transporter: a critical review of current multidrug resistant cell lines. *Biochim Biophys A*cta, 1992, 1139, 169-183.

Niitsu N; Honma Y. Adenosine analogs as possible differentiation-inducing agents against acute myeloid leukemia. *Leuk Lymphoma*, 1999, 34, 261-271.

Ninkovic, J; Gotz, M. Signaling in adult neurogenesis: from stem cell niche to neuronal networks. *Curr Opin Neurobiol*, 2007, in press.

Niwa, H; Miyazaki, J; Smith, AG. Quantitative expression of Oct-3/4 defines differentiation, dedifferentiation or self-renewal of ES cells. *Nature Genet*, 2000, 24, 372–376.

O'Carroll, D; Erhardt, S; Pagani, M; et al. The Polycomb-group gene Ezh2 is required for early mouse development. *Mol Cell Biol*, 2001, 21, 4330-4336.

O'Donoghue, K; Choolani, M; Chan, J; et al. Identification of fetal mesenchymal stem cells in maternal blood: implications for non-invasive prenatal diagnosis. *Mol Hum Reprod*, 2003, 9, 497-502.

Odorico, JP; Kaufman, DS; Thomson, JA. Multilineage differentiation from human embryonic stem cell lines. *Stem Cells*, 2001, 19, 193-204.

Oh, S; Witek, R; Bae, S; et al. Bone marrow-derived hepatic oval cells differentiate into hepatocytes in 2-acetylaminofluorene/partial hepatectomy-induced liver regeneration. *Gastroenterology*, 2007, 132, 1077-1087.

Ohyashiki, JH; Sashida, G; Tauchi, T; Ohyashiki, K. Telomeres and telomerase in hematologic neoplasia. *Oncogene*, 2002, 21, 680-687.

Orlic, D; Kajstura, J; Chimenti, S; et al. Bone marrow cells regenerate infarcted myocardium. *Nature*, 2001a, 410, 701-705.

Orlic, D; Kajstura, J; Chimenti, S; et al. Mobilized bone marrow cells repair the infarcted heart, improving function and survival. *Proc Natl Acad Sci USA*, 2001b, 98, 10344-10349.

Ortiz, LA; Gambelli, F; McBride, C; et al. Mesenchymal stem cell engraftment in lung is enhanced in response to bleomycin exposure and ameliorates its fibrotic effects. *Proc Natl Acad Sci USA*, 2003, 100, 8407-8411.

Otto TC and Lane MD. Adipose development: from stem cell to adipocyte. *Crit Rev Biochem Mol Biol,* 2005, 40, 229-242.

Oyama, T; Kagawa, N; Kunugita, N; et al. Expression of cytochrome P450 in tumor tissues and its association with cancer development. *Front Biosci*, 2004, 9, 1967-1976.

Park, Y; Gerson, SL. DNA repair defects in stem cell function and aging. *Annu Rev Med*, 2005, 56, 495-508.

Parker WB; Cheng YC. Metabolism and mechanism of action of 5-fluorouracil. *Pharmacol Ther*, 1990, 48, 381-395.

Pasini, D; Bracken, AP; Jensen, MR; et al. Suz12 is essential for mouse development and for EZH2 histone methyltransferase activity. *EMBO J*, 2004, 23, 4061-4071.

Pedemonte, E; Benvenuto, F; Casazza, S; et al. The molecular signature of therapeutic mesenchymal stem cells exposes the architecture of the hematopoietic stem cell niche synapse. *BMC Genomics*, 2007, 6, 8-65.

Pinto, A; Zagonel, V. 5-Aza-2′-deoxicidine in the treatment of acute myeloid leukaemias and myelodisplasic syndromes: past, present and future trends. *Leukaemia*, 1993, 7, 51-60.

Pittenger, MF; Mackay, AM; Beck, SC; et al. Multilineage potential of adult human mesenchymal stem cells. *Science*, 1999, 284, 143-147.

Pla P; Soloveva O; Moore R et al. Dct:lacZ ES cells: a novel cellular model to study melanocyte determination and differentiation. *Pigment Cell Res*, 2004, 17, 142-149.

Pomerantz, J; Blau, HM. Nuclear reprogramming: a key to stem cell function in regenerative medicine. *Nat Cell Biol*, 2004, 6, 810-816.

Prados, J; Melguizo, C; Carrillo, E; et al. Cancer mechanism of multidrug resistance mediated by P-glycoprotein In: Marchal, JA; Gaforio, JA; Aránega, A (eds). *Trends in basic and applied oncology*. Granada: Editorial Universidad de Granada, 2006, 266-283.

Prados, J; Melguizo, C; Fernandez, JE; et al. Inverse expression of mdr 1 and c-myc genes in a human rhabdomyosarcoma cell line resistant to actinomycin D. *J Pathol*, 1996, 180, 85-89.

Prados, J; Melguizo, C; Marchal JA; et al. Therapeutic differentiation in a human rhabdomyosarcoma cell line selected for resistance to actinomycin D. *Int J Cancer*, 1998, 75, 379-383.

Preston, SL; Alison, MR; Forbes, SJ; et al. The new stem cell biology: something for everyone. *Mol Pathol*, 2003, 56, 86-96.

Qi, X; Li, TG; Hao, J; et al. BMP4 supports self-renewal of embryonic stem cells by inhibiting mitogen-activated protein kinase pathways. *Proc Natl Acad Sci USA*, 2004, 101, 6027-6032.

Quarto, R; Mastrogiacomo, M; Cancedda, R; et al. Repair of large bone defects with the use of autologous bone marrow stromal cells. *N Engl J Med*, 2001, 344, 385-396.

Radford, IR; Lobachevsky, PN. An enteroendocrine cell-based model for a quiescent intestinal stem cell niche. *Cell Prolif*, 2006, 39, 403-414.

Rambhatla, I; Ram-Mohan, S; Cheng, JJ; Sherley, JL. Immortal DNA strang congregation requires p53/IMPDH-dependent asymmetric self-renewal associated with adult stem cells. *Cancer Res*, 2005, 65, 3155-3161.

Rangappa, S; Entwistle, JW; Wechsler, AS; Kresh, JY. Cardiomyocyte-mediated contact programs human mesenchymal stem cells to express cardiogenic phenotype. *J Thorac Cardiovasc Surg*, 2003, 126, 124-132.

Reya, T; Morrison, SJ; Clarke, MF; Weissman, IL. Stem cells, cancer, and cancer stem cells. *Nature*, 2001, 414, 105-111.

Reyes, M; Dudek, A; Jahagirdar, B; et al. Origin of endothelial progenitors in human postnatal bone marrow. *J Clin Invest*, 2002, 109, 337-46.

Reyes, M; Lund, T; Lenvik, T; et al. Purification and ex vivo expansion of postnatal human marrow mesodermal progenitor cells. *Blood*, 2001, 98, 2615-2625.

Rhodes, NP; Srivastava, JK; Smith, RF; Longinotti, C. Heterogeneity in proliferative potential of ovine mesenchymal stem cell colonies. *J Mater Sci Mater Med*, 2004, 15, 397-402.

Ribatti, D. The discovery of endothelial progenitor cells. An historical review. *Leukemia Res*, 2007, 31, 439-444.

Ribatti, D. The involvement of endothelial progenitor cells in tumor angiogenesis. *J Cell Mol Med*, 2004, 8, 294-300.

Ringe, J; Haupl, T; Sittinger, M. Mesenchymal stem cells for tissue engineering of bone and cartilage. *Med Klin*, 2003, 98, 35-40.

Ringrose, L; Paro, R. Epigenetic regulation of cellular memory by the Polycomb and Trithorax group proteins. *Annu Rev Genet*, 2004, 38, 413-443.

Rodda, DJ; Chew, JL; Lim, LH; et al. Transcriptional regulation of nanog by OCT4 and SOX2. *J Biol Chem*, 2005, 280, 24731–24737.

Rojas, M; Xu, J; Woods, CR; et al. Bone marrow-derived mesenchymal stem cells in repair of the injured lung. *Am J Respir Cell Mol Biol*, 2005, 33, 145-152.

Romagnani, P; Annunziato, F; Liotta, F; et al. CD14+CD34low cells with stem cell phenotypic and functional features are the major source of circulating endothelial progenitors. *Cir Res*, 2005, 97, 314-322.

Rosner, MH; Vigano, MA; Ozato, K; et al. A POU-domain transcription factor in early stem cells and germ cells of the mammalian embryo. *Nature*, 1990, 345, 686-692.

Ross, SE; Hemati, N; Longo, KA; et al. Inhibition of adipogenesis by Wnt signaling. *Science*, 2000, 289, 950-953.

Rossant, J. Stem cells and lineage development in the mammalian blastocyst. *Reprod Fertil Dev*, 2007, 19, 111-118.

Rustum, YM; Radel, S; Campbell, J; Mayhew, E. Approaches to overcome in vivo anti-cancer drug resistance. *Prog Clin Biol Res*, 1986, 223, 187-202.

Saito, T; Kuang, JQ; Lin, CC; Chiu, RC. Transcoronary implantation of bone marrow stromal cells ameliorates cardiac function after myocardial infarction. *J Thorac Cardiovasc Surg*, 2003, 126, 114-123.

Sanai, N; Tramontin, AD; Quinones-Hinojosa, A; et al. Unique astrocyte ribbon in adult human brain contains neural stem cells but lacks chain migration. *Nature*, 2004, 427, 740-744.

Sancho, E; Batlle, E; Clevers, H. Live and let die in the intestinal epithelium. *Curr Opin Cell Biol*, 2003, 15, 763-770.

Satake, K; Lou, J; Lenke, LG. Migration of mesenchymal stem cells through cerebrospinal fluid into injured spinal cord tissue. *Spine*, 2004, 29, 1971-1979.

Satija, NK; Gurudutta, GU; Sharma, S; et al. Mesenchymal stem cells: molecular targets for tissue engineering. *Stem Cell Dev*, 2007, 16, 7-23.

Sato, N; Meijer, L; Skaltsounis, L; et al. Maintenance of pluripotency in human and mouse embryonic stem cells through activation of Wnt signaling by a pharmacological GSK-3-specific inhibitor. *Nature Med*, 2004, 10, 55-63.

Sato, Y; Araki, H; Kato, J; et al. Human mesenchymal stem cells xenografted directly to rat liver differentiated into human hepatocytes without fusion. *Blood*, 2005, 106, 756-763.

Sausville EA; Johnson J; Alley M et al. Inhibition of CDKs as a therapeutic modality. *Ann N Y Acad Sci*, 2000, 910, 207-222.

Scadden, DT. The stem cell niche in health and leukemic disease. *Best Pract Res Clin Haematol*, 2007, 20, 19-27.

Scadden, DT. The stem-cell niche as an entity of action. *Nature*, 2006, 441, 1075-1079.

Scheper, RJ; Broxterman, HJ; Sheffer, GL; et al. Overexpression of a M(r) 110,000 vesicular protein in non-P-glycoprotein-mediated multidrug resistance. *Cancer Res*, 1993, 53, 1475-1479.

Schiller, PC; D'Ippolito, G; Brambilla, R; et al. Inhibition of Gap-Junctional Communication Induces the Trans-differentiation of Osteoblasts to an Adipocytic Phenotype in Vitro. *Biol Chem*, 2001, 276, 14133-14138.

Scott RE. Differentiation, differentiation/gene therapy and cancer. *Pharmacol Ther*, 1997, 73, 51-65.

Sell, S. Stem cell origin of cancer and differentiation therapy. *Crit Rev Oncol Hematol*, 2004, 51, 1-28.

Sell, S; Pierce, GB. Maturation arrest of stem cell differentiation is a common patway for the cellular origin of teratocarcinomas and epithelial cancer. *Lab Invest*, 1994, 70, 6-22.

Serakinci, N; Guldberg, P; Burns, JS; et al. Adult human mesenchymal stem cell as a target for neoplastic transformation. *Oncogene*, 2004, 23, 5095-5098.

Serakinci, N; Keith, NW. Therapeutic potential of adult stem cells. *Eur J Cancer*, 2006, 42, 1243-1246.

Serup, P; Madsen, OD; Mandrup-Poulsen T. Islet and stem cell transplantation for treating diabetes. *Br Med J*, 2001, 322, 29-32.

Shantsila, E; Watson, T; Lip, GYH. Endothelial progenitor cells in cardiovascular disorders. *J Am Coll Cardiol*, 2007, 49, 741-752.

Sherley, JL. Asymmetric cell kinetics genes: the key to expansion of adult stem cells in culture. *Stem Cells*, 2002, 20, 561-572.

Shi, Y; Hou, L; Tang, F; et al. Inducing embryonic stem cells to differentiate into pancreatic beta cells by a novel three-step approach with activin A and all-trans retinoic acid. *Stem Cells*, 2005, 23, 656-662.

Shintani, S; Murohara, T; Ikeda, H; et al. Mobilization of endothelial progenitor cells in patients with acute myocardial infarction. *Circulation*, 2001, 103, 2776-2779.

Shufaro, Y; Reubinoff, BE. Therapeutic applications of embryonic stem cells. *Best Pract Res Clin Obstet Gynaecol*, 2004, 18, 909-927.

Silva, GV; Litovsky, S; Assad, JA; et al. Mesenchymal stem cells differentiate into an endothelial phenotype, enhance vascular density and improve heart function in a canine chronic ischemia model. *Circulation*, 2005, 111, 150-156.

Silverman LR; Holland JF; Weinberg RS et al. Effects of treatment with 5-azacytidine on the in vivo and in vitro hematopoiesis in patients with myelodysplastic sydromes. *Leukemia*, 1993, 7, 21-29.

Singh, SK; Clarke, ID; Hide, T; Dirks, PB. Cancer stem cells in nervous system tumors. *Oncogene*, 2004, 23, 7267-7273.

Singh, SK; Clarke, ID; Terasaki, T; et al. Identification of a cancer stem cell in human brain tumors. *Cancer Res*, 2003, 63, 5821-5828.

Slack, JMW; Tosh, D. Transdifferentiation and metaplassia: switching of cell types. *Curr Opin Genet Dev*, 2001, 11, 581-586.

Smith LN; Anderson JR; Coffin CM. Cytodifferentiation and clinical outcome after chemotherapy and radiation therapy for rhabdomyosarcoma (RMS). *Med Pediatr Oncol* 2002, 38, 398-404.

Smith, AG. Embryo-derived stem cells: of mice and men. *Annu Rev Cell Dev Biol*, 2001, 17, 435-462.

Snykers, S; Vinken, M; Rogiers, V; Vanhaecke, T. Differential role of epigenetic modulators in malignant and normal stem cells: a novel tool in preclinical in vitro toxicology and clinical therapy. *Arch Toxicol*, 2007, in press.

Song, LL; Miele, L. Cancer stem cells an old idea that's new again: implications for the diagnosis and treatment of breast cancer. *Expert Opin Biol Ther*, 2007, 7, 431-438.

Song, X; Xie, T. Wingless signalling regulates the maintenance of ovarian somatic stem cells in Drosophila. *Development*, 2003, 130, 3259-3268.

Sottile, V; Li, M; Scotting, PJ. Stem cell marker expression in the Bergmann glia population of the adult mouse brain. *Brain Res*, 2006, 1099, 8-17.

Stafford, D; Prince, VE. Retinoic acid signaling is required for a critical early step in Zebrafish pancreatic development. *Curr Biol*, 2002, 12, 1215-1220.

Steiner, E; Holzmann, K; Elbling, L; Berger, W. Cellular functions of vaults and their involvement in multidrug resistance. *Curr Drug Targets*, 2006, 7, 923-934.

Sun, B; Jeong, YH; Jung, JW; et al. Regulation of human umbilical cord blood-derived multipotent stem cells by autogenic osteoclast-based niche-like structure. *Biochem Biophys Res Commun*, 2007, 357, 92-98.

Taipale, J; Beachy, PA. The Hedgehog and Wnt signalling pathways in cancer. *Nature*, 2001, 411, 349-354.

Tan, B; Piwnica-Worms, D; Ratner, L. Multidrug resistance transporters and modulation. *Curr Opin Oncol*, 2000, 12, 450-458.

Tan, BT; Park, CY; Ailles, LE; Weissman, IL. The cancer stem cell hypothesis: a work in progress. *Lab Invest*, 2006, 86, 1203-1207.

Taylor, SM. 5-Aza-2′-deoxicidine: cell differentiation and DNA methylation. *Leukaemia*, 1993, 7, 3-8.

Terada, N; Hamazaki, T; Oka, M; et al. Bone marrow cells adopt the phenotype of other cells by spontaneous cell fusion. *Nature*, 2002, 416, 542-545.

Thomson JA; Itskovitz-Eldor, J; Shapiro, SS; Waknitz, MA; Swiergiel, JJ; Marshall, VS; Jones, JM. Embryonic stem cell lines derived from human blastocysts. *Science*, 1998, 282, 1145-1147.

Thomson, JA; Odorico, JS. Human embryonic stem cell and embryonic germ cell lines. *Trends Biotechnol*, 2000, 18, 53-57.

Tian, XC; Smith, SL; Zhan, SQ; et al. Nuclear reprogramming by somatic cell nuclear transfer the cattle story. *Soc Reprod Fertil Suppl*, 2007, 64, 327-339.

Tobita T; Takeshita A; Kitamura K; et al. Treatment with a new synthetic retinoid Am80, of acute promyelocytic leukemia relapsed from complet remission induced by all-*trans* retinoic acid. *Blood*, 1997, 90, 967-973.

Togel, F; Hu, Z; Weiss, K; et al. Amelioration of Acute Renal Failure by Stem Cell Therapy–Paracrine Secretion Versus Transdifferentiation into Resident Cells: Administered Mesenchymal Stem Cells Protect against Ischemic Acute Renal Failure through Differentiation-Independent Mechanisms. *Am J Physiol Renal Physiol*, 2005, 16, 1153-1163.

Tomita, S; Li, RK; Weisel, RD; et al. Autologous transplantation of bone marrow cells improves damaged heart function. *Circulation*, 1999, 100, Suppl II, II247-256.

Tomita, S; Mickle, DA; Weisel, RD; et al. Improved heart function with myogenesis and angiogenesis after autologous porcine bone marrow stromal cell transplantation. *J Thorac Cardiovasc Surg*, 2002, 123, 1132-1140.

Tonini, T; Rossi, F; Claudio, PP. Molecular basis of angiogenesis and cancer. *Oncogene*, 2003, 22, 6549-6556.

Toogood PL. Progress toward the development of agents to modulate the cell cycle. *Curr Op Chem Biol*, 2002, 6, 472-478.

Tosh, D; Slack, JM. How cells change their phenotype. *Nat Rev Mol Cell Biol*, 2002, 3, 187-194.

Tsuji-Takayama K; Inoue T; Ijiri Y et al. Demethylating agent, 5-azacytidine, reverses differentiation of embryonic stem cells. *Biochem Bioph Res Co*, 2004, 323, 86-90.

Tsurtuo, T. Molecular cancer therapeutics: recent progress and targets in drug resistance. *Inter Med*, 2003, 42, 237-243.

Uccelli, A; Moretta, L; Pistoia, V. Immunoregulatory function of mesenchymal stem cells. *Eur J Immunol*, 2006, 36, 2566-2573.

Udani, VM. The continuum of stem cell transdifferentiation: possibility of hematopoietic stem cell plasticity with concurrent CD45 expression. *Stem Cells Dev*, 2006, 15, 1-3.

Valk-Lingbeek, ME; Bruggeman, SW; van Lohuizen, M. Stem cells and cancer, the polycomb connection. *Cell*, 2004, 118, 409-418.

Vallier, L; Alexander, M; Pedersen, RA. Activin/Nodal and FGF pathways cooperate to maintain pluripotency of human embryonic stem cells. *J Cell Sci*, 2005, 118, 4495-4509.

Vast, A; Bielby, RC; Tolley, NS; et al. Stem cells. *Lancet*, 2005, 366, 592-602.

Venezia, TA; Merchant, AA; Ramos, CA; et al. Molecular signatures of proliferation and quiescence in hematopoietic stem cells. *PlotS Biol*, 2004, 2, 301.

Venezia, TA; Merchant, AA; Ramos, CA; et al. Molecular signatures of proliferation and quiescence in hematopoietic stem cells. *PLoS Biol*, 2004, 2, 301.

Ventura, C; Cantoni, S; Bianchi, F; et al. Hyaluronan mixed esters of butyric and retinoi acid drive cardiac and endothelial fate in term placenta human mesenchymal stem cells and enhance cardiac repair in infarcted rat hearts. *J Biol Chem*, 2007, 282, 14243-14252.

Ventura, C; Maioli, M; Asara, Y; et al. Butyric and retinoic mixed ester of hyaluronan. A novel differentiating glycoconjugate affording a high throughput of cardiogenesis in embryonic stem cells. *J Biol Chem*, 2004, 279, 23574-23579.

Villadsen, R. In search of a stem cell hierarchy in the human breast and its relevance to breast cancer evolution. *Apmis*, 2005, 113, 903-921.

Villadsen, R; Fridriksdottir, AJ; Ronnov-Jessen, L; et al. Evidence for a stem cell hierarchy in the adult human breast. *J Cell Biol*, 2007, 177, 87-101.

Wakamaya T; Tabar V; Rodriguez I; Perry, AC; Studer, L; Mombaerts, P. Differentiation of embryonic stem cells lines generated from adult somatic cells by nuclear transfer. *Science*, 2001, 292, 740-743.

Wakitani, S; Imoto, K; Yamamoto, T; et al. Human autologous culture expanded bone marrow mesenchymal cell transplantation for repair of cartilage defects in osteoarthritic knees. *Osteoarthritis Cartilage*, 2002, 10, 199-206.

Wang, S; Yang, D; Lippman, ME. Targeting Bcl-2 and Bcl-XL with nonpeptidic small-molecule antagonists. *Semin Oncol*, 2003, 16,133-142.

Waters, JM; Richardson, GD; Jahoda, CA. Hair follicle stem cells. *Semin Cell Dev B*iol, 2007, 18, 245-254.

Watkins, DN; Berman, DM; Burkholder, SG; et al. Hedgehog signalling within airway spithelial progenitors and in small-cell lung cancer. *Nature*, 2003, 422, 313-317.

Weimann, JM; Charlton, CA; Brazelton, TR; Hackman, RC; et al. Contribution of transplanted bone marrow cells to Purkinje neurons in human adult brains. *Proc Natl Acad Sci USA*, 2003, 100, 2088-2093.

Werner, N; Kosiol, S; Schiegl, T; et al. Circulating endothelial progenitor cells and cardiovascular outcomes. *N Engl J Med*, 2005, 353, 999-1007.

Werner, N; Nickenig, G. Clinical and therapeutical implications of EPC biology in atherosclerosis. *J Cell Mol Med*, 2006, 10, 318-332.

Wicha, MS; Liu, S; Dontu, G. Cancer stem cells: An old idea-A paradigm shift. *Cancer Res*, 2006, 66, 1883-1890.

Wijermans P; Lubbert M; Verhoef G et al. Low-dose 5-Aza-2′-deoxicytidine, a DNA hypomethylating agent, for the treatment of high-risk myelodisplastic syndrome: A multicenter phase II study in elderley patinets. *J Clin Oncol*, 2000, 18, 956-962.

Wijermans, P; Lubbert, M; Verhoef, G; et al. Low-dose 5-Aza-2′-deoxicidine, a DNA hypomethylating agent, for the treatment of high-risk Myelodisplasic syndrome: a multicenter phase II study in elderly patients. *J Clin Oncol*, 2000, 18, 956-962.

Wislet-Gendebien, S; Wautier, F; Leprince, P; Rogister, B. Astrocytic and neuronal fate of mesenchymal stem cells expressing nestin. *Brain Res Bull*, 2005, 68, 95-102.

Wobus, AM; Boheler, KR. Embryonic stem cells: Prospects for developmental biology and cell therapy. *Physiol Rev*, 2005, 85, 635–678.

Wollert, KC; Meyer, GP; Lotz, J; et al. Intracoronary autologous bone-marrow cell transfer after myocardial infarction: The BOOST randomized controlled clinical trial. *Lancet*, 2004, 364, 141-148.

Wu, DC; Byod, AS; Wood, KJ. Embryonic stem cell transplantation: potential applicability in cell replacement therapy and regenerative medicine. *Front Biosci*, 2007, 12, 4525-35.

Wurmser, AE; Gage, FH. Cell fusion causes confusion. *Nature*, 2002, 416, 485-487.

Xiong-Zhi, W; Dan, C; Guang-Ru, X. Bone marrow-derived cells: roles in solid tumor. *Neoplasma*, 2007, 54, 1-6.

Xu, M; Uemura, R; Dai, Y; et al. In vitro and in vivo effects of bone marrow stem cells on cardiac structure and function. *J Mol Cell Cardiol*, 2007, 42, 441-448.

Yamashita, YM; Jones, DL; Fuller, MT. Orientation of asymmetric stem cell division by the APC tumor suppressor and centrosome. *Science*, 2003, 301, 1547-1550.

Yang Y-W; Chang Y-H. Induction of erytroid differentiation by 5-fluorouracil in K562 leukemia cells. *Jpn J Cancer Res*, 1995, 86, 948-955.

Yang, L; Li, S; Hatch, H; et al. In vitro trans-differentiation of adult hepatic stem cells into pancreatic endocrine hormone-producing cells. *Proc Natl Acad Sci USA*, 2002, 99, 8078-8083.

Yang, ZJ; Wechsler-Reya, RJ. Hit 'em where they live: targeting the cancer stem cell niche. *Cancer Cell*, 2007, 11, 3-5.

Yilmaz, OH; Valdez, R; Theisen, BK; et al. Pten dependence distinguishes haematopoietic stem cells from leukaemia-initiating cells. *Nature*, 2006, 441, 475-482.

Ying, QL; Nichols, J; Chambers, I; Smith, A. BMP induction of Id proteins suppresses differentiation and sustains embryonic stem cell self-renewal in collaboration with STAT3. *Cell*, 2003, 115, 281-292.

Yoon, YS; Park, JS; Tkebuchava T; et al. Unexpected severe calcification after transplantation of bone marrow cells in acute myocardial infarction. *Circulation*, 2004, 109, 3154-3157.

Zhang, DZ; Gai, LY, Liu, HW; Zhu, XY. Adipose-derived stem cells differentiate into endothelial cells in vitro. *Chin J Clin Rehabil*, 2005a, 9, 14-17.

Zhang, DZ; Gai, LY; Liu, HW; Zhu, XY. Transformation of adipose derived stem cells into cardiomyocytes. *Chin Heart* J, 2005b, 17, 405-408.

Zhang, Y; Kalderon, D. Hedgehog acts as a somatic stem cell factor in the Drosophila evary. *Nature*, 2001, 410, 599-604.

Zhou, S; Morris, JJ; Barnes, Y; et al. Bcrpl gene expression is required for normal numbers of side population stem cells in mice, and confers relative protection to mitoxantrone in hematopoietic cells in vivo. *Proc Natl Acad Sci USA*, 2002, 99, 12339-12344.

In: New Cell Differentiation Research Topics
Editor: Hitoshi Saitama, pp. 79-116
ISBN: 978-1-60021-937-5

Chapter II

Antioxidants in Differentiation of Leukemic Cells

Guillermo López-Lluch*, Emilio Siendones*,
Daniel J. Moreno Fernández-Ayala, and Plácido Navas
Centro Andaluz de Medicina del Desarrollo, CABD-CSIC, Universidad Pablo de Olavide, 41013 Sevilla, Spain.

Abstract

The therapy of leukaemia based on the use of compounds able to reinitiate the differentiation program has been extensively studied. The use of calcitriol ($1,25D_3$) or trans-retinoic acid (ATRA) in myeloid or monocytic leukemic cell differentiation has been also clinically taken into consideration. Both are natural compounds accepted at relatively high doses by the organism, however, several other secondary effects take place because their other respective main functions different than regulating the differentiation program in cancerous cells. To avoid the secondary effects due to overloading of both, ATRA or $1,25D_3$, therapies based on the co-administration of other natural compounds such as vitamin C, vitamin E, polyphenols, carotenoids or other antioxidants acting as anti-cancer active compounds have been taking into consideration. In several cases, these natural compounds do not affect by themselves the differentiation program in leukemic cells but they are able to enhance the effect of the differentiation-active compounds such as ATRA or $1,25D_3$. The present chapter reviews the effect of the combination of these natural antioxidant compounds on the treatment of leukemic cells and will try to define the probable common mechanism of action for this plethora of substances.

---

* These two authors contribute equally to this work.

INTRODUCTION

A Complex Picture in Acute Myeloid Leukaemia Therapy

Leukaemia is defined as a haematopoietic stem cells malignant neoplasia initiated in the bone marrow but that invades blood and other organs. Leukaemia is classified depending on the lineage of the compromised cell (myeloid or lymphoid) and of the mature state of the leukemic cell (acute or chronic). When the leukemic cell involved an early state in the development of the haematopoiesis, we are treating an acute leukaemia but when leukemic cells are at the latest steps in the differentiation program, it is a chronic leukaemia.

Between the different kinds of leukaemia, acute myeloid leukaemia (AML) is at the moment, the most common leukaemia form and the most frequent cause of death by leukaemia disease. It can be defined as a clonal disorder by the accumulation of abnormally differentiated myeloid cells that are not mature. Due to the different steps in the maturation of blood cells and the different lineages into the myeloid group, several different types of AML can be found (Table I).

Table I. FAB Classification of AML

Morphological Features of AML	
TYPE, FAB, classification	**Morphological characteristics**
Acute myelocytic leukemia (M1)	Very undifferentiated cells, very low cytoplasmic granules, some promyelocytes
Acute myelocytic leukemia (M2)	Predominance of granulated blasts, small number of monocytoid cells, differentiation beyond promyelocytic stage evident
Acute promyelocytic leukemia (M3)	Predominance of typically, hypergranular promyelocytes, cells have large basophilic and esoinophilic granules
Acute myelomonocytic leukemia (M4)	Precursors of monocytes and granulocytes, serum lysozyme elevated
Acute monoblastic leukemia (M5)	Large monoblasts showing abundant agranular cytoplasm that may be vacuolated and basophilic
Erythroleukemia (M6)	Predominance of megaloblastoid red cell precursors, some myeloid blasts can also be seen, presence of multinucleated red cell precursors
Megakaryocytic leukemia (M7)	Variable morphology, megakaryocytic features may not be seen with light microscopy

The broadly most well known causes in the development of AML are multiple genetic defects, which often involve either translocations or mutations within oncogenes. These genes can be critical components of signalling pathways that regulate proliferation, differentiation, and apoptosis.

An unresolved problem is how to know the proteins involved in leukemogenesis and to understand the mechanisms by which the leukaemia develops. So that, a new drug therapy could be developed against this disease. The rational combination of different strategies based on regimens defined by prospective randomised studies will help in future therapies in order to conquer the disease.

Treatment of AML has not changed in the last years. Anthracyclines and cytarabine are still the mainstays of first-line therapy. Novel chemotherapeutic agents, such as temozolomide, an oral alkylating agent, and troxacitabine, a novel nucleotide analogue, have been studied, but it is unclear whether they are better than conventional therapy. In addition to this, most AML patients are intrinsically resistant to chemotherapy. In fact, most patients die either after relapse or from the side effects associated with the treatment. The advances made in the elucidation of the mechanisms involved in leukemogenesis and the molecular alterations underlying the pathogenesis of AML have enabled the development and use of new therapies directed to specific molecular targets. One important think to understand is that a unique genetic abnormality may not be sufficient to promote leukemogenesis. The cooperation or the interaction between various regulatory pathways is frequent events in the development of leukaemia (Delgado, *et al* 2000, Hayakawa, *et al* 2000, Mizuki, *et al* 2000, Stirewalt, *et al* 2003). Therefore, if multiple defects are required, targeting one of these multiple genetic abnormalities may not be sufficient to reverse leukemogenesis, but this reversion may be enhanced by simultaneously targeting several different pathways within the same AML clone.

Small-molecule inhibitors of signalling pathways are being currently developed for the treatment of AML. These compounds are typically targeted against pathways that have been activated by mutation. However, only modest clinical results have been shown thus far. The causes are diverse: in some cases, the target molecule is inhibited for an inadequate duration, and in other cases, the clinical effect does not show a relationship with the inhibition of the target. Although small-molecule inhibitors are a promise in the treatment of AML, this field of pharmacotherapy is still quite new and they are several obstacles to overcome, and in several cases, when administered as monotherapy, the inhibitors seem unlikely to be curative. In addition, pharmacokinetic problems must be overcome with many of them. However, it is important to believe that the discovery and characterization of new molecular alterations in AML will help to develop novel therapies that will cure the majority of patients. The prototype for this approach, is the treatment of acute promyelocytic leukaemia (APL, M3 status in table I) with all-trans retinoic acid (ATRA) (Tallman, *et al* 2002). In this type of AML, the retinoic acid receptor fused to the promyelocytic leukaemia (PML) gene product. The appropriate treatment with ATRA has turned a poor-prognosis disease into one in which most cases are cured. AML, like APL, could be cured with the appropriate combination of chemotherapy with targeted therapy. Numerous investigations have been performed to elucidate the mechanisms and search for effective agents in the treatment of other types of AML. Now, we know that the molecular abnormalities in AML can be found both at the epigenetic level and at the genetic level. Many of these molecular pathways result in changes in gene transcription, which result in a blockade in differentiation of leukemic cells. Following to ATRA, several new agents have emerged, although their clinical effectiveness remains to be confirmed. Genetic therapy and a combination of different strategies will improve the treatment of AML.

Therapeutic Targets in Leukaemia

The therapeutic targets can be grouped in three levels that distinguish AML cells from the normal haematopoietic stem cell: mutated proteins, adaptive non-mutational changes to survive after the initial mutational event, and cell surface antigens.

Therapy against Mutated Proteins

It has become evident that tyrosine kinases are common and frequent targets for mutations in human leukaemia. In fact, the FMS-like tyrosine kinase (Flt3), a cellular receptor involved in cell survival and differentiation, is commonly overexpressed in AML cells. Furthermore, two distinct groups of Flt3 mutations (10-30% of cases), which confer ligand Flt3-independent activation and lead to a deregulated protein tyrosine kinase activity, have been described (Kiyoi, *et al* 1998, Yamamoto, *et al* 2001). Such activation also induces factor-independent proliferation of human haematopoietic cell lines, and in transgenic mice produces a myeloproliferative syndrome (Kelly, *et al* 2002), although does not result in full-phenotype AML. In fact, recently it has been shown that constitutive activation of signalling pathways was not correlated to Flt3 activity in some patient's AML cells (Siendones, *et al* 2007). This suggests that other genetic events, such as mutations in genes affecting differentiation, are also required for the development of AML (Deguchi and Gilliland 2002). Flt3 is now recognized to be a critical mediator in the pathogenesis of myeloid and some lymphoid leukaemia. Thus, Flt3 is nowadays a promising target for therapeutic intervention. The efficacy of new small molecule Flt3 tyrosine kinase inhibitors has been shown in clinical trials (O'Farrell, *et al* 2003), but there is insufficient information about the long-term effects of these drugs. Several broad classes of Flt3 protein tyrosine kinase inhibitors are undergoing evaluation in clinical trials. Although the agents are well tolerated by patients, clinical responses in relapsed or refractory AML are limited and transient. Nowadays it is known that the acquisition of mutations in the kinase catalytic domain of Flt3 can confer resistance to inhibitors (Clark, *et al* 2004, Weisberg, *et al* 2002). These data enhance the idea that the most appropriate treatments for most AML patients may require the combination of various specific agents. A reasonable approach is to target not only the oncogenic tyrosine kinase, but also the additional pathways that may be required for transformation and blockade of differentiation.

RAS molecules are a group of signalling proteins downstream of many tyrosine kinase receptors that are involved in cell proliferation and survival (FIGURE 1). Mutations in members of the RAS family are found in 15-20% of cases, most commonly in N-RAS and less frequently in K-RAS and H-RAS (Radich, *et al* 1990). The RAS proteins are synthesized as cytosolic precursors and require the addition of a farnesyl group to be attached to the plasma membrane, so that they are able to transmit signals to the rest of the cell. Thus the blockade of farnesylation of RAS with farnesyltransferase inhibitors could be a target for therapeutic intervention in leukaemia harbouring a mutation in RAS. Among the more prominent proteins activated by RAS are the MAP/ERK kinases, which are involved in cell proliferation, Rac and Rho, which influence cytoskeletal organization, and phosphatidyl inositol (PI)-3/Akt kinases, and STAT proteins which may protect against apoptosis.

The enzyme 3-hydroxy-3-methylglutaryl-coenzyme A (HMGCoA) reductase is the rate-limiting enzyme of the mevalonate pathway and catalyses the reduction of HMG-CoA to mevalonate, which is an intermediate in the synthesis of various substances including cholesterol and the isoprenoid molecules farnesylpyrophosphate (FPP) and geranylgeranylpyrophosphate (GGPP) (Goldstein and Brown 1990). These isoprenoids are used for posttranslational modification of a variety of proteins, including the above-indicated RAS and also the proteins of the RHO family. Thus, the mevalonate pathway can be also a therapeutic target through the inhibition of HMG-CoA reductase by inhibition of cell proliferation or induction of apoptosis and chemosensitivity elicited by RAS.

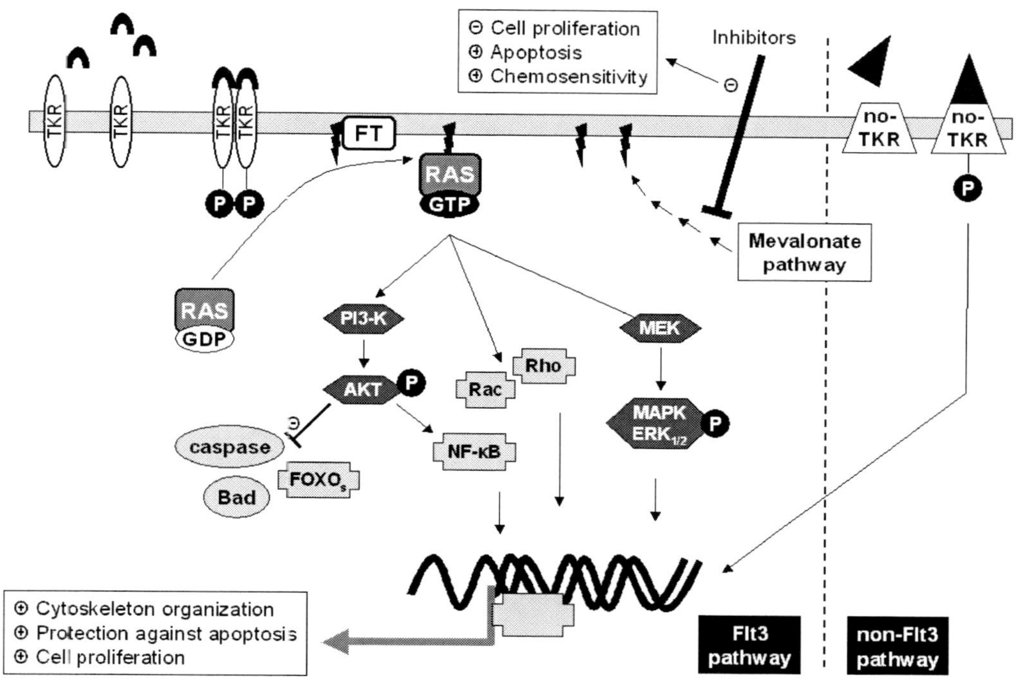

Figure 1. Ras-dependent mechanism of activation of leukemogenesis in AML. Mutations in members of the RAS family are found in a significative population of AML cases. Ras proteins transmit signals to several intracellular intermediatres that affect the PI3K/Akt kinases MAPK, and STAT pathways.

Therapy against Adaptive Non-mutational Changes

AML may also require the cell to make other adaptive changes in order to survive after leukemogenesis. In that case, AML cells may become more dependent than normal cells on the function of certain pathways and, thus, blocking them may specifically sensitise malignant cells to conventional therapy. A known example is Bcl-2, an antiapoptotic gene, which is overexpressed in almost all AML samples. Moreover, Bcl-2 expression can be enhanced after treatment with antileukemic agents. Thus, the interference of Bcl-2 by RNAi sensitises AML cells to the exposure to cytotoxic drugs (Banker, *et al* 1997, Campos, *et al* 1993), providing the basis for ongoing clinical studies.

Oxidative stress has been suggested to play an important role in the carcinogenic process (Klaunig and Kamendulis 2004). Reactive oxygen species (ROS) can elicit multiple effects by mediating carcinogen activation, causing DNA damage, and interfering with the repair machinery of the damaged DNA (Jackson and Loeb 2001). In addition, ROS are increased in preneoplastic cells during their promotion stage. This allows for the selective growth of initiated cells, and they may also play a role during cell cycle progression either by further damage the unstable genetic material or by altering the growth of tumour cells (Behrend, *et al* 2003, Denda, *et al* 1993). Thus, tumour cells continually undergo high and persistent oxidative stress by ROS; however, this stress does not appear to be enough to induce cell death. One reason why preneoplastic cells have increased cell resistance to ROS may be because they have shown high levels of antioxidants such as coenzyme Q (CoQ). These levels induced by persistent oxidative stress may increase the chemotherapeutic resistance of the cells. Recently we have shown that anticancer drugs such as camptothecin (CPT), a topoisomerase I inhibitor that induces oxidative stress, affects cellular CoQ levels (Brea-Calvo, *et al* 2006). CPT but also other anticancer drugs -such as etoposide, doxorubicin, and methotrexate- induces a significant increase in ROS production (Sanchez-Alcazar, *et al* 2003). Thus, cells induce the up-regulation of CoQ levels in an attempt to protect them against the oxidative stress generated after exposure to anticancer drugs. Another adaptive response that occurs is the increase of cholesterol synthesis. Following exposure to radiation or chemotherapy, surviving cells greatly increase cholesterol synthesis acquiring a state of cytoresistance. CoQ and cholesterol share the first part of their respective synthesis chain being FPP a common substrate for CoQ and cholesterol biosynthesis. If CoQ and/or cholesterol synthesis are blocked, the acquired cytoresistance is diminished and the cells will remain sensitive to the damaging agents. Therefore, statins, a class of hypolipidemic agents, which inhibit the HMG-CoA reductase, could be appropriate candidates for the treatment of AML. In fact, it has already being showed that exposure to lovastatin sensitise both AML cell lines and primary AML cell cultures to apoptosis (Dimitroulakos, *et al* 1999, Wong, *et al* 2002).

Cell Surface Antigens

Cell surface antigens can be also potential targets for antibodies and cellular immunotherapy. For example, CD33 is a glycoprotein found on blasts in more than 90% of cases of AML. It is also expressed on almost all normal early myeloid and erythroid progenitors, but is not expressed on the haematopoietic stem cell or by nonhematopoietic tissue. Antibodies against cell surface antigens are being made to elicit antitumor responses by immunizing patients with vaccine (Pinilla-Ibarz, *et al* 2000).

Treatment with Apoptosis-inducing Agents

The imbalance in the delicate dance between cell proliferation and apoptotic cell death occurs in disease scenarios, including leukaemia, where cells acquire a selective survival advantage that promotes neoplasia and malignancy. Therefore, a great number of new apoptosis inducers are being studied. Although the knowledge of the majority of these agents is based on in vitro studies, they permit to discover antiapoptotic proteins and proapoptotic proteins as targets for possible drug design. According to the mechanism of induction, the apoptotic-inducing agents can be grouped as follow (Wang 2003):

activator agents of:

- death pathway receptors, such as FAS, or TNF, (*GM-CSF, Tryptanthrin*)
- calcium-dependent apoptosis signal pathway, (*calphostin C*)
- caspases, (*methyl selenocysteine, the retinoid analog CD437, curcumin, diallyl disulfide and resveratrol*)
- ROS production or ROS-dependent apoptosis, (*falconensone A dioxime, β2 microglobulin, baicalin, curcumin*)
- disruption or modification of mitochondrial transmembrane potential, (*potassium antimonyl tartrate, homoharringtonine, theasinensin A, artepillin C carnosol, curcumin, baicalin*, sophoranone, *dolichyl monophosphate, farnesylpyridinium, lovastatin*, bis(4,7-dimethyl-1,10 phenanthroline) sulfatooxovanadium)

inhibitors agents of (Wang 2003) :

- Bcl-2,(*carnosol, erianin, momordin I*)
- cyclin D kinase (CDK) (*arginine deiminase, monensin*)
- apoptosis activation gene or protein, (*STI571 + TRAIL, arsenic trioxide + interferon (IFN)-α*)
- caspase-3 activation (*squamocin*)

One of the main pathways found in AML, the Flt3 tyrosine kinase receptor and PI3K/AKT protein kinase pathway, are involved in cell growth, proliferation, and apoptosis. The functional activation of Flt3 and PI3K/Akt provides survival signals and, then, blockade of this pathway may facilitate cell death. Downstream targets of PI3K-AKT include the proapoptotic protein BAD, caspase-9 (Zhao, *et al* 2004b), NF-kB (Birkenkamp, *et al* 2004, Kordes, *et al* 2000), and Forkhead Box class O (FOXO) transcription factors (Carter and Brunet 2007, Nakamura, *et al* 2000, Scheijen, *et al* 2004). It is now known that FOXO transcription factors play an important role in repression of tumour development by their ability to induce apoptosis. In fact, the FOXOs are non-functional in a broad spectrum of human tumours. Akt, which is overexpressed at high frequency in many tumour types, directly phosphorylates FOXO family members at three key regulatory sites (Thr32, Ser253, and Ser315), leading to their nuclear export and consequently the inhibition of the proapoptotic FOXO-dependent transcription (Brunet, *et al* 1999, Kops, *et al* 1999). Recently, it has been demonstrated that oncogenic tyrosine kinase FLT3 can negatively regulate FOXO transcription factors in AML cells, through Akt activation, thereby promoting cell survival and proliferation (Brandts, *et al* 2005, Scheijen, *et al* 2004). Thus the FLT3/Akt/FOXO pathway has become an attractive target for therapeutic intervention in AML.

Treatment with Differentiation-inducing Agents

From the first treatment with retinoids in patients with acute promyelocytic leukaemia (APL) in 1983, it has been discovered that ATRA improved the outcome of APL at a magnitude that is unparalleled in any other cancer. ATRA induces a complete remission by cell differentiation in nearly every newly diagnosed APL patient. Moreover, when combined with chemotherapy, ATRA doubles the cure rate to 75% compared to chemotherapy alone. Further, ATRA is the only transcription factor therapy that has become standard in clinical practice, by specifically targeting the aberrant transcription factor that underlies the pathogenesis of this disease.

ATRA has been transformed into the most common differentiation agent used as therapeutic drug. The classic chromosome abnormality of APL is a translocation t(15;17)(q22;q21) resulting in the generation of fusion gene and protein PML-RARα, which plays a central role in APL pathogenesis. PML-RARα fusion protein forms a complex with retinoid x receptor (RXR), nuclear corepressors (N-CoR), Sin3A, and histone deacetylase (HDAC). This complex repress the transcriptional expression of target genes (Collins 1998) and acts as a dominant negative factor on the retinoic acid-signalling pathway (Melnick and Licht 1999) blocking the differentiation of myeloid cells. The action mechanism of ATRA has been broadly described (FIGURE 2). When ATRA (pharmacological doses) binds to the RAR receptors it causes degradation of PML-RARα protein through the ubiquitin-proteosome and caspases (Nervi, *et al* 1998, Zhu, *et al* 1999) leading to restoration of terminal differentiation of promyelocytes. Moreover, APL cells treated with ATRA induces relocalization of PML (Melnick and Licht 1999), and a pharmacological dose (1 μM), CoR is dissociated from the repressive complex, whereas CoA (coactivator) is recruited to the complex (Collins 1998). As a result, the repression of transcriptional activation of target genes is relieved and the differentiation of promyelocytic process is restored.

A synthetic retinoid, 4-[(5,6,7,8-tetrahydro-5.5.8.8-tetramethyl-2-naphthyl)-carbamoyl] benzoic acid, AM80 that is an specific RAR-α agonist, is effective in treating APL appearing after relapse from ATRA-induced complete remission. Am80 has several potential advantages over ATRA: it is 10 times more potent than ATRA in inducing differentiation in vitro, and is markedly more stable. Among 24 APL patients who relapsed after ATRA therapy, 14 (58%) achieved a second CR with Am80 alone (Takeuchi, *et al* 1998).

Other differentiation-inducing agents have been studied as potential differentiation therapy of acute leukaemia. They can be divided into the following groups: *vitamin analogs* (other retinoids and vitamin D derivatives), *cytokines* (granulocyte-macrophage colony-stimulating factor, granulocyte colony stimulating factor, interferons, tumour necrosis factor), *Polar-aplanar compounds* (hexamethylene bisacetate), *inhibitors of DNA methylation* (5-aza-2′- deoxycytidine64), *cyclic AMP analogs* (8-Cl-cAMP, dibutyl cAMP), *chemotherapeutic agents* (aclarubicin, cytosine arabinoside, hydroxyurea), *medicinal plant-derived products and plant growth regulators* (meisoindigo, flavonoids, cotylenin A, tanshinone IIA, pyranocoumarins, intermedeol, and magnolialide). Most of these agents have only been studied in vitro on only leukaemia cell lines, such as NB4, H-60, U937, K562, Kasumi-2, or in a primary leukaemia cells culture assay. Therefore, the results of clinical trials of their effectiveness are limited.

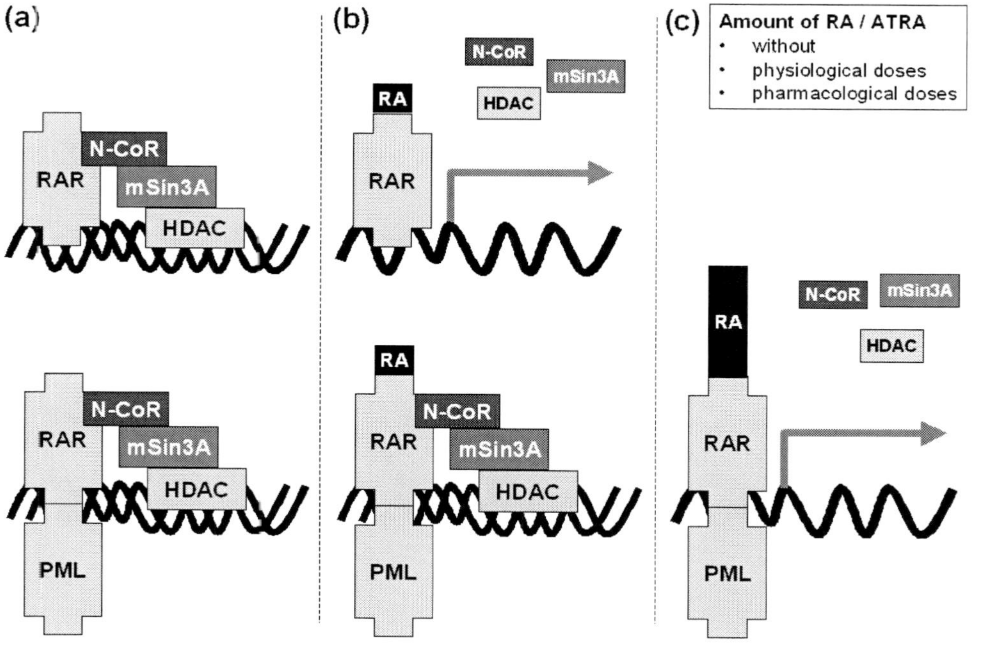

Figure 2. Mechanism of action of ATRA in differentiation of APL cells. The receptor of retinoids is complexed with several other proteins in DNA. In the case of the fusion protein between RAR and PML found in APL cells this complex is also found in the absence of retinoids. Physiological doses of ATRA induce the gene expression directed by RAR in normal cells but not in APL. However, the addition of pharmacological doses of ATRA can unblock the effect of PML in APL and induce gene expression restoring the differentiation pathway.

Combination of Differentiation Treatment, Apoptosis-inducing Treatment, and Chemotherapy

Several objectives are pursued to design and develop a treatment to revert leukemogenesis, the main one is the total cure of the disease or, at least, a longer survival of the patient and to reduce both, the risk of the side effects and the resistance to the treatment. Thus, cooperation between various treatments may improve their individual effectiveness in order to reverse leukemogenesis. In this sense, a combined treatment with chemotherapy and differentiation/apoptosis inducers and has already been proved in the treatment of in vitro AML:

- the combination of KH 1060 (a 1,25D_3 analogue) with ATRA can irreversibly inhibit clonal growth, decrease Bcl-2 expression, and induce apoptosis in HL- 60 cells (Elstner, *et al* 1996).
- the sensibility to receptor-mediated apoptosis induced by IFN-γ is enhanced by doxorubicin in some primary AML cells (Varela, *et al* 2001, Wuchter, *et al* 2001).

- arsenic trioxide is capable of inducing apoptosis in multidrug-resistant AML cells that express BCR-ABL or overexpress MDR, MRP, Bcl-2, or Bcl-x(L) (Perkins, *et al* 2000).
- treatment of BCR-ABL-positive leukaemia cells with arsenic combined with STI-571 can potentially induce apoptosis and differentiation (Porosnicu, *et al* 2001).
- histone deacetylase inhibitors (trichostatin, phenylbutyrate, apicidin, depsipeptide) markedly influence gene transcription in APL cells in response to ATRA restoring ATRA responsiveness in ATRA-resistant APL cell lines, allowing ATRA to activate gene transcription via the RAR moiety (Lin, *et al* 1998)

Differentiation induction therapy has been already tested in myelodisplastic syndromes to ameliorate maturation defects and to restore normal haematopoietic function. In a pilot study, the combination of ATRA, granulocyte colony-stimulating factor (G-CSF), erythropoietin (EPO) and α-tocopherol demonstrated the well tolerance, the normalization of neutrophils counts and improvement of platelets and red blood cells in a significant subgroup of patients (Ganser, *et al* 1996). This study demonstrates the importance of a combined therapy in the treatment of leukaemia.

On the other hand, differentiation therapy of cancer remains an only partially attained goal. During the last years, several studies have demonstrated that the combination of different natural compounds with antioxidant activity with leukaemia differentiation inducers such as ATRA or 1,25D_3 enhance the effectiveness of these inducers. These combinations permit to decrease the dose of the inducers avoiding side effect such as the calcemic effect of 1,25D_3 that preclude their use in the clinic. Even today, an active investigation on the effect of derivatives 1,25D_3, on differentiation is being carried out. Then, another approach that may obviate this problem is to combine 1,25D_3 or its derivatives with other agents that increase the antineoplastic effects of low, non-toxic concentrations of vitamin D compounds. The following part of the chapter will review the experiments with these antioxidants and their mechanism of action on leukaemia. However, most of these studies have been developed in vitro and the results only indicate good perspectives as promising agents in the combinatory therapy of leukaemia.

Plant Extracts in the Treatment of Leukaemia

The use of vegetables as source of preventive agents against cancer and the use high doses of vitamins between which we can found several antioxidants such as vitamin E or vitamin C or combinations of several of them in the treatment of several different cancers has been proposed (Potter and Steinmetz 1996, Prasad, *et al* 1999).

In fact, plant foods show several potentially anticarcinogenic substances such as carotenoids, vitamin C, vitamin E, selenium, and dietary fibre. Moreover, other animal and in vitro studies have demonstrate anticarcinogenic effect of other substances from plants such as phytoestrogens and other phytochemicals such as polyphenols. Most of the phytochemicals such as glucosinolates and indoles, thiocyanates and isothiocyanates, phenols, and coumarins can induce a faster elimination of carcinogenic compounds by inducing detoxifying phase II enzymes; other as ascorbate and phenols can block the formation of carcinogens such as nitrosamines; other as flavonoids and carotenoids can disable the oxidative stress produced by some carcinogens acting as antioxidants; carotenoids and other compounds can affect DNA synthesis and induce differentiation; etc...Thus, it has been proposed that a low consumption of plant foods in the diet results in an increase in cancer risk due to the deficiency in these compounds (Potter and Steinmetz 1996)

On the other hand, not all the studies have shown cooperative activity of antioxidants on leukaemia therapy. In fact, several antioxidants have shown inhibitory effect of differentiation of mouse myeloid cells (M1) induced by the glucocorticoid synthetic compound dexamethasone (Takenaga, *et al* 1981). In this case, the effect of the antioxidants was higher as higher the inhibition of PGE2 synthesis. This effect was also found in the case of ascorbate enhancing effect on vitamin D3 differentiation although in this case, the antioxidant effect increased differentiation (Lopez-Lluch, *et al* 2005).

Differentiation Inducers Plus Antioxidants in Combinatory Therapy of Leukaemia

Ascorbate

Due to the absence of the last enzyme of its biosynthesis pathway, ascorbate is an essential nutrient in several animal species including humans (Chatterjee 1978). Ascorbate can affect several essential processes in the organism being scurvy the most severe manifestation of ascorbate deprivation (Padh 1990). Moreover, potential benefits of ascorbate in different pathological processes in humans such as prevention of cancer, cardiovascular risk and cataracts, protection against damage of sperm cells or cigarette smoke-dependent damage have demonstrated that the influence of ascorbate on human health is very important (reviewed in (Bendich and Langseth 1995)). In fact, ascorbate is involved in many cellular functions including modulation of cell growth and differentiation (Alcain and Buron 1994), collagen, carnitine and norepinephrine biosynthesis, activation of hormones, oxygen radical scavenging, and regulation of transcription factors (Alcain and Buron 1994, Frei, *et al* 1989, Gershoff 1993, Lopez-Lluch, *et al* 2001, Munoz, *et al* 1997, Padh 1990).

The use of ascorbate or some of its derivatives has been suggested for cancer therapy, especially against chemotherapy resistant tumour cells (Osmak, *et al* 1997). Actually, ascorbate prolongs survival in cancer patients (Cameron and Pauling 1976) probably by reducing the growth rate of tumoral cells as has been shown in cell cultures (Osmak, *et al* 1997, Park, *et al* 1980, Venugopal, *et al* 1996) and animal models (Gruber, *et al* 1980, Pauling, *et al* 1985). However, although ascorbate is considered a model of antioxidant compound, the anticancerous effect of this antioxidant has been truly related to its prooxidant activity through the production of H2O2 since catalase avoids its growth inhibitory and even its toxic effect (Arakawa, *et al* 1994, Nemoto, *et al* 1996, Sestili, *et al* 1996). In vitro studies have demonstrate that near physiologic levels of ascorbate are able to suppress growth of bone marrow cells from patients with AML (Park, *et al* 1980) while normal haematopoietic cells were not affected.

On the other hand, ascorbate is also able to stimulate the differentiation of several mesenchyme-derived cell types including osteoblasts (Franceschi and Iyer 1992) also other cell types (Farquharson, *et al* 1998, Lee, *et al* 2003, Ragab, *et al* 1998, Savini, *et al* 2002, Takahashi, *et al* 2003). For example, ascorbate increases the in vitro differentiation of epidermal keratinocytes (Savini, *et al* 2002), it also differentiates embryonic stem cells to cardiac myocytes (Takahashi, *et al* 2003) or affects the differentiation of embryonic brain cortical precursors into neurons and astrocytes (Lee, *et al* 2003). The role of ascorbate in collagen synthesis has been associated to its essential activity on mesenchymal differentiation (Franceschi and Iyer 1992). However, not all the effect of ascorbate on differentiation can be explained by its essential activity on collagen synthesis. Moreover, we have demonstrated that ascorbate has also shown cooperative activity with 1,25D_3 in myelomonocytic cell differentiation to mature monocytes (Lopez-Lluch, *et al* 2001, Lopez-Lluch, *et al* 1998, Lopez-Lluch, *et al* 2005, Quesada, *et al* 1996).

Addition of physiological concentrations of ascorbate to HL-60 cells increased specific differentiation markers induced by 1,25D_3 (Quesada, *et al* 1996). Although ascorbate, as several other antioxidants, were ineffective in the absence of any other inducer, it increased by about a 50% the number of cells displaying differentiation-associated enzymes or cell surface markers in the presence of 1,25D_3. On the other hand, Kang et al., have shown that millimolar concentrations of L-ascorbic acid (LAA) induce the expression of the granulocytic-specific cell surface marker CD 66b in HL-60 cells (Kang, *et al* 2003). In this case, catalase abolished and superoxide dismutase promoted LAA-induced differentiation of HL-60 indicating a role of H2O2 in this pro-differentiating activity of ascorbate. However, our studies have demonstrated that such these high doses of ascorbate induce apoptosis in the cells in agreement with other studies that report the same apoptotic effect at millimolar concentrations of ascorbate (Sakagami and Satoh 1997). In that case, we consider that high concentrations of ascorbate do not increase differentiation in AML cells but induce cell death by apoptosis.

Our results indicate that ascorbate affects this differentiation pathway by modulating the activity of the plasma membrane redox system (PMRS) that

The mechanism by which ascorbate enhances the differentiation activity of calcitriol in leukemic cells remains to be completely elucidated. To date, we have found that ascorbate induces a long-term increase in cAMP levels in HL-60 cells (Lopez-Lluch, *et al* 1998) and changes the early response to 1,25D_3 by modulating AP-1 transcription factor (Lopez-Lluch, *et al* 2001). In most cases, ascorbate-dependent mechanism seems to be dependent on its activity as external acceptor or electrons through the plasma membrane redox system (Navas, *et al* 1994). This system is modulated during TPA-induced differentiation in HL-60 cells (Burón, *et al* 1993). Moreover, this system is able to regulate the response to several hormonal factors such as growth factors or insulin (Brightman, *et al* 1992, Crane, *et al* 1982, Krieger-Brauer and Kather 1995). Then, it can regulate the cytosolic redox balance and then, control intracellular signalling mechanisms during the early steps of the differentiation program. Thus, PMRS can be related to a localized rise of intracellular levels of H2O2 during the first times of incubation that would modulate the response to 1,25D_3 and by this mechanism increase the differentiation of the cells. This increase if H2O2 can be accompanied by a decrease in superoxide ($O_2^{\bullet -}$) as we have already found (Lopez-Lluch, *et al*

2001) and similarly to the effect of other compounds with higher depletion activity of $O_2^{\bullet -}$ that also induce differentiation of HL.60 cells (Satoh, *et al* 1997).

But ascorbate not only affects the differentiation program induced by $1,25D_3$ in leukemic cells. Recently, imatinib, a BRC/ABL tyrosine kinase inhibitor have been shown as a differentiation inducers with remarkable clinical effects in Chronic Myelogenous Leukaemia (CML) although resistance to this compound appears in most of the blast crisis cases. In this case of resistance, addition of ascorbate to these cells together with imatinib partly restores sensitivity to imatinib. The mechanism of action of ascorbate in this case seems to be by the inhibition of the Nrf2-dependent transcription of detoxifying limiting enzymes such as gamma-glutamylcysteine synthetase (gamma-CGS) that is involved in GSH synthesis (Tarumoto, *et al* 2004). Ascorbate seems to inhibit the Nrf2-DNA binding found in imatinib-treated cells.

Vitamin E

Vitamin E lumps together a family of eight isomers consisting of four tocopherols and four tocotrienols: α, β, γ, and δ in each case. These compounds are essential components of the human diet since they are synthesized exclusively by photosynthetic organisms. Recently, these components were found to exert important and unique biological effects on nutrition and health other than antioxidant properties and are, therefore, now receiving increased attention.

The effect of vitamin E on haematopoiesis and blood cell differentiation has been studied from long time. Early studies demonstrate abnormal erythropoiesis in vitamin E deficient monkeys (Fitch, *et al* 1980) and contradictory results in mouse myeloid leukaemia cells (M1) were vitamin E induces the appearance of several differentiation markers whereas , as other antioxidants, it inhibited the dexamethasone-induced differentiation process (Sakagami, *et al* 1981, Takenaga, *et al* 1981).

Like ascorbate, vitamin E, by itself does not strongly affect differentiation in cells. In vitro studies have shown that only very high doses of α-tocopherol induce the appearance of myelomonocytic differentiation markers such as NBT-reducing activity and the expression of the integrin CD11b. This modest effect is probably due to the lipidic nature of this compound. In our hands, a-tocopherol was less effective than ascorbate on $1,25D_3$–induced differentiation (unpublished results). However, more soluble forms such as vitamin E-succinate (VES) are able to partially induce monocytic differentiation in HL-60 cells. Already in 1982, VES was established as the most effective form of vitamin E against cancer. Several studies have demonstrate that alone or in combination with other dietary micronutrients, VES can be useful as an adjunct to standard cancer therapy to increase tumour response and even to decrease the toxicity of chemotherapics in normal cells (Prasad, *et al* 2003). In fact, VES stimulates HL-60 cells to produce superoxide radicals and to show non-specific esterase activity although does not change the expression of cell surface antigens (Turley, *et al* 1992). Moreover, other experiments have demonstrate that VES added to low levels of $1,25D_3$ increased the differentiation induced by this last compound (Sokoloski, *et al* 1997a).

More recently, hybrid compounds between α-tocopherol and retinoic acid (ATRA) have been designed and used in the treatment of leukaemia. Tretinoin tocoferil is a α-tocopherol ester of ATRA and is able to induce the granulocytic differentiation of human promyelocytic leukaemia HL-60 cells. The inactivity of tretinoin tocoferil as teratogenic agent and its very low toxicity have suggested its use on use on myelomonocytic leukaemia (Makishima, *et al* 1996). Furthermore, this compound also enhances cellular differentiation induced by sub-optimal concentrations of ATRA, dimethyl sulfoxide, phorbol esters and 1,25D_3. In fact, the combination of tretinoin tocoferil and 1,25D_3 may be useful for treating myelomonocytic leukaemia since it have enhanced the 1,25D_3–dependent differentiation of several different human myelomonocytic cell lines such as HL-60, U937, ML-1, THP-1, P39/TSU, and P31/FUJ cells. (Makishima and Honma 1997, Makishima, *et al* 1996, Makishima, *et al* 1998).

Another hybrid compound, in this case between 9-cis retinoic acid and α-tocopherol (9CTT), have shown properties as granulocytic differentiation inducer in the APL cells NB4 and HT-93 cells whereas its effect was weaker in non-APL leukemic cells such as HL-60 and U937 cells. In this case, its combination with 1,25D_3 also produced a more than additive induction of differentiation in APL cells (Makishima, *et al* 1998).

Recently, the other members of the vitamin E, tocotrienols have been taking into consideration due to their powerful neuroprotective, anticancer and cholesterol-lowering properties often not exhibited by tocopherols. Their use in the prevention and treatment of different cancer types are currently under study (Wada, *et al* 2005). Tocotrienols are found in seed endosperm and are rarely found in vegetative tissues of plants. The structural difference between these compounds and tocopherols is the presence of three *trans* double bonds in the hydrocarbon tail in tocotrienols. On the other hand, other vitamin E derived compounds, tocopherylquinones (TQ) that are the product of vitamin E oxidation have been suggested as a new class of alkylating electrophilic quinones that can function as highly cytotoxic agents able to escape multidrug resistance in acute lymphoblastic leukaemia cell lines specially in the case of γ-TQ and δ-TQ whereas α-TQ does not induce cytotoxicity (Cornwell, *et al* 1998).

Polyphenols

Polyphenols are a big group of compounds found in plants and uniquely characterize by the presence of more than one phenol group in the molecule. They are considerer the most abundant antioxidant in the diet although the diversity of their structures makes them different from other antioxidants. Moreover, the amount of different polyphenols found to date in plants and plant foods, several thousands, make them a complex family of compounds with very interesting therapeutic properties against cancer, cardiovascular diseases, inflammation and other diseases. Polyphenols are classified according to the nature of their carbon skeleton in: phenolic acids, flavonoids, stilbenes and lignans. Several compounds in each subgroup have been already studied in cancer and leukaemia therapy. In fact, the effect of polyphenols in leukemic cell differentiation is known from the late 90's and from there, several different compounds have been studied to determine the in vitro effect on several models of leukemic cell differentiation.

Phenolic acids are abundant in foods being caffeic acid and ferulic acid the most present. It seems that caffeic acid, ferulic acid and other phenolic acid such as chlorogenic acid do not increase the differentiation of HL-60 cells produced by 1,25D_3. However, derived compounds such as the more lipophilic ethyl of ferulic and caffeic acid are capable of inducing the differentiation of HL-60 cells and enhance the maturation produced by vitamin 1,25D_3 (Sokoloski, *et al* 1997b).

One of the most known phenolic acids in leukaemia treatment is carnosic acid, a diterpene derived from rosemary known to be a strong dietary antioxidant showing also antimutagenic and anticarcinogenic properties (Steiner, *et al* 2001). This compound inhibits proliferation in leukemic cell lines such as HL-60 and U937 but cause only marginal induction of differentiation. However, when added to ATRA, TPA or 1,25D_3-induced cells, carnosic acid substantially augments the differentiating effect of these compounds (Danilenko, *et al* 2001). This compound also synergistically potentiate the activity of 1,25D_3 or its derivative Ro25-4020 in Balb/c mice bearing WEHI-3B D(-) murine myelomonocytic leukaemia delaying tumour appearance and reducing tumour size. Rosemary dry extract mixed with food produced a similar effect (Sharabani, *et al* 2006). These results suggest that carnosic acid and derivatives of 1,25D_3 can cooperate not only in vitro assays but also show antileukemic activity in vivo and then, their combination should be evaluated in the treatment of myeloid leukaemia (Danilenko, *et al* 2003).

Flavonoids are the most abundant polyphenols in the diet because they are present ubiquitously in plants. These compounds can be classified in several different classes. According to the degree of oxidation of the oxygen heterocycle we can find: flavones (*luteolin, apigenin*), flavonols (*quercetin*), isoflavones (*genistein and daidzein*), anthocyanins (*red fruids pigments*), flavanols (*catechins*), proanthocyanidins (*polymeric flavanols*) and flavanones (*hesperidin, narigin*) (Scalbert and Williamson 2000).

Most of these compounds have been also shown to present cancer preventive properties. In human colon carcinoma cells HT-29, flavonoids affects proliferation, differentiation and apoptosis with selectivity toward transformed cells (Wenzel, *et al* 2000). Specific catechins have been proposed to optimise current chemotherapeutic protocols in leukaemia (Annabi, *et al* 2006). Flavonoid-rich polyphenol fraction from pomegranate (*Punica granatum*) fruit exert antiproliferative and prodifferentiation activity on HL-60 cells in a similar way than in breast and prostate cancer cells (Kawaii and Lansky 2004). The anthocyanin cyanidin-3-O-beta-glucopyranoside (Cy-g) is able to induce apoptosis in T-lymphoblastoid whereas in HL-60 promyelocytic cells Cy-g also induces differentiation to macrophage-like cells and granulocytes (Fimognari, *et al* 2004). Furthermore, proanthocyanidins extracted from barley bran, prodelphinidin B-3, T1, T2, and T3, induce low levels of cell differentiation in HL-60 cells but potentiate ATRA-induced granulocytic and also butyrate-induced monocytic differentiation in HL-60 cells (Tamagawa, *et al* 1998). Moreover, complex mixture of bioflavanoids and proanthocyanidins and organic acid already found in more that 500 dietary supplements, foods and beverages as is Pycnogenol, also have shown differentiation activities in leukemic cells (Huang, *et al* 2005b).

Stilbenes, are phenol-based plant metabolites widely represented in nature. Among the stilbenes, the phytoalexin resveratrol have recently attracted attention for its different effects at the cell, tissue and physiology level. Resveratrol inhibits growth of several myeloid leukaemia cell lines (HL-60, NB4, U937, THP-1, ML-1, Kasumi-1) and fresh samples from 17 patients with acute myeloid leukaemia. Moreover, resveratrol also induce the appearance of several differentiation markers depending on the cell line. Thus, in erithroleukemic K562 cells, resveratrol also induces a remarkable γ-globin synthesis an erythroid differentiation marker (Ragione, *et al* 2003, Rodrigue, *et al* 2001). It also enhances the expression of adhesion molecules (CD11a, CD11b, CD18, CD54) in each of the cell lines except in Kasumi-1. Moreover, resveratrol induces the production of superoxide in U937 cells. Resveratrol also synergistically increased the differentiation induced by ATRA (Asou, *et al* 2002). Resveratrol also have shown inhibitory effect on the growth of AML cells suggesting that this compound can have a future role in the therapy of AML (Asou, *et al* 2002, Estrov, *et al* 2003).

Lignans are polyphenolic substances derived from phenylalanine via dimerization of substituted cinnamic alcohols. Lignans are present in foods such as flax, pumpkin and sesame seeds, rye, soybeans, broccoli, beans and some berries. However, although they have been identified in human plasma and urine, the exact dietary origin is unknown and probably, the dietary lignans are metabolised by the gut microflora and derived metabolites absorbed through the gut epithelium. Between lignans, magnolol (MG) and honokiol (HK), abundantly available in the medicinal plants *Magnolia officinalis* and *M. obovata*, have shown anti-inflammatory and anti-oxidant properties and also are able to synergistically enhance HL-60 cell differentiation initiated by low doses of 1,25D_3 and ATRA (Fong, *et al* 2005). Previously, other lignans such as Saucernetin-8 and –7 have shown antiproliferative and pro-differentiating activity in leukemic cells (Seo, *et al* 2004a, Seo, *et al* 2004b). Moreover, several other lignans and terpenoids found in extracts from different plants have shown activity as inducers of differentiation in HL-60 (Luyengi, *et al* 1996, Umehara, *et al* 1996).

Carotenoids

As precursors of retinoids, carotenoids also show activity as differentiation agents in leukemic cells. However, the induction of cell differentiation by carotenoids without (lutein) and with (b-carotene) ATRA activity suggests that these agents induce differentiation in a mode of action independent of ATRA (Gross, *et al* 1997). The carotenoid lycopene also functions as a very potent antioxidant that has been shown to affect several human cancer cells. As in other cases, the combination of low concentrations of lycopene with 1,25-dihydroxyvitamin D3 synergistically affects cell proliferation and differentiation in the HL-60 promyelocytic leukaemia cell line (Amir, *et al* 1999, Heber and Lu 2002). Recent studies have demonstrated that the oxidation products of acyclic carotenoids other than lycopene are the responsible agents of apoptosis induction in leukemic cells (Nara, *et al* 2001).

Moreover, several different carotenoids have been considered as agents able to revert the multidrug resistance found in several leukaemia and cancer (Gyemant, *et al* 2006), then, the importance of these compounds in chemotherapy resistance is clear.

Sesquiterpene Lactones

Sesquiterpenes are a class of terpenes consisting in three isoprene units. They can be acyclic of can contain one, two or three cycles. Terpenes and terpenoids are the primary constituents in the essential oils of many types of plants and flowers. This is the reason why terpenes and sesquiterpene lactones are found in several extracts from plants. Several of these compounds have received considerable attention in pharmacological research in the treatment of inflammation or neoplasia.

The use of sesquiterpene lactones in the treatment of leukaemia is known from early 70s and several different compounds have shown anti-leukemic activity both by itself or in combination with ATRA or 1,25 D_3. For example, sesquiterpene lactones from extracts from Tanacetum parthenium (L.) or other Mexican-Indian medical plants are responsible of anti-inflammatory effects in vitro and enhance differentiation induced by 1,25D3 in HL-60 cells (Kang, *et al* 2002). In this extract, parthenolide was the most active compound whereas other sesquiterpenes were inactive depending on the modification of signal pathways during differentiation (see below). Costunolide, is another germacranolide sesquiterpene lactone that exists in several medicinal plants that also increase the degree of differentiation when combined with 1,25D3 *(Kim, et al 2002)*. Another example can be found with helenalin, a cell-permeable pseudoguainolide sesquiterpene lactone, extracted from several plant species of the Asteraceae family. In this case, this compound is able to induce differentiation in HL-60 cells by itself (Kim, *et al* 2005).

Signalling Pathways Affected by Antioxidant in Differentiation Therapy

The different nature of the above-described plethora of compounds and other antioxidant compounds able to modulate differentiation and even apoptosis in leukemic cells does not permit to obtain a clear picture of the molecular mechanism involved in their effect. In most cases, the antioxidant capacity of these compounds has been associated to its pro-differentiating activity, but, recently, most of these compounds have shown pro-oxidant activity in cell cultures. However, it seems that there are a few common pathways that are affected by most of these compounds.

MAPK Pathway

The effect of VES on leukemic differentiation seems to be dependent on the induction of ERK during the first moments of treatment. In fact, VES-induced ERK phosphorylation is abolished by the ERK inhibitor, PD98059, which resulted in a remarkable prevention of VES-induced monocytic differentiation. However, the ERK is not the only factor affected by VES since inhibition of the ERK activity by PD98059 also diminished the VES-induced p21WAF1 protein expression, but is unable to change the phosphorylation state of the retinoblastoma protein (Lee, *et al* 2002). This same effect has been found with several other antioxidants such as resveratrol, pyrrolidine dithiocarbamate or hydroxityrosol (Della Ragione, *et al* 2002).

The effect induced by the combination of the polyphenol carnosic acid in the presence of 1,25D_3 is associated with the increase in the activation of MAPK pathway affecting the Raf-mitogen-activated protein/ERK. This pathway also enhance the activation of and the binding of AP1 to its cognate DNA element in the promoter of the vitamin D receptor gene. Then, the potentiation due to carnosic acid can be due to the activation of this MAPK pathway (Danilenko, *et al* 2003).

Although the mechanism of action of these compounds in MAPK activation is not clear several mechanisms are being proposed. Most of these phenolic compounds generally appear to be electrophiles. Then, they can activate MAPK pathway via an electrophilic-mediated stress response. This response will activate the bZIP transcription factor NRF2 that dimerizes with Mafs and binds to the antioxidant response element found in promoters (ARE). By this mechanism, most of these compounds would induce many of the phase II drug metabolic enzymes such as thioredoxins, γGCS, HO-1 (Rushmore and Kong 2002).

However, not all the polyphenols could affect in a similar way. In fact, it has been described that different polyphenols can affect antagonistic pathways and show different effect in the same cells. Hence, EGCG and curcumin produce opposite response in keratinocyte differentiation. Like carnosic acid in leukemic cells, the catechin EGCG alters MAPK cascade function activating involucrin gene transcription via Ras, MEKK1, MEK3, ERK1/2-p38delta cascade and then, targets AP1 and CAATT enhancer binding protein transcription factors. However, the turmeric curcumin, also able to potentiate differentiation in leukemic cells, antagonizes the EGCG-dependent response by interfering in this signalling pathway (Eckert, *et al* 2006). Moreover, in pancreatic stellate cells, curcumin also inhibit the effect of PDGF by inhibiting MAP kinase pathway (ERK. JNK and p·38) but not affecting NF-kB (Masamune, *et al* 2006). Then, these results indicate that polyphenols can affect the same signal pathway in opposite ways in different cells.

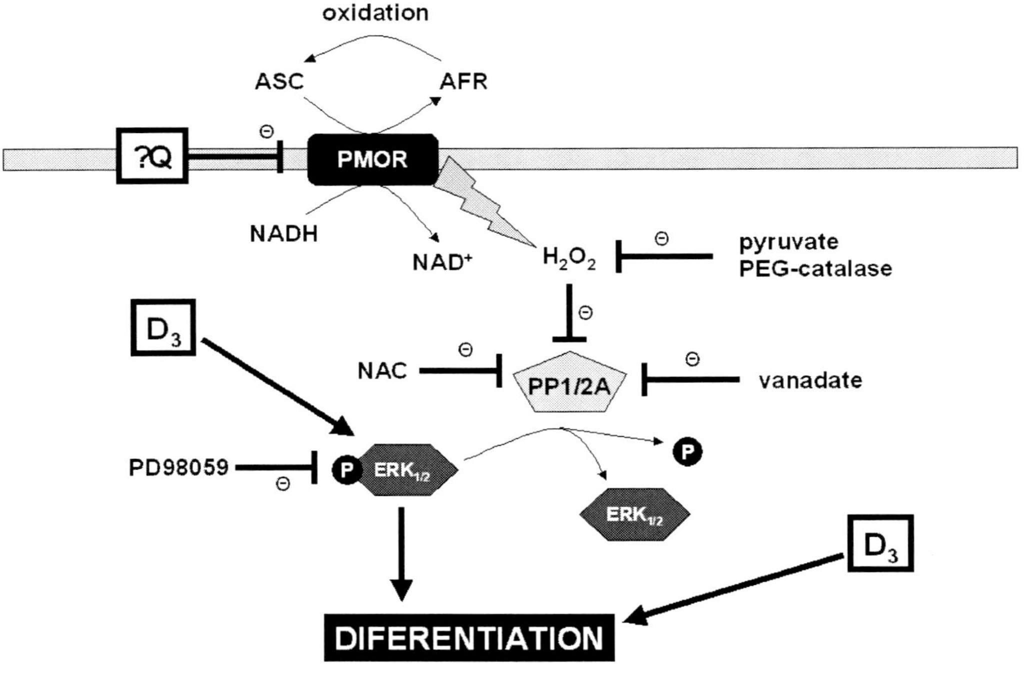

Figure 3. A putative role of plasma membrane redox system (PMRS) activated by ascorbate in the stimulation of differentiation by $1,25D_3$. The prooxidant activity of ascorbate or even other antioxidants with prooxidant capacity will increase H_2O_2 levels that will inhibit the phosphatase PP1/2A as other compounds do. This inhibition will permit the maintenance of the phosphorylation of ERK and the enhancement of the differentiation induced by $1,25D_3$.

The same ERK-dependent mechanism has been reported for sesquiterpene lactones such as costunolide that potentiates $1,25D_3$-induced cell differentiation predominantly into monocytes. In this case, inhibitors for PKC, PI3-K, and ERK markedly inhibit differentiation induced by costunolide and 1,25D_3 (Kim, *et al* 2002).

Ascorbic acid enhances the activation of the Rac1 and p38 induced by arsenic trioxide in leukemic cells. The effect of ascorbate in this pathway can be related to its antiapoptotic effect in this treatment since inhibition of p38 potentiates arsenic-dependent apoptosis (Verma, *et al* 2002). Then, p38 in this case acts and a negative regulator of arsenic trioxide-induced apoptosis and will increase malignant cell growth. Other authors indicate that ascorbate can induce the ERK MAPK pathway in AML cell lines by a Raf1 dependent mechanism (Park, *et al* 2005). As downstream effector, ascorbate increases the activity of MAPK-activated protein kinase 2. On the other hand, inhibition of ERK pathway reduces ascorbate-dependent apoptosis and inhibition of cell growth in AML cell lines. Our work demonstrate that ascorbate can increase differentiation in 1,25D_3-induced HL-60 cells. Recent results demonstrate that the mechanism of ascorbate depends on H_2O_2 since catalase or pyruvate, known scavengers of H_2O_2, abrogate the effect of ascorbate (unpublished results). Probably, the activity of ascorbate as electron acceptor in the PMRS will induce the production of H_2O_2 in cells. In fact, PMRS have been linked to the regulation of hormone-dependent intracellular signalling in most cases depending on the H_2O_2 levels (Brightman, *et al* 1992, Crane, *et al* 1982, Goldenberg, *et al* 1978). This activity will inhibit the activity of the PP1/2A phosphatase and, by this mechanism, will maintain the activity of the ERK pathway (FIGURE 3). However, this mechanism must not be implicated in VES-induced activation since α-tocopherol seems to act by activating PPA_2 causing dephosphorylation of PKC and, probably, dephosphorylation and then, activation of other proteins.

The Role of Cyclooxygenase and Inflammatory Pathway

Most of the antioxidants studied to far show anti-inflammatory properties. Ascorbate and polyphenols such as resveratrol, curcumin, or quercetin and α-tocopherol affects COX activity. Moreover, known COX inhibitors such as aspirin or indomethacin also enhance differentiation induced by 1,25D_3 in HL-60 cells (Lopez-Lluch, *et al* 2005).

In most cases, these compounds inhibit the inflammatory mechanisms of cells affecting both, enzymatic activity and pro-inflammatory gene transcription. We have shown that ascorbate increases cAMP levels in 1,25D_3-induced cells. Our work has demonstrate that the increase in cAMP levels in the cells during the differentiation process is a key component in the enhancing effect of ascorbate (Lopez-Lluch, *et al* 1998). Although part of the effect of ascorbate can be explained by its activity on the PMRS (Lopez-Lluch, *et al* 1998), its known effect as anti-inflammatory agent and the cAMP increase found with other known COX inhibitors, suggest that inhibition of COX during early phases of differentiation induction can increase the response to, at least, 1,25D_3 (Lopez-Lluch, *et al* 2005). Our results agree with the ascorbate-dependent downregulation of the expression of COX-2, probably by inhibition of NF-kB activity, found in HL-60 cells (Han, *et al* 2004)

On the other hand, not all the studies have shown cooperative activity of antioxidants on leukaemia therapy. In fact, several antioxidants have shown inhibitory effect of differentiation of mouse myeloid cells (M1) induced by the glucocorticoid synthetic compound dexamethasone (Takenaga, *et al* 1981). In this case, the effect of the antioxidants was higher as higher the inhibition of PGE2 synthesis. This effect was also found in the case of ascorbate enhancing effect on vitamin D3 differentiation although in this case, the antioxidant effect increased differentiation (Lopez-Lluch, *et al* 2005).

The Role of NFkB

Another important common mechanism in antioxidants is the inhibition of NF-kB. Near all the above-indicated antioxidant compounds show inhibitory activity against the NF-kB pathway. However, the role of NF-kB in differentiation of leukemic cells seems to be controversial since there are contradictory studies about the role of this factor in differentiation. For example, the inhibition of NF-kB was present as a key factor in differentiation since antisense treatment against the RelA subunit of NF-kB in HL-60 cells increases differentiation in these cells (Sokoloski, *et al* 1997a). However, the differentiation induced by 1,25D3 in NB4 cells was shown to be mediated by phosphorylation and proteolytic cleavage of IkB followed by translocation of NF-kB to the nucleus (Berry, *et al* 2002). Although 1,25D_3 inhibited NF-kB activation and binding to the promoter of Il-8 and IL-12 in human melanoma cells (Harant, *et al* 1997). Also, the differentiation induced by ATRA is linked to the activation of NF-kB although the inactivation of the transcription factor does not affect granulocytic differentiation (Mathieu, *et al* 2005).

Besides that, it seems that all the antioxidants that have shown cooperation with ATRA or 1,25D3 are inhibitors of NF-kB translocation to the nucleus although they seems to act through different mechanisms. VES and other antioxidants such as butylated hydroxianisole, beta-carotene and lipoic acid reduces the nuclear content of NF-kB in both, the presence or the absence of 1,25D_3. The same effect has been found when curcumin, a polyphenol with potent antitumour properties and enhancer of differentiation in leukemic cells, is added to 1,25D_3 (Sokoloski, *et al* 1997b). Curcumin also reduced NF-kB activity in nuclear extracts of HL-60 cells in both, the presence or absence of 1,25D_3, indicating that this regulation is not important by itself in non-stimulated cells but when 1,25D_3 is present, the inhibition of NF-kB activity can enhance the differentiation program. In fact, curcumin inhibits IkB kinase, (IkBK), then inhibiting NFkB activation and several oncogenes responding to this factor (Lin 2004).

Other polyphenol, resveratrol, also affect NFkB in AML cells. Fresh AML cells and AML cell lines OCIM2 and OCI/AML3 proliferate after incubation with IL-1b, however, the addition of resveratrol inhibits the proliferation of these cells in a dose-dependent fashion arresting then at the S phase (Estrov, *et al* 2003). The mechanism of action of IL-1b is dependent on NFkB activation and the presence of resveratrol suppressed the activation of nuclear NF-kB. Moreover, exposure of U937 to resveratrol suppress the binding activity of NFkB (Asou, *et al* 2002).

Parthenolide, a sesquiterpene lactone, also enhances 1,25D3-induced differentiation in HL-60 cells via inhibition of NF-kB activity although partenolide is unable to activate cell differentiation by itself (Kang, *et al* 2002). Very importantly, santonin, another sesquiterpene lactone without inhibitory activity on NF-kB is unable to enhance 1,25D_3-induced cell differentiation. In the case of the sesquiterpene lactone costunolide, the enhanced levels of cell differentiation found when combined with 1,25D_3 closely correlate with the inhibitory levels of NF-kB-binding activity. (Kim, *et al* 2002)..

Although it has been not demonstrated in the differentiation effect of ascorbate, NFkB can be also affected by ascorbate since it also suppress the activation of NF-kB under some circumstances such as TNFa activation. In this case, intracellular ascorbate inhibits TNF-a-induced activation of NF-kB in several cell lines including leukemic cells U937 and HL-60. The mechanism of action of ascorbate was inhibition of nuclear translocation of NF-kB, the reporter transcription of this expression factor and phosphorylation of its inhibitor IKB. It seems that ascorbate can affect NIK (NF-kB inducing kinase) and IkB-kinase (IKK) activity (Carcamo, *et al* 2002). Moreover, L-ascorbic acid also represses the constitutive activation of NF-kB in HL-60 cells by suppression of the binding of the p65/p50 heterodimer NF-kB to DBA through inhibition of the degradation of IKB-a and prevention of the nuclear translocation of p65: This inhibitory effect was dependent on GSH levels and generation of H2O2 levels, indicating a prooxidative role of ascorbate in this effect (Han, *et al* 2004).

Then, it seems clear that, at least in the monocytic differentiation induced by 1,25D3, inhibition of NF-kB seems to be a key factor. Most of the leukemic cells show high levels of NF-kB activity that would be involved in the inhibition of the genomic pathway of cellular differentiation when induced by 1,25D3. In fact, tumour necrosis factor-a (TNF-a) is a known activator of NF-kB and is known to suppress the vitamin D response element (VDRE)-promoter activity through the activation of NF-kB. These agents, as parthenolide, are able to suppress the inhibitory activity of TNF-a and permit the activation of the differentiation process even in the presence of TNF-a.

The Role of cAMP

The role of cAMP in myeloid differentiation is known from early 70s (Chaplinski and Niedel 1982) and the potentiation between intracellular cAMP-elevating agents and inducers of leukemic cell differentiation seems to be clear (Fontana, *et al* 1985). However, experiments performed in murine erithroleukemic cells have shown that cAMP-increasing agents inhibits cell differentiation probably indicating differences between murine and human haematopoiesis (Sherman, *et al* 1986).

We have found that the enhancement of differentiation induced by ascorbate in HL-60 cells treated with $1,25D_3$ depends on the increase of cAMP (Lopez-Lluch, *et al* 1998). The importance of cAMP in monocytic differentiation induced by $1,25D_3$ was highlighted by the potentiation found with known inducers of cAMP pathway such as isobutylmethylxanthine, forskolin and prostaglandin E_2. Moreover, we also found that ascorbate inhibits activity of COX in these cells and that inhibitors of this enzyme such as ibuprofen and aspirin also increase cAMP and enhance differentiation in $1,25D_3$-induced cells (Lopez-Lluch, *et al* 2005). Then, cAMP is another important intermediate in the monocytic differentiation program.

The same role of cAMP have been found in ATRA induction of granulocytic differentiation (Yang, *et al* 1998). In fact, ATRA induces a marked increase in cAMP within few minutes after treatment (Zhao, *et al* 2004a). In APL cells, addition of cAMP derivatives or the increase of its levels in cells by treatment with theophilline or other agents also enhances the granulocytic pathway induced by ATRA (Guillemin, *et al* 2002, Zhu, *et al* 2002). The activation of the differentiation program by cAMP is related to the release of the RAR subunit of the RAR/RXR heterodimer or the PML-RAR complex found in APL depending on the PKA activity. It is known that RXR can dimerize with VDR and activate the 1,25D_3–induced genomic activation and then, activate the differentiation of leukemic cells.

A similar mechanism in the potentiating effect of other phytochemicals has been suggested. For example, the polyphenol carnosic acid is able to increase VDR and RXR expression and then, greatly enhance the differentiation induced by both 1,25D_3 and ATRA (Danilenko, *et al* 2003, Danilenko, *et al* 2001). However, apart of the papers about ascorbate, no information is found on the role of antioxidants in cAMP-depending pathway in leukemic cells.

The Role of the Cellular Redox State

It is clear that one of the main activities of antioxidants compounds in differentiation may be based on the activity of these compounds in the scavenging of reactive oxygen species (ROS). In fact, the very different chemical structure of polyphenols, carotenoids, VES, lipoic acid, ascorbate, hydroxyurea, dithiocarbamates, or ebselen that enhance the differentiation induced by ATRA or 1,25D_3 in leukemic cells, indicate that the only common property between them is the modulation of the intracellular redox state. However, it is not completely clear that the antioxidant activity of these compounds is the responsible factor in their potentiating effect but the prooxidant activity found in most of these compounds must be also considered. This last section will be focussed on the role of these "antioxidants" in differentiation and apoptosis in leukemic cells.

Several works have suggested that ROS can play a role in the differentiation program of different types of AML (Nagy, *et al* 1993, Nagy, *et al* 1995). On the other hand, decrease of GSH levels in cells have been related to the differentiation of the monocytic THP-1 cells induced by deferoxamine (Seo, *et al* 2006).

The topoisomerase inhibitor β-lapachone induces differentiation and apoptosis in various human cancer cells. This compound increases hydrogen peroxide (H_2O_2) in human leukaemia HL-60 cells, an increase blocked by N-acetyl-L-cysteine (NAC), ascorbic acid, and α-tocopherol. This H_2O_2 increase found in HL-60, U937, and Molt-4 is due to the low level of GSH found in these cells. Then, it is not rare to think that the increase in ROS levels in these cells is responsible of the monocytic differentiation induction found with β-Lapachone at lower concentrations. Moreover because this monocytic differentiation is suppressed by NAC (Chau, *et al* 1998).

On the contrary, deregulation of ROS levels has been related to the leukemogenesis in blood cells. In fact, granulocyte colony-stimulating factor (G-CSF) stimulates the production of ROS in a mechanism dependent on the activation of Lyn, PI3K and Akt (Zhu, *et al* 2006). Most of the acute myeloid leukaemia (AML) express the truncated G-CSF receptor and they produce more ROS than those cells showing a wild type receptor and addition of NAC diminish G-CSF induced ROS production and cell proliferation by inhibiting Akt activation. Furthermore, leukocyte ascorbic acid is reduced in abnormal leukocyte states such as leukaemia (Barton and Roath 1976). Intracellular ascorbate levels were reduced in CML and CLL and in a significant population of patients with AML and ALL. Then, the presence of this antioxidant may be important in the regulation of intracellular pathways in these cells. Studies performed in mice, able to synthesize ascorbate whereas humans are not, demonstrate that vitamin C deficiency in leukaemia occurs during malignancy due to an enhanced rate of utilization of this compound (Bhattacharjee, *et al* 1985). Then, the leukemogenesis is accompanied by the increase in ROS production. Therefore, ROS levels can have an important role in both leukemogenesis and normal haematopoiesis program.

Several works demonstrate that cells, including the leukemic cell line HL-60 cells, incorporate very high amounts of reduced ascorbic acid by the uptake of the oxidized form, dehydroascorbate through the facilitate hexose transporters (glucose transporters) and in a mechanism independent of intracellular GSH content (Guaiquil, *et al* 1997, Vera, *et al* 1995, Vera, *et al* 1994, Wolf 1996). Reduced ascorbate levels in HL-60 cells protects cells against oxidative damage induced by high levels of H_2O_2, serum or glucose-free cultures or X radiation, then, intracellular levels of ascorbate are used by these cells mainly in protection against oxidative damage since ascorbate levels decrease after oxidative injury (Witenberg, *et al* 1999a, Witenberg, *et al* 1999b). Thus, ascorbate may be an important component of the cell machinery against oxidative injury and subsequent leukemogenesis.

However, the role of ROS and antioxidants in these processes seems to be contradictory. In fact, ROS are related with differentiation and antioxidants are scavengers of ROS. Polyphenols are considered as antioxidants because in most cases, these compounds induce the antioxidant response of cells increasing enzymes that protects cells against oxidative injury. However, some reports have found that treatment of human myeloid leukemia cells with the polyphenols carnosic acid (Danilenko, *et al* 2001) or caffeic acid (Nardini, *et al* 2001) results in the decrease of intracellular levels of ROS. On the contrary, our results on HL-60 cells have shown that carnosic acid, resveratrol, curcumin and even high levels of ascorbate increases H_2O_2 levels in HL-60 cells indicating that these compounds increase ROS during early phases of differentiation. In agreement with our results, Strasser et al., (Strasser, *et al* 2005) have shown that curcumin 10 μM increases ROS species just after 1 h after administration in undifferentiated U937 cells. The induction of ROS levels just after treatment produces an increase in GSH levels and further decrease in ROS levels after 24 h of incubation. This study demonstrate that the antioxidative effect of curcumin is preceded by an oxidative stimulus in a time and dose-dependent effect being the GSH induction a response of the cell to the oxidative insult (Strasser, *et al* 2005). This cell response have been denominated as the hormetic response where mild oxidative injury causes an antioxidant response that increases the protection of the cell (Rattan 1998).

This hormetic effect has been also found with other classic antioxidant compounds. In fact, α-tocopherol and hydrocoumarins have been also proposed as antimelanogenic compounds by affecting melanosome differentiation by interrupting melanisation by increasing the intracellular GSH content (Yamamura, *et al* 2002). Also, resveratrol, that have shown antiproliferative and apoptotic effect on several myeloid cell lines, increases GSH level after 24-48 h of exposure affecting also to the activity of glutathione reductase and peroxidase (Luzi, *et al* 2004).

Another important effect of ROS in leukemic cells is the induction of apoptosis. However, the literature is not clear about the role of ROS in apoptosis in these cells. In some cases, the induction of the apoptotic pathway seems to be dependent on the antioxidant role of the compounds, although in other cases, the reduction of ROS levels avoids apoptosis in these cells. Our experience demonstrates that leukemic cell lines respond in a different way to antioxidant compounds depending mainly on cell concentration. In fact, in HL-60 cells, high concentrations of cells prevent the apoptosis found with high levels of ascorbate (0.5 mM). Moreover, the cell line also affects the *in vitro* experiment. For example, NB4 cells are less sensible to ascorbate-induced apoptosis than HL-60 cells. These *in vitro* experiments can highlight the importance of ROS on the apoptosis program in leukemic cells but do not explain the exact role of these compounds into the organism and the mechanism by which antioxidants affect apoptosis in these cells. Moreover, and in agreement with our results, a recent work have demonstrated that ascorbic acid at moderated concentrations can reduce cell growth and induce apoptosis in three different AML cell lines and also in leukemic cells from peripheral blood from patients (Park, *et al* 2004). In this mechanism, the effect of ascorbate was prooxidant since it induced oxidation of GSH to GSSG, and accumulation of H_2O_2 in a dose-dependent fashion. This prooxidant activity of ascorbate was directly related to the induction of apoptosis since catalase completely abrogate this effect.

On the other hand, specific natural polyphenols such as the anthocyanin, cyanidin-3-rutinoside, have been shown to selectively kill leukemic cells by induction of oxidative stress (Feng, *et al* 2007). Also, ethyl acetate extracts from three marine algae significantly increase ROS levels and induce apoptosis in U937 and HL-60 cells and addition of NAC to these cells blocks apoptosis suggesting that ROS are key mediators in apoptosis in these cells (Huang, *et al* 2005a). New compounds in study for leukaemia treatment such as the farnesyltransferase inhibitor manumycin A, also induce apoptosis and increase nitric oxide and other ROS production in these cells, an effect that also was inhibited by the addition of NAC to these cells (She, *et al* 2005). The same effect can be found when cells are treated with homocysteine thiolactone where NAC, ascorbate, α-tocopherol and folate reduced the apoptosis induced by this agent (Huang, *et al* 2002).

On the other hand, quercetin is also an antioxidant that acts as anticancer drug. This compound exerts its role by either diminishing or promoting ROS generation (antioxidant or pro-oxidant role). A recent study have demonstrated that addition of soluble antioxidants such as ascorbate, NAC or GSH increase whereas H_2O_2 attenuates the antiproliferative and apoptotic effect of quercetin on HL-60 cells indicating a role of the antioxidant function of this compound on growth arrest and cell death (Chen, *et al* 2004). Then, the reduction of ROS levels is the factor involved in cell arrest and cell death (Chen, *et al* 2005). However, in this work, the levels of ascorbate were too high (1 mM). At these levels, ascorbate exerts different effects on ROS since our results demonstrate that whereas cells show a dramatic reduction in superoxide anion (Lopez-Lluch, *et al* 2001), H_2O_2 levels increases. Moreover, other agents that induce apoptosis in APL and also in AML such as arsenic trioxide are enhanced in their apoptotic effect by the addition of ascorbate suggesting a putative future role of both compounds in the treatment of AML (Bachleitner-Hofmann, *et al* 2001).

In other cases, the use of nutritional antioxidants such as ascorbate and α-tocopherol has been proposed to avoid the appearance of secondary acute myelogenous leukaemia after etoposite (VP-16) treatment of other cancer (Kagan, *et al* 1999). In this case, the antioxidants prevent the pro-oxidative activity of etoposide phenoxyl radical after one-electron oxidation of etoposide mainly by myeloperoxidase that is highly active in haematopoietic stem cells and other cell lines such as HL-60. Then, dietary antioxidants can minimize the genotoxic effect of radicals in bone marrow myelogenous progenitor cells after chemotherapy.

Independently of the antioxidant or prooxidant mechanism used by these compounds, it is clear that their activity in cell differentiation or apoptosis open an important field of research to find a common mechanism able to increase the effectiveness of antileukemic pharmacological compounds. In fact, currently, phase I to III trials are demonstrating that the use of these phytochemicals with antioxidant activity is important in the treatment of leukaemia and other cancers (as published in the National Cancer Institute web page (www.cancer.gov). For example, curcumin and ubiquinone are currently studies as agents in the treatment of Myelodysplastic Syndromes, known as preleukemic syndromes. Long time ago Prasad et al., (Prasad, *et al* 1999), already proposed that supplements with multiple antioxidant vitamin together with diet and lifestyle modifications may improve the efficacy of standard and experimental cancer therapies. The present work has highlighted the different and complex mechanisms used by these compounds in differentiation induction and enhancement in leukemic cells. The identification of the main pathway implicated in their reinforcement of differentiation mechanism will help in the design of more specific and effective compounds in the therapy of leukaemia by reestablishment of the differentiation pathway.

References

Alcain, F.J. & Buron, M.I. (1994) Ascorbate on cell growth and differentiation. *J Bioenerg Biomembr, 26*, 393-398.

Amir, H., Karas, M., Giat, J., Danilenko, M., Levy, R., Yermiahu, T., Levy, J. & Sharoni, Y. (1999) Lycopene and 1,25-dihydroxyvitamin D3 cooperate in the inhibition of cell cycle progression and induction of differentiation in HL-60 leukemic cells. *Nutr Cancer, 33*, 105-112.

Annabi, B., Currie, J.C., Moghrabi, A. & Beliveau, R. (2006) Inhibition of HuR and MMP-9 expression in macrophage-differentiated HL-60 myeloid leukemia cells by green tea polyphenol EGCg. *Leuk Res.*

Arakawa, N., Nemoto, S., Suzuki, E. & Otsuka, M. (1994) Role of hydrogen peroxide in the inhibitory effect of ascorbate on cell growth. *J Nutr Sci Vitaminol (Tokyo), 40*, 219-227.

Asou, H., Koshizuka, K., Kyo, T., Takata, N., Kamada, N. & Koeffier, H.P. (2002) Resveratrol, a natural product derived from grapes, is a new inducer of differentiation in human myeloid leukemias. *Int J Hematol, 75*, 528-533.

Bachleitner-Hofmann, T., Gisslinger, B., Grumbeck, E. & Gisslinger, H. (2001) Arsenic trioxide and ascorbic acid: synergy with potential implications for the treatment of acute myeloid leukaemia? *Br J Haematol, 112*, 783-786.

Banker, D.E., Groudine, M., Norwood, T. & Appelbaum, F.R. (1997) Measurement of spontaneous and therapeutic agent-induced apoptosis with BCL-2 protein expression in acute myeloid leukemia. *Blood, 89*, 243-255.

Barton, G.M. & Roath, O.S. (1976) Leucocyte ascorbic acid in abnormal leucocyte states. *Int J Vitam Nutr Res, 46*, 271-274.

Behrend, L., Henderson, G. & Zwacka, R.M. (2003) Reactive oxygen species in oncogenic transformation. *Biochem Soc Trans, 31*, 1441-1444.

Bendich, A. & Langseth, L. (1995) The health effects of vitamin C supplementation: a review. *J Am Coll Nutr, 14,* 124-136.

Berry, D.M., Clark, C.S. & Meckling-Gill, K.A. (2002) 1alpha,25-dihydroxyvitamin D3 stimulates phosphorylation of IkappaBalpha and synergizes with TPA to induce nuclear translocation of NFkappaB during monocytic differentiation of NB4 leukemia cells. *Exp Cell Res, 272*, 176-184.

Bhattacharjee, J., Chakraborty, A.S., Sarkar, N.K., Basu, A. & Mitra, S. (1985) Study of ascorbate status in murine and human leukaemias. *J Comp Pathol, 95*, 87-91.

Birkenkamp, K.U., Geugien, M., Schepers, H., Westra, J., Lemmink, H.H. & Vellenga, E. (2004) Constitutive NF-kappaB DNA-binding activity in AML is frequently mediated by a Ras/PI3-K/PKB-dependent pathway. *Leukemia, 18*, 103-112.

Brandts, C.H., Sargin, B., Rode, M., Biermann, C., Lindtner, B., Schwable, J., Buerger, H., Muller-Tidow, C., Choudhary, C., McMahon, M., Berdel, W.E. & Serve, H. (2005) Constitutive activation of Akt by Flt3 internal tandem duplications is necessary for increased survival, proliferation, and myeloid transformation. *Cancer Res, 65*, 9643-9650.

Brea-Calvo, G., Rodriguez-Hernandez, A., Fernandez-Ayala, D.J., Navas, P. & Sanchez-Alcazar, J.A. (2006) Chemotherapy induces an increase in coenzyme Q10 levels in cancer cell lines. *Free Radic Biol Med, 40*, 1293-1302.

Brightman, A.O., Wang, J., Miu, R.K., Sun, I.L., Barr, R., Crane, F.L. & Morre, D.J. (1992) A growth factor- and hormone-stimulated NADH oxidase from rat liver plasma membrane. *Biochim Biophys Acta, 1105*, 109-117.

Brunet, A., Bonni, A., Zigmond, M.J., Lin, M.Z., Juo, P., Hu, L.S., Anderson, M.J., Arden, K.C., Blenis, J. & Greenberg, M.E. (1999) Akt promotes cell survival by phosphorylating and inhibiting a Forkhead transcription factor. *Cell, 96*, 857-868.

Burón, M.I., Rodríguez-Aguilera, J.C., Alcaín, F.J. & Navas, P. (1993) Transplasma membrane redox system in HL-60 cells is modulated during TPA-induced differentiation. *Biochem. Biophys. Res. Commun., 192*, 439-445.

Cameron, E. & Pauling, L. (1976) Supplemental ascorbate in the supportive treatment of cancer: Prolongation of survival times in terminal human cancer. *Proc Natl Acad Sci U S A, 73*, 3685-3689.

Campos, L., Rouault, J.P., Sabido, O., Oriol, P., Roubi, N., Vasselon, C., Archimbaud, E., Magaud, J.P. & Guyotat, D. (1993) High expression of bcl-2 protein in acute myeloid leukemia cells is associated with poor response to chemotherapy. *Blood, 81*, 3091-3096.

Carcamo, J.M., Pedraza, A., Borquez-Ojeda, O. & Golde, D.W. (2002) Vitamin C suppresses TNF alpha-induced NF kappa B activation by inhibiting I kappa B alpha phosphorylation. *Biochemistry, 41*, 12995-13002.

Carter, M.E. & Brunet, A. (2007) FOXO transcription factors. *Curr Biol, 17*, R113-114.

Clark, J.J., Cools, J., Curley, D.P., Yu, J.C., Lokker, N.A., Giese, N.A. & Gilliland, D.G. (2004) Variable sensitivity of FLT3 activation loop mutations to the small molecule tyrosine kinase inhibitor MLN518. *Blood, 104*, 2867-2872.

Collins, S.J. (1998) Acute promyelocytic leukemia: relieving repression induces remission. *Blood, 91*, 2631-2633.

Cornwell, D.G., Jones, K.H., Jiang, Z., Lantry, L.E., Southwell-Keely, P., Kohar, I. & Thornton, D.E. (1998) Cytotoxicity of tocopherols and their quinones in drug-sensitive and multidrug-resistant leukemia cells. *Lipids, 33*, 295-301.

Crane, F.L., Crane, H.E., Sun, I.L., MacKellar, W.C., Grebing, C. & Low, H. (1982) Insulin control of a transplasma membrane NADH dehydrogenase in erythrocyte membranes. *J Bioenerg Biomembr, 14*, 425-433.

Chaplinski, T.J. & Niedel, J.E. (1982) Cyclic nucleotide-induced maturation of human promyelocytic leukemia cells. *J Clin Invest, 70*, 953-964.

Chatterjee, I.B. (1978) Ascorbic acid metabolism. *World Rev Nutr Diet, 30*, 69-87.

Chau, Y.P., Shiah, S.G., Don, M.J. & Kuo, M.L. (1998) Involvement of hydrogen peroxide in topoisomerase inhibitor beta-lapachone-induced apoptosis and differentiation in human leukemia cells. *Free Radic Biol Med, 24*, 660-670.

Chen, J., Kang, J., Da, W. & Ou, Y. (2004) Combination with water-soluble antioxidants increases the anticancer activity of quercetin in human leukemia cells. *Pharmazie, 59*, 859-863.

Chen, J., Wanming, D., Zhang, D., Liu, Q. & Kang, J. (2005) Water-soluble antioxidants improve the antioxidant and anticancer activity of low concentrations of curcumin in human leukemia cells. *Pharmazie, 60*, 57-61.

Danilenko, M., Wang, Q., Wang, X., Levy, J., Sharoni, Y. & Studzinski, G.P. (2003) Carnosic acid potentiates the antioxidant and prodifferentiation effects of 1alpha,25-dihydroxyvitamin D3 in leukemia cells but does not promote elevation of basal levels of intracellular calcium. *Cancer Res, 63*, 1325-1332.

Danilenko, M., Wang, X. & Studzinski, G.P. (2001) Carnosic acid and promotion of monocytic differentiation of HL60-G cells initiated by other agents. *J Natl Cancer Inst, 93*, 1224-1233.

Deguchi, K. & Gilliland, D.G. (2002) Cooperativity between mutations in tyrosine kinases and in hematopoietic transcription factors in AML. *Leukemia, 16*, 740-744.

Delgado, M.D., Vaque, J.P., Arozarena, I., Lopez-Ilasaca, M.A., Martinez, C., Crespo, P. & Leon, J. (2000) H-, K- and N-Ras inhibit myeloid leukemia cell proliferation by a p21WAF1-dependent mechanism. *Oncogene, 19*, 783-790.

Della Ragione, F., Cucciolla, V., Criniti, V., Indaco, S., Borriello, A. & Zappia, V. (2002) Antioxidants induce different phenotypes by a distinct modulation of signal transduction. *FEBS Lett, 532*, 289-294.

Denda, A., Tang, Q., Tsujiuchi, T., Tsutsumi, M., Amanuma, T., Murata, Y., Tamura, K., Horiguchi, K., Nakae, D. & Konishi, Y. (1993) Effects of oxidative stress induced by redox-enzyme modulation on the progression stage of rat hepatocarcinogenesis. *Carcinogenesis, 14*, 95-101.

Dimitroulakos, J., Nohynek, D., Backway, K.L., Hedley, D.W., Yeger, H., Freedman, M.H., Minden, M.D. & Penn, L.Z. (1999) Increased sensitivity of acute myeloid leukemias to lovastatin-induced apoptosis: A potential therapeutic approach. *Blood, 93*, 1308-1318.

Eckert, R.L., Crish, J.F., Efimova, T. & Balasubramanian, S. (2006) Opposing action of curcumin and green tea polyphenol in human keratinocytes. *Mol Nutr Food Res, 50*, 123-129.

Elstner, E., Linker-Israeli, M., Umiel, T., Le, J., Grillier, I., Said, J., Shintaku, I.P., Krajewski, S., Reed, J.C., Binderup, L. & Koeffler, H.P. (1996) Combination of a potent 20-epi-vitamin D3 analogue (KH 1060) with 9-cis-retinoic acid irreversibly inhibits clonal growth, decreases bcl-2 expression, and induces apoptosis in HL-60 leukemic cells. *Cancer Res, 56*, 3570-3576.

Estrov, Z., Shishodia, S., Faderl, S., Harris, D., Van, Q., Kantarjian, H.M., Talpaz, M. & Aggarwal, B.B. (2003) Resveratrol blocks interleukin-1beta-induced activation of the nuclear transcription factor NF-kappaB, inhibits proliferation, causes S-phase arrest, and induces apoptosis of acute myeloid leukemia cells. *Blood, 102*, 987-995.

Farquharson, C., Berry, J.L., Mawer, E.B., Seawright, E. & Whitehead, C.C. (1998) Ascorbic acid-induced chondrocyte terminal differentiation: the role of the extracellular matrix and 1,25-dihydroxyvitamin D. *Eur J Cell Biol, 76*, 110-118.

Feng, R., Ni, H.M., Wang, S.Y., Tourkova, I.L., Shurin, M.R., Harada, H. & Yin, X.M. (2007) Cyanidin-3-rutinoside, a natural polyphenol antioxidant, selectively kills leukemic cells by induction of oxidative stress. *J Biol Chem, 282*, 13468-13476.

Fimognari, C., Berti, F., Nusse, M., Cantelli-Forti, G. & Hrelia, P. (2004) Induction of apoptosis in two human leukemia cell lines as well as differentiation in human promyelocytic cells by cyanidin-3-O-beta-glucopyranoside. *Biochem Pharmacol, 67*, 2047-2056.

Fitch, C.D., Broun, G.O., Jr., Chou, A.C. & Gallagher, N.I. (1980) Abnormal erythropoiesis in vitamin E-deficient monkeys. *Am J Clin Nutr, 33*, 1251-1258.

Fong, W.-F., Tse, A.K.-W., Poon, K.-H. & Wang, C. (2005) Magnolol and honokiol enhance HL-60 human leukemia cell differentiation induced by 1,25-dihydroxyvitamin D3 and retinoic acid. *The International Journal of Biochemistry & Cell Biology, 37*, 427-441.

Fontana, J., Munoz, M. & Durham, J. (1985) Potentiation between intracellular cyclic-AMP-elevating agents and inducers of leukemic cell differentiation. *Leuk Res, 9*, 1127-1132.

Franceschi, R.T. & Iyer, B.S. (1992) Relationship between collagen synthesis and expression of the osteoblast phenotype in MC3T3-E1 cells. *J Bone Miner Res, 7*, 235-246.

Frei, B., England, L. & Ames, B.N. (1989) Ascorbate is an outstanding antioxidant in human blood plasma. *Proc Natl Acad Sci U S A, 86*, 6377-6381.

Ganser, A., Maurer, A., Contzen, C., Seipelt, G., Ottmann, O.G., Schadeck-Gressel, C., Kolbe, K., Haas, R., Zander, C., Reutzel, R. & Hoelzer, D. (1996) Improved multilineage response of hematopoiesis in patients with myelodysplastic syndromes to a combination therapy with all-trans-retinoic acid, granulocyte colony-stimulating factor, erythropoietin and alpha-tocopherol. *Ann Hematol, 72*, 237-244.

Gershoff, S.N. (1993) Ascorbic acid: new roles, new requirements? *Nutr Rev, 51*, 313-326.

Goldenberg, H., Crane, F.L. & Morre, D.J. (1978) Influence of hormones on NADH-dehydrogenase in mouse liver plasma membrane. *Biochem Biophys Res Commun, 83*, 234-240.

Goldstein, J.L. & Brown, M.S. (1990) Regulation of the mevalonate pathway. *Nature, 343*, 425-430.

Gross, M.D., Bishop, T.D., Belcher, J.D. & Jacobs, D.R., Jr. (1997) Induction of HL-60 cell differentiation by carotenoids. *Nutr Cancer, 27*, 169-173.

Gruber, H.E., Tewfik, H.H. & Tewfik, F.A. (1980) Cytoarchitecture of Ehrlich ascites carcinoma implanted in the hind limb of ascorbic acid-supplemented mice. *Eur J Cancer, 16*, 441-448.

Guaiquil, V.H., Farber, C.M., Golde, D.W. & Vera, J.C. (1997) Efficient transport and accumulation of vitamin C in HL-60 cells depleted of glutathione. *J Biol Chem, 272*, 9915-9921.

Guillemin, M.C., Raffoux, E., Vitoux, D., Kogan, S., Soilihi, H., Lallemand-Breitenbach, V., Zhu, J., Janin, A., Daniel, M.T., Gourmel, B., Degos, L., Dombret, H., Lanotte, M. & De The, H. (2002) In vivo activation of cAMP signaling induces growth arrest and differentiation in acute promyelocytic leukemia. *J Exp Med, 196*, 1373-1380.

Gyemant, N., Tanaka, M., Molnar, P., Deli, J., Mandoky, L. & Molnar, J. (2006) Reversal of multidrug resistance of cancer cells in vitro: modification of drug resistance by selected carotenoids. *Anticancer Res, 26*, 367-374.

Han, S.S., Kim, K., Hahm, E.R., Lee, S.J., Surh, Y.J., Park, H.K., Kim, W.S., Jung, C.W., Lee, M.H., Park, K., Yang, J.H., Yoon, S.S., Riordan, N.H., Riordan, H.D., Kimler, B.F., Park, C.H., Lee, J.H. & Park, S. (2004) L-ascorbic acid represses constitutive activation of NF-kappaB and COX-2 expression in human acute myeloid leukemia, HL-60. *J Cell Biochem, 93*, 257-270.

Harant, H., Andrew, P.J., Reddy, G.S., Foglar, E. & Lindley, I.J. (1997) 1alpha,25-dihydroxyvitamin D3 and a variety of its natural metabolites transcriptionally repress nuclear-factor-kappaB-mediated interleukin-8 gene expression. *Eur J Biochem, 250*, 63-71.

Hayakawa, F., Towatari, M., Kiyoi, H., Tanimoto, M., Kitamura, T., Saito, H. & Naoe, T. (2000) Tandem-duplicated Flt3 constitutively activates STAT5 and MAP kinase and introduces autonomous cell growth in IL-3-dependent cell lines. *Oncogene, 19*, 624-631.

Heber, D. & Lu, Q.Y. (2002) Overview of mechanisms of action of lycopene. *Exp Biol Med (Maywood)*, 227, 920-923.

Huang, H.L., Wu, S.L., Liao, H.F., Jiang, C.M., Huang, R.L., Chen, Y.Y., Yang, Y.C. & Chen, Y.J. (2005a) Induction of apoptosis by three marine algae through generation of reactive oxygen species in human leukemic cell lines. *J Agric Food Chem, 53*, 1776-1781.

Huang, R.F., Huang, S.M., Lin, B.S., Hung, C.Y. & Lu, H.T. (2002) N-Acetylcysteine, vitamin C and vitamin E diminish homocysteine thiolactone-induced apoptosis in human promyeloid HL-60 cells. *J Nutr, 132*, 2151-2156.

Huang, W.W., Yang, J.S., Lin, C.F., Ho, W.J. & Lee, M.R. (2005b) Pycnogenol induces differentiation and apoptosis in human promyeloid leukemia HL-60 cells. *Leukemia Res., 29*, 685-692.

Jackson, A.L. & Loeb, L.A (2001) The contribution of endogenous sources of DNA damage to the multiple mutations in cancer. *Mutat Res, 477*, 7-21.

Kagan, V.E., Yalowich, J.C., Borisenko, G.G., Tyurina, Y.Y., Tyurin, V.A., Thampatty, P. & Fabisiak, J.P. (1999) Mechanism-based chemopreventive strategies against etoposide-induced acute myeloid leukemia: free radical/antioxidant approach. *Mol Pharmacol, 56*, 494-506.

Kang, H.K., Suh, J.H., Lee, J.J., Yoon, S.H., Hyun, J.W., Choi, S.W., Choi, J.Y., Ryu, K.H. & Chung, M.H. (2003) Induction of the differentiation of HL-60 promyelocytic leukemia cells by L-ascorbic acid. *Free Radic Res, 37*, 773-779.

Kang, S.N., Kim, S.H., Chung, S.W., Lee, M.H., Kim, H.J. & Kim, T.S. (2002) Enhancement of 1 alpha,25-dihydroxyvitamin D(3)-induced differentiation of human leukaemia HL-60 cells into monocytes by parthenolide via inhibition of NF-kappa B activity. *Br J Pharmacol, 135*, 1235-1244.

Kawaii, S. & Lansky, E.P. (2004) Differentiation-promoting activity of pomegranate (Punica granatum) fruit extracts in HL-60 human promyelocytic leukemia cells. *J Med Food, 7*, 13-18.

Kelly, L.M., Liu, Q., Kutok, J.L., Williams, I.R., Boulton, C.L. & Gilliland, D.G. (2002) FLT3 internal tandem duplication mutations associated with human acute myeloid leukemias induce myeloproliferative disease in a murine bone marrow transplant model. *Blood, 99*, 310-318.

Kim, S.H., Kang, S.N., Kim, H.J. & Kim, T.S. (2002) Potentiation of 1,25-dihydroxyvitamin D3-induced differentiation of human promyelocytic leukemia cells into monocytes by costunolide, a germacranolide sesquiterpene lactone. *Biochemical Pharmacology, 64*, 1233-1242.

Kim, S.H., Oh, S.M. & Kim, T.S. (2005) Induction of human leukemia HL-60 cell differentiation via a PKC/ERK pathway by helenalin, a pseudoguainolide sesquiterpene lactone. *Eur J Pharmacol, 511*, 89-97.

Kiyoi, H., Towatari, M., Yokota, S., Hamaguchi, M., Ohno, R., Saito, H. & Naoe, T. (1998) Internal tandem duplication of the FLT3 gene is a novel modality of elongation mutation which causes constitutive activation of the product. *Leukemia, 12*, 1333-1337.

Klaunig, J.E. & Kamendulis, L.M. (2004) The role of oxidative stress in carcinogenesis. *Annu Rev Pharmacol Toxicol,* 44, 239-267.

Kops, G.J., de Ruiter, N.D., De Vries-Smits, A.M., Powell, D.R., Bos, J.L. & Burgering, B.M. (1999) Direct control of the Forkhead transcription factor AFX by protein kinase B. *Nature, 398*, 630-634.

Kordes, U., Krappmann, D., Heissmeyer, V., Ludwig, W.D. & Scheidereit, C. (2000) Transcription factor NF-kappaB is constitutively activated in acute lymphoblastic leukemia cells. *Leukemia, 14*, 399-402.

Krieger-Brauer, H.I. & Kather, H. (1995) The stimulus-sensitive H2O2-generating system present in human fat-cell plasma membranes is multireceptor-linked and under antagonistic control by hormones and cytokines. *Biochem J, 307* (Pt 2), 543-548.

Lee, J.K., Jung, J.C., Chun, J.S., Kang, S.S. & Bang, O.S. (2002) Expression of p21WAF1 is dependent on the activation of ERK during vitamin E-succinate-induced monocytic differentiation. *Mol Cells, 13*, 125-129.

Lee, J.Y., Chang, M.Y., Park, C.H., Kim, H.Y., Kim, J.H., Son, H., Lee, Y.S. & Lee, S.H. (2003) Ascorbate-induced differentiation of embryonic cortical precursors into neurons and astrocytes. *J Neurosci Res, 73*, 156-165.

Lin, J.K. (2004) Suppression of protein kinase C and nuclear oncogene expression as possible action mechanisms of cancer chemoprevention by Curcumin. *Arch Pharm Res, 27*, 683-692.

Lin, R.J., Nagy, L., Inoue, S., Shao, W., Miller, W.H., Jr. & Evans, R.M. (1998) Role of the histone deacetylase complex in acute promyelocytic leukaemia. *Nature, 391*, 811-814.

Lopez-Lluch, G., Blazquez, M.V., Perez-Vicente, R., Macho, A., Buron, M.I., Alcain, F.J., Munoz, E. & Navas, P. (2001) Cellular redox state and activating protein-1 are involved in ascorbate effect on calcitriol-induced differentiation. *Protoplasma, 217*, 129-136.

Lopez-Lluch, G., Buron, M.I., Alcain, F.J., Quesada, J.M. & Navas, P. (1998) Redox regulation of cAMP levels by ascorbate in 1,25-dihydroxy- vitamin D3-induced differentiation of HL-60 cells. *Biochem J, 331* (Pt 1), 21-27.

Lopez-Lluch, G., Fernandez-Ayala, D.J.M., Alcain, F.J., Buron, M.I., Quesada, J.M. & Navas, P. (2005) Inhibition of COX activity by NSAIDs or ascorbate increases cAMP levels and enhances differentiation in 1[alpha],25-dihydroxyvitamin D3-induced HL-60 cells. *Archives of Biochemistry and Biophysics, 436*, 32-39.

Luyengi, L., Suh, N., Fong, H.H., Pezzuto, J.M. & Kinghorn, A.D. (1996) A lignan and four terpenoids from Brucea javanica that induce differentiation with cultured HL-60 promyelocytic leukemia cells. *Phytochemistry, 43*, 409-412.

Luzi, C., Brisdelli, F., Cinque, B., Cifone, G. & Bozzi, A. (2004) Differential sensitivity to resveratrol-induced apoptosis of human chronic myeloid (K562) and acute lymphoblastic (HSB-2) leukemia cells. *Biochem Pharmacol, 68*, 2019-2030.

Makishima, M. & Honma, Y. (1997) Tretinoin tocoferil as a possible differentiation-inducing agent against myelomonocytic leukemia. *Leuk Lymphoma, 26*, 43-48.

Makishima, M., Kanatani, Y., Yamamoto-Yamaguchi, Y. & Honma, Y. (1996) Enhancement of activity of 1alpha, 25-dihydroxyvitamin D3 for growth inhibition and differentiation induction of human myelomonocytic leukemia cells by tretinoin tocoferil, an alpha-tocopherol ester of all-trans retinoic acid. *Blood, 87*, 3384-3394.

Makishima, M., Umesono, K., Shudo, K., Naoe, T., Kishi, K. & Honma, Y. (1998) Induction of differentiation in acute promyelocytic leukemia cells by 9-cis retinoic acid alpha-tocopherol ester (9-cis tretinoin tocoferil). *Blood, 91*, 4715-4726.

Masamune, A., Suzuki, N., Kikuta, K., Satoh, M., Satoh, K. & Shimosegawa, T. (2006) Curcumin blocks activation of pancreatic stellate cells. *J Cell Biochem, 97*, 1080-1093.

Mathieu, J., Giraudier, S., Lanotte, M. & Besancon, F. (2005) Retinoid-induced activation of NF-kappaB in APL cells is not essential for granulocytic differentiation, but prolongs the life span of mature cells. *Oncogene, 24*, 7145-7155.

Melnick, A. & Licht, J.D. (1999) Deconstructing a disease: RARalpha, its fusion partners, and their roles in the pathogenesis of acute promyelocytic leukemia. *Blood, 93*, 3167-3215.

Mizuki, M., Fenski, R., Halfter, H., Matsumura, I., Schmidt, R., Muller, C., Gruning, W., Kratz-Albers, K., Serve, S., Steur, C., Buchner, T., Kienast, J., Kanakura, Y., Berdel, W.E. & Serve, H. (2000) Flt3 mutations from patients with acute myeloid leukemia induce transformation of 32D cells mediated by the Ras and STAT5 pathways. *Blood, 96*, 3907-3914.

Munoz, E., Blazquez, M.V., Ortiz, C., Gomez-Diaz, C. & Navas, P. (1997) Role of ascorbate in the activation of NF-kappaB by tumour necrosis factor-alpha in T-cells. *Biochem J, 325* (Pt 1), 23-28.

Nagy, K., Pasti, G., Bene, L. & Zs-Nagy, I. (1993) Induction of granulocytic maturation in HL-60 human leukemia cells by free radicals: a hypothesis of cell differentiation involving hydroxyl radicals. *Free Radic Res Commun, 19,* 1-15.

Nagy, K., Pásti, G., Bene, L. & Zs-Nagy, I. (1995) Involvement of fenton reaction products in differentiation induction of K562 human leukemia cells. *Leukemia Res., 19*, 203-212.

Nakamura, N., Ramaswamy, S., Vazquez, F., Signoretti, S., Loda, M. & Sellers, W.R. (2000) Forkhead transcription factors are critical effectors of cell death and cell cycle arrest downstream of PTEN. *Mol Cell Biol, 20*, 8969-8982.

Nara, E., Hayashi, H., Kotake, M., Miyashita, K. & Nagao, A. (2001) Acyclic carotenoids and their oxidation mixtures inhibit the growth of HL-60 human promyelocytic leukemia cells. *Nutr Cancer, 39*, 273-283.

Nardini, M., Leonardi, F., Scaccini, C. & Virgili, F. (2001) Modulation of ceramide-induced NF-kappaB binding activity and apoptotic response by caffeic acid in U937 cells: comparison with other antioxidants. *Free Radic Biol Med, 30*, 722-733.

Navas, P., Villalba, J.M. & Cordoba, F. (1994) Ascorbate function at the plasma membrane. *Biochim Biophys Acta, 1197*, 1-13.

Nemoto, S., Otsuka, M. & Arakawa, N. (1996) Inhibitory effect of ascorbate on cell growth: relation to catalase activity. *J Nutr Sci Vitaminol (Tokyo), 42*, 77-85.

Nervi, C., Ferrara, F.F., Fanelli, M., Rippo, M.R., Tomassini, B., Ferrucci, P.F., Ruthardt, M., Gelmetti, V., Gambacorti-Passerini, C., Diverio, D., Grignani, F., Pelicci, P.G. & Testi, R. (1998) Caspases mediate retinoic acid-induced degradation of the acute promyelocytic leukemia PML/RARalpha fusion protein. *Blood, 92*, 2244-2251.

O'Farrell, A.M., Foran, J.M., Fiedler, W., Serve, H., Paquette, R.L., Cooper, M.A., Yuen, H.A., Louie, S.G., Kim, H., Nicholas, S., Heinrich, M.C., Berdel, W.E., Bello, C., Jacobs, M., Scigalla, P., Manning, W.C., Kelsey, S. & Cherrington, J.M. (2003) An innovative phase I clinical study demonstrates inhibition of FLT3 phosphorylation by SU11248 in acute myeloid leukemia patients. *Clin Cancer Res, 9*, 5465-5476.

Osmak, M., Kovacek, I., Ljubenkov, I., Spaventi, R. & Eckert-Maksic, M. (1997) Ascorbic acid and 6-deoxy-6-chloro-ascorbic acid: potential anticancer drugs. *Neoplasma, 44*, 101-107.

Padh, H. (1990) Cellular functions of ascorbic acid. *Biochem Cell Biol, 68*, 1166-1173.

Park, C.H., Amare, M., Savin, M.A. & Hoogstraten, B. (1980) Growth suppression of human leukemic cells in vitro by L-ascorbic acid. *Cancer Res, 40*, 1062-1065.

Park, S., Han, S.S., Park, C.H., Hahm, E.R., Lee, S.J., Park, H.K., Lee, S.H., Kim, W.S., Jung, C.W., Park, K., Riordan, H.D., Kimler, B.F., Kim, K. & Lee, J.H. (2004) L-Ascorbic acid induces apoptosis in acute myeloid leukemia cells via hydrogen peroxide-mediated mechanisms. *Int J Biochem Cell Biol, 36*, 2180-2195.

Park, S., Park, C.H., Hahm, E.R., Kim, K., Kimler, B.F., Lee, S.J., Park, H.K., Lee, S.H., Kim, W.S., Jung, C.W., Park, K., Riordan, H.D. & Lee, J.H. (2005) Activation of Raf1 and the ERK pathway in response to l-ascorbic acid in acute myeloid leukemia cells. *Cell Signal, 17*, 111-119.

Pauling, L., Nixon, J.C., Stitt, F., Marcuson, R., Dunham, W.B., Barth, R., Bensch, K., Herman, Z.S., Blaisdell, B.E., Tsao, C. & et al. (1985) Effect of dietary ascorbic acid on the incidence of spontaneous mammary tumors in RIII mice. *Proc Natl Acad Sci U S A, 82*, 5185-5189.

Perkins, C., Kim, C.N., Fang, G. & Bhalla, K.N. (2000) Arsenic induces apoptosis of multidrug-resistant human myeloid leukemia cells that express Bcr-Abl or overexpress MDR, MRP, Bcl-2, or Bcl-x(L). *Blood, 95*, 1014-1022.

Pinilla-Ibarz, J., Cathcart, K., Korontsvit, T., Soignet, S., Bocchia, M., Caggiano, J., Lai, L., Jimenez, J., Kolitz, J. & Scheinberg, D.A. (2000) Vaccination of patients with chronic myelogenous leukemia with bcr-abl oncogene breakpoint fusion peptides generates specific immune responses. *Blood, 95*, 1781-1787.

Porosnicu, M., Nimmanapalli, R., Nguyen, D., Worthington, E., Perkins, C. & Bhalla, K.N. (2001) Co-treatment with As2O3 enhances selective cytotoxic effects of STI-571 against Brc-Abl-positive acute leukemia cells. *Leukemia, 15*, 772-778.

Potter, J.D. & Steinmetz, K. (1996) Vegetables, fruit and phytoestrogens as preventive agents. *IARC Sci Publ*, 61-90.

Prasad, K.N., Kumar, A., Kochupillai, V. & Cole, W.C. (1999) High doses of multiple antioxidant vitamins: essential ingredients in improving the efficacy of standard cancer therapy. *J Am Coll Nutr, 18*, 13-25.

Prasad, K.N., Kumar, B., Yan, X.D., Hanson, A.J. & Cole, W.C. (2003) Alpha-tocopheryl succinate, the most effective form of vitamin E for adjuvant cancer treatment: a review. *J Am Coll Nutr, 22*, 108-117.

Quesada, J.M., Lopez, L.G., Buron, M.I., Alcain, F.J., Borrego, F., Velde, J.P., Blanco, I., Bouillon, R. & Navas, P. (1996) Ascorbate increases the 1,25 dihydroxyvitamin D3-induced monocytic differentiation of HL-60 cells. *Calcif Tissue Int, 59*, 277-282.

Radich, J.P., Kopecky, K.J., Willman, C.L., Weick, J., Head, D., Appelbaum, F. & Collins, S.J. (1990) N-ras mutations in adult de novo acute myelogenous leukemia: prevalence and clinical significance. *Blood, 76*, 801-807.

Ragab, A.A., Lavish, S.A., Banks, M.A., Goldberg, V.M. & Greenfield, E.M. (1998) Osteoclast differentiation requires ascorbic acid. *J Bone Miner Res, 13*, 970-977.

Ragione, F.D., Cucciolla, V., Criniti, V., Indaco, S., Borriello, A. & Zappia, V. (2003) p21Cip1 Gene Expression Is Modulated by Egr1: A NOVEL REGULATORY MECHANISM INVOLVED IN THE RESVERATROL ANTIPROLIFERATIVE EFFECT. *J. Biol. Chem., 278*, 23360-23368.

Rattan, S.I. (1998) Repeated mild heat shock delays ageing in cultured human skin fibroblasts. *Biochem Mol Biol Int, 45*, 753-759.

Rodrigue, C.M., Arous, N., Bachir, D., Smith-Ravin, J., Romeo, P.H., Galacteros, F. & Garel, M.C. (2001) Resveratrol, a natural dietary phytoalexin, possesses similar properties to hydroxyurea towards erythroid differentiation. *Br J Haematol*, 113, 500-507.

Rushmore, T.H. & Kong, A.N. (2002) Pharmacogenomics, regulation and signaling pathways of phase I and II drug metabolizing enzymes. *Curr Drug Metab, 3*, 481-490.

Sakagami, H., Asaka, K., Abe, E., Miyaura, C., Suda, T. & Konno, K. (1981) Effect of DL-alpha-tocopherol (vitamin E) on the differentiation of mouse myeloid leukemia cell. *J Nutr Sci Vitaminol (Tokyo), 27*, 291-300.

Sakagami, H. & Satoh, K. (1997) Modulating factors of radical intensity and cytotoxic activity of ascorbate (review). *Anticancer Res, 17*, 3513-3520.

Sanchez-Alcazar, J.A., Bradbury, D.A., Brea-Calvo, G., Navas, P. & Knox, A.J. (2003) Camptothecin-induced apoptosis in non-small cell lung cancer is independent of cyclooxygenase expression. *Apoptosis, 8*, 639-647.

Satoh, K., Sakagami, H. & Motohashi, N. (1997) Radical modulation activity of benzo[a]phenothiazine. *Anticancer Res, 17*, 2539-2543.

Savini, I., Catani, M.V., Rossi, A., Duranti, G., Melino, G. & Avigliano, L. (2002) Characterization of keratinocyte differentiation induced by ascorbic acid: protein kinase C involvement and vitamin C homeostasis. *J Invest Dermatol, 118*, 372-379.

Scalbert, A. & Williamson, G. (2000) Dietary intake and bioavailability of polyphenols. *J Nutr, 130*, 2073S-2085S.

Scheijen, B., Ngo, H.T., Kang, H. & Griffin, J.D. (2004) FLT3 receptors with internal tandem duplications promote cell viability and proliferation by signaling through Foxo proteins. *Oncogene, 23*, 3338-3349.

Seo, B.R., Lee, K.W., Ha, J., Park, H.J., Choi, J.W. & Lee, K.T. (2004a) Saucernetin-7 isolated from Saururus chinensis inhibits proliferation of human promyelocytic HL-60 leukemia cells via G0/G1 phase arrest and induction of differentiation. *Carcinogenesis, 25*, 1387-1394.

Seo, B.R., Yoo, C.B., Park, H.J., Choi, J.W., Seo, K., Choi, S.K. & Lee, K.T. (2004b) Saucernetin-8 isolated from Saururus chinensis induced the differentiation of human acute promyelocytic leukemia HL-60 cells. *Biol Pharm Bull, 27*, 1594-1598.

Seo, G.-S., Lee, S.-H., Choi, S.-C., Choi, E.-Y., Oh, H.-M., Choi, E.-J., Park, D.-S., Kim, S.-W., Kim, T.-H. & Nah, Y.-H. (2006) Iron chelator induces THP-1 cell differentiation potentially by modulating intracellular glutathione levels. *Free Radical Biology and Medicine, 40,* 1502-1512.

Sestili, P., Brandi, G., Brambilla, L., Cattabeni, F. & Cantoni, O. (1996) Hydrogen peroxide mediates the killing of U937 tumor cells elicited by pharmacologically attainable concentrations of ascorbic acid: cell death prevention by extracellular catalase or catalase from cocultured erythrocytes or fibroblasts. *J Pharmacol Exp Ther, 277*, 1719-1725.

Sharabani, H., Izumchenko, E., Wang, Q., Kreinin, R., Steiner, M., Barvish, Z., Kafka, M., Sharoni, Y., Levy, J., Uskokovic, M., Studzinski, G.P. & Danilenko, M. (2006) Cooperative antitumor effects of vitamin D3 derivatives and rosemary preparations in a mouse model of myeloid leukemia. *Int J Cancer, 118*, 3012-3021.

She, M., Pan, I., Sun, L. & Yeung, S.C. (2005) Enhancement of manumycin A-induced apoptosis by methoxyamine in myeloid leukemia cells. *Leukemia, 19*, 595-602.

Sherman, M.L., Shafman, T.D. & Kufe, D.W. (1986) Inhibitory effects of theophylline and dibutyryl cAMP on murine erythroleukemia cell differentiation. *Biochem Biophys Res Commun, 134*, 845-851.

Siendones, E., Barbarroja, N., Torres, L.A., Buendia, P., Velasco, F., Dorado, G., Torres, A. & Lopez-Pedrera, C. (2007) Inhibition of Flt3-activating mutations does not prevent constitutive activation of ERK/Akt/STAT pathways in some AML cells: a possible cause for the limited effectiveness of monotherapy with small-molecule inhibitors. *Hematol Oncol, 25*, 30-37.

Sokoloski, J.A., Hodnick, W.F., Mayne, S.T., Cinquina, C., Kim, C.S. & Sartorelli, A.C. (1997a) Induction of the differentiation of HL-60 promyelocytic leukemia cells by vitamin E and other antioxidants in combination with low levels of vitamin D3: possible relationship to NF-kappaB. *Leukemia, 11*, 1546-1553.

Sokoloski, J.A., Shyam, K. & Sartorelli, A.C. (1997b) Induction of the differentiation of HL-60 promyelocytic leukemia cells by curcumin in combination with low levels of vitamin D3. *Oncol Res, 9*, 31-39.

Steiner, M., Priel, I., Giat, J., Levy, J., Sharoni, Y. & Danilenko, M. (2001) Carnosic acid inhibits proliferation and augments differentiation of human leukemic cells induced by 1,25-dihydroxyvitamin D3 and retinoic acid. *Nutr Cancer, 41*, 135-144.

Stirewalt, D.L., Meshinchi, S. & Radich, J.P. (2003) Molecular targets in acute myelogenous leukemia. *Blood Rev, 17*, 15-23.

Strasser, E.M., Wessner, B., Manhart, N. & Roth, E. (2005) The relationship between the anti-inflammatory effects of curcumin and cellular glutathione content in myelomonocytic cells. *Biochem Pharmacol, 70*, 552-559.

Takahashi, T., Lord, B., Schulze, P.C., Fryer, R.M., Sarang, S.S., Gullans, S.R. & Lee, R.T. (2003) Ascorbic acid enhances differentiation of embryonic stem cells into cardiac myocytes. *Circulation, 107*, 1912-1916.

Takenaga, K., Honma, Y. & Hozumi, M. (1981) Inhibition of differentiation of mouse myeloid leukemia cells by phenolic antioxidants and alpha-tocopherol. *Gann, 72*, 104-112.

Takeuchi, M., Yano, T., Omoto, E., Takahashi, K., Kibata, M., Shudo, K., Harada, M., Ueda, R. & Ohno, R. (1998) Relapsed acute promyelocytic leukemia previously treated with all-trans retinoic acid: clinical experience with a new synthetic retinoid, Am-80. *Leuk Lymphoma, 31*, 441-451.

Tallman, M.S., Nabhan, C., Feusner, J.H. & Rowe, J.M. (2002) Acute promyelocytic leukemia: evolving therapeutic strategies. *Blood, 99*, 759-767.

Tamagawa, K., Fukushima, S., Kobori, M., Shinmoto, H. & Tsushida, T. (1998) Proanthocyanidins from barley bran potentiate retinoic acid-induced granulocytic and sodium butyrate-induced monocytic differentiation of HL60 cells. *Biosci Biotechnol Biochem, 62*, 1483-1487.

Tarumoto, T., Nagai, T., Ohmine, K., Miyoshi, T., Nakamura, M., Kondo, T., Mitsugi, K., Nakano, S., Muroi, K., Komatsu, N. & Ozawa, K. (2004) Ascorbic acid restores sensitivity to imatinib via suppression of Nrf2-dependent gene expression in the imatinib-resistant cell line. *Exp Hematol, 32*, 375-381.

Turley, J.M., Sanders, B.G. & Kline, K. (1992) RRR-alpha-tocopheryl succinate modulation of human promyelocytic leukemia (HL-60) cell proliferation and differentiation. *Nutr Cancer, 18*, 201-213.

Umehara, K., Nakamura, M., Miyase, T., Kuroyanagi, M. & Ueno, A. (1996) Studies on differentiation inducers. VI. Lignan derivatives from Arctium fructus. (2). *Chem Pharm Bull (Tokyo), 44*, 2300-2304.

Varela, N., Munoz-Pinedo, C., Ruiz-Ruiz, C., Robledo, G., Pedroso, M. & Lopez-Rivas, A. (2001) Interferon-gamma sensitizes human myeloid leukemia cells to death receptor-mediated apoptosis by a pleiotropic mechanism. *J Biol Chem, 276*, 17779-17787.

Venugopal, M., Jamison, J.M., Gilloteaux, J., Koch, J.A., Summers, M., Hoke, J., Sowick, C. & Summers, J.L. (1996) Synergistic antitumour activity of vitamins C and K3 against human prostate carcinoma cell lines. *Cell Biol Int, 20*, 787-797.

Vera, J.C., Rivas, C.I., Velasquez, F.V., Zhang, R.H., Concha, II & Golde, D.W. (1995) Resolution of the facilitated transport of dehydroascorbic acid from its intracellular accumulation as ascorbic acid. *J Biol Chem, 270*, 23706-23712.

Vera, J.C., Rivas, C.I., Zhang, R.H., Farber, C.M. & Golde, D.W. (1994) Human HL-60 myeloid leukemia cells transport dehydroascorbic acid via the glucose transporters and accumulate reduced ascorbic acid. *Blood, 84*, 1628-1634.

Verma, A., Mohindru, M., Deb, D.K., Sassano, A., Kambhampati, S., Ravandi, F., Minucci, S., Kalvakolanu, D.V. & Platanias, L.C. (2002) Activation of Rac1 and the p38 mitogen-activated protein kinase pathway in response to arsenic trioxide. *J Biol Chem, 277*, 44988-44995.

Wada, S., Satomi, Y., Murakoshi, M., Noguchi, N., Yoshikawa, T. & Nishino, H. (2005) Tumor suppressive effects of tocotrienol in vivo and in vitro. *Cancer Lett, 229*, 181-191.

Wang, Z.Y. (2003) Ham-Wasserman lecture: treatment of acute leukemia by inducing differentiation and apoptosis. *Hematology Am Soc Hematol Educ Program,* 1-13.

Weisberg, E., Boulton, C., Kelly, L.M., Manley, P., Fabbro, D., Meyer, T., Gilliland, D.G. & Griffin, J.D. (2002) Inhibition of mutant FLT3 receptors in leukemia cells by the small molecule tyrosine kinase inhibitor PKC412. *Cancer Cell, 1*, 433-443.

Wenzel, U., Kuntz, S., Brendel, M.D. & Daniel, H. (2000) Dietary flavone is a potent apoptosis inducer in human colon carcinoma cells. *Cancer Res, 60*, 3823-3831.

Witenberg, B., Kalir, H.H., Raviv, Z., Kletter, Y., Kravtsov, V. & Fabian, I. (1999a) Inhibition by ascorbic acid of apoptosis induced by oxidative stress in HL-60 myeloid leukemia cells. *Biochem Pharmacol, 57*, 823-832.

Witenberg, B., Kletter, Y., Kalir, H.H., Raviv, Z., Fenig, E., Nagler, A., Halperin, D. & Fabian, I. (1999b) Ascorbic acid inhibits apoptosis induced by X irradiation in HL60 myeloid leukemia cells. *Radiat Res, 152*, 468-478.

Wolf, G. (1996) The mechanism of uptake of ascorbic acid into osteoblasts and leukocytes. *Nutr Rev, 54*, 150-152.

Wong, W.W., Dimitroulakos, J., Minden, M.D. & Penn, L.Z. (2002) HMG-CoA reductase inhibitors and the malignant cell: the statin family of drugs as triggers of tumor-specific apoptosis. *Leukemia, 16*, 508-519.

Wuchter, C., Krappmann, D., Cai, Z., Ruppert, V., Scheidereit, C., Dorken, B., Ludwig, W.D. & Karawajew, L. (2001) In vitro susceptibility to TRAIL-induced apoptosis of acute leukemia cells in the context of TRAIL receptor gene expression and constitutive NF-kappa B activity. *Leukemia, 15*, 921-928.

Yamamoto, Y., Kiyoi, H., Nakano, Y., Suzuki, R., Kodera, Y., Miyawaki, S., Asou, N., Kuriyama, K., Yagasaki, F., Shimazaki, C., Akiyama, H., Saito, K., Nishimura, M., Motoji, T., Shinagawa, K., Takeshita, A., Saito, H., Ueda, R., Ohno, R. & Naoe, T. (2001) Activating mutation of D835 within the activation loop of FLT3 in human hematologic malignancies. *Blood, 97*, 2434-2439.

Yamamura, T., Onishi, J. & Nishiyama, T. (2002) Antimelanogenic activity of hydrocoumarins in cultured normal human melanocytes by stimulating intracellular glutathione synthesis. *Arch Dermatol Res, 294*, 349-354.

Yang, K.D., Chao, C.Y. & Shaio, M.F. (1998) Pentoxifylline synergizes with all-trans retinoic acid to induce differentiation of HL-60 myelocytic cells, but suppresses tRA-augmented clonal growth of normal CFU-GM. *Acta Haematol, 99*, 191-199.

Zhao, Q., Tao, J., Zhu, Q., Jia, P.M., Dou, A.X., Li, X., Cheng, F., Waxman, S., Chen, G.Q., Chen, S.J., Lanotte, M., Chen, Z. & Tong, J.H. (2004a) Rapid induction of cAMP/PKA pathway during retinoic acid-induced acute promyelocytic leukemia cell differentiation. *Leukemia, 18*, 285-292.

Zhao, S., Konopleva, M., Cabreira-Hansen, M., Xie, Z., Hu, W., Milella, M., Estrov, Z., Mills, G.B. & Andreeff, M. (2004b) Inhibition of phosphatidylinositol 3-kinase dephosphorylates BAD and promotes apoptosis in myeloid leukemias. *Leukemia, 18*, 267-275.

Zhu, J., Gianni, M., Kopf, E., Honore, N., Chelbi-Alix, M., Koken, M., Quignon, F., Rochette-Egly, C. & de The, H. (1999) Retinoic acid induces proteasome-dependent degradation of retinoic acid receptor alpha (RARalpha) and oncogenic RARalpha fusion proteins. *Proc Natl Acad Sci U S A, 96*, 14807-14812.

Zhu, Q., Zhang, J.W., Zhu, H.Q., Shen, Y.L., Flexor, M., Jia, P.M., Yu, Y., Cai, X., Waxman, S., Lanotte, M., Chen, S.J., Chen, Z. & Tong, J.H. (2002) Synergic effects of arsenic trioxide and cAMP during acute promyelocytic leukemia cell maturation subtends a novel signaling cross-talk. *Blood, 99,* 1014-1022.

Zhu, Q.S., Xia, L., Mills, G.B., Lowell, C.A., Touw, I.P. & Corey, S.J. (2006) G-CSF induced reactive oxygen species involves Lyn-PI3-kinase-Akt and contributes to myeloid cell growth. *Blood, 107*, 1847-1856.

In: New Cell Differentiation Research Topics
Editor: Hitoshi Saitama, pp. 117-151

ISBN: 978-1-60021-937-5

Chapter III

CANCER CELL + STEM CELL = CANCER STEM CELL?

Jeanette Seidel, Emilio Batistin, Sarah Schwitalla, Bernd Niggemann, Kurt S. Zänker and Thomas Dittmar*

Institute of Immunology, Witten/ Herdecke University, 58448 Witten, Germany.

ABSTRACT

Bone marrow-derived stem cells (BMDCs) gather considerable attention for their ability to trans-differentiate into various dedicated cell types, e.g. liver, heart and skeletal muscle, and brain. Thereby, BMDCs trans-differentiate in accordance to the specific tissue blueprint, which could be the local milieu (soluble factors, cell adhesion molecules and/or nanotubes), a single cell (cell fusion) or a combination of both. The local tissue blueprint has to guide the changes from a stem cell, by definition endowed with the ability for self-renewal, to a differentiated, dedicated cell exhibiting specific tissue functions. A similar process has to occur during tissue stem cell differentiation. Again, the local tissue milieu is responsible for providing a correct blueprint guaranteeing an accurate differentiation process.

The question arises what will be the result of BMDC trans- and tissue stem cell differentiation if stem cells will receive a faulty set of instructions, which might be provided by chronic inflamed tissue or tumor tissue. Both environments have lost many of the characteristics of the tissue of origin. Tumor tissue *per se* is heterogeneous and resembles chronic inflamed tissue. Chronic inflamed tissue itself is characterized by a chaotic interplay of tissue cells, immune competent cells, cytokines, chemokines, growth factors and reactive oxygen species.

An incomplete set of instructions concomitant with the combination of proliferative stimuli and DNA damaging agents represents an explosive mixture with a fatal outcome for (trans-) differentiating stem cells, especially for those stem cells adopting tissue

* Correspondence concerning this article should be addressed to: Dr. Thomas Dittmar, Institute of Immunology, Witten/ Herdecke University, Stockumer Str. 10, 58448 Witten, Germany; Phone: +49 2302 926 165; FAX: +49 2302 926 158; E-mail: thomasd@uni-wh.de.

function by cell fusion. Aneuploidy is frequently found in tumor cells and has been postulated to be the driving force in the pathogenesis of cancer. Fusion of stem cells and tumor cells will result in an increased aneuploidy in hybrid cells. Conjointly, the proliferative and DNA damaging environment will lead to the evolution of hybrid cells exhibiting new properties as e.g. a higher proliferation rate, the ability to metastasize and/ or to be drug resistant. In particular the latter property has been associated with so-called cancer stem cells, which apparently play a crucial role in metastasis formation and tumor relapse. Thus, it might be hypothesized that cancer stem cells might be the product of cell fusion between stem cells and tumor cells. The understanding of cell fusion and its consequences for the pathogenesis of cancer, possibly driving the evolution of cancer stem cells, is therefore crucial for the development of novel anti-cancer strategies.

INTRODUCTION

Cancer as a disease has already been described in the earliest medical records found in the history of mankind, dating back to ancient Egypt. The Greek physician Hipporcrates defined the term "cancer" by describing the bizarre "crablike" growth of tumors ("karkinoma" [καρκινομα] is the Greek word for "crab"). To date, cancer is still the second most prevalent cause of death after cardiovascular diseases in the industrialized world. Thereby, the primary cause of cancer is not attributed to primary tumor formation, but rather to the growth of metastases at distant organ sites [1]. Multiple cell-cell and cell-matrix interactions are necessary to allow tumor cells to detach from the primary tumor, get accession to the lymphatic or blood circulatory system, survive in the circulation, arrest at distant sites, transmigrate across the endothelial lining into the parenchymal tissue, and form secondary tumors.

For several decades it was considered that the primary cause of cancer is attributed to an accumulation of point mutations in somatic cells. Based on the knowledge about caretakers, gatekeepers, oncogenes, and tumor suppressors Fearon and Vogelstein developed the genetic multistep model of colon cancer [2]. Thereby, each step in the pathogenesis of colon cancer is defined by a genetic alteration (either activation or loss) of tumor suppressor genes and oncogenes. For instance, the transition from an early adenoma to an intermediate adenoma is characterized by *k-ras* activation [2]. Unfortunately, this model does not provide clues into later stages of tumorigenesis including invasiveness and metastasis [3], and tumor tissue heterogeneity [4,5].

Additionally, the multistep model of cancer does not give a valid answer to the initial step of carcinogenesis, which must involve the "immortilization of a normal somatic cell", so that it can survive long enough to accrue all of the other "hallmarks of cancer" including invasiveness and metastatic behavior [6]. Till now it remains unclear how a single mutation event converts a mortal cell to be immortal. Moreover, the "conventional" models of carcinogenesis fail to explain the many similarities of phenotypic traits of cancer cells and embryonic cells such as immortality, telomerase activity, drug resistance, responsiveness to chemokines, migratory activity, and a graduate un-differentiated phenotype [7,8].

The stem cell theory of carcinogenesis provided elegant answers to these dilemmas by postulating that a tumor is derived from adult tissue stem cells [9,10]. Tissue stem cells are long-lived, slow-dividing, self-renewing cells and therefore highly susceptible for an accumulation of genetic alterations [10]. Subsequent genetic alterations in single cells will give rise to various subclones exhibiting different properties, e.g. escape of normal growth control.

In addition to tissue stem cells, another type of stem cells has become of interest in cancer research. In a recent work Houghton and colleagues demonstrated impressively that bone marrow-derived stem cells (BMDCs), which have been recruited to chronically inflamed tissue, gave rise to metaplasia formation, which is considered to be the direct precursor of gastric adenocarcinoma [11]. Similar findings were reported by Rizivi and colleagues by showing fusion events between BMDCs and normal and neoplastic intestinal epithelium [12]. Liu and colleagues showed that transformed BMDCs could not only give rise to multiple tumor types but also to multipotent cancer stem cells (CSCs) [13]. CSCs have been described as a rare-population of cancer cells with self-renewing and differentiation capacity [14,15]. Moreover, due to their capacity to initiate tumor growth and to be drug-resistant CSCs have further been linked to metastatic spreading and recurrence of disease after conventional therapy [14,15].

In the present book chapter we will give a brief overview about the role of tissue stem cells and BMDCs in tumor initiation, progression, and metastasis formation by being the precursors of CSCs.

THE CAUSE OF CANCER – MUTATIONS?

The question, what is the primary cause of cellular transformation is still subject to controversial discussions. Although it is well established that most carcinomas arise from the stepwise accumulation of genetic alterations [16], the nature and temporal sequence of such changes is less clear [17]. Based on experiments of the German scientist Hermann J. Müller showing that radiation, of that type derived from radioactive substances or X-ray machines, does produce permanent changes, mutations, in the hereditary constitution of living beings of all kind [18-20]. Thus it was postulated that mutations are the driving force in tumorigenesis. Mutations are commonly found in tumors and most of them implicated in tumorigenesis have been associated with a dysfunction of the so-called gatekeepers and caretakers [21]. Gatekeepers are genes that control cell growth and death and mutated forms of these genes are frequently found in a variety of human cancers such as pRb and retinoblastoma [22,23], von Hippel-Lindau (vHL) protein and renal cell carcinoma [24,25], and adenomatous polyposis coli (APC) gene and colon cancer [26,27]. In contrast, caretaker genes do not directly regulate proliferation, but act to prevent genomic instability [28]. Mutations of caretaker genes that are associated with cancer-prone clinical syndromes include the Werner protein (WRN) and Werner's syndrome [29], BLM and Bloom' syndrome [30], breast cancer susceptible genes BRCA1/2, and ovarian cancer and hereditary breast cancer [31,32]. Gatekeepers and caretaker genes could either act as oncogenes if activated by gain of function mutations, e.g. ras [33], Flt-3 receptor tyrosine kinase [34], c-kit [35], or amplification such as c-erbB-2 (HER2/*neu*) [36,37], as well as tumor suppressors if inactivated by loss of function mutations, e.g. pRb [22,23] and APC [26,27]. Based on the knowledge about caretakers, gatekeepers, oncogenes and tumor suppressors Fearon and Vogelstein developed the genetic multistep model of colon cancer [2]. Thereby, each step in the pathogenesis of colon cancer is defined by a genetic alteration (either activation or loss) of tumor suppressor genes and oncogenes. For instance, the transition from an early adenoma to an intermediate adenoma is characterized by *k-ras* activation [2].

However, while this model nicely supports the multistep nature of tumorigenesis (and the mutation hypothesis), unfortunately it does not provide clues into later stages of tumorigenesis including invasiveness and metastasis [3]. Additionally, this model does not explain sufficiently the well-known phenomenon of tumor tissue heterogeneity [4,5]. To explain these "hallmarks of cancer" it was postulated that cancer cells must exhibit a "mutator phenotype" early during their evolution [38-40]. This hypothesis was based on the assumption that there was an increase in the rate of errors in DNA synthesis, e.g. due to mutations in DNA polymerases that render them error-prone, or from mutations in DNA repair that render them defective, during tumor cell division [39]. This is in accordance with Nowell who postulated that an accumulation of mutations occur during successive waves of clonal selection [41]. It is well recognized that several inherited cancer predisposing syndromes such as the Li-Fraumeni [42], hereditary non-polyposis colon cancer (HNPCC) [43], or familial adenomatous polyposis [26,27] are based on inherited mutation(s) in one allele of the genes involved in DNA integrity and mismatch repair. Moreover, the loss of the second wild-type allele occurs with high frequency thus explaining the extremely high tumor incidence rate of afflicted patients. For instance, familial adenomatous polyposis afflicted patients have an almost 100% chance of developing colorectal cancer after having multiple (typically 100s to 1000s) of adenomatous polyps [44].

The Cause of Cancer – Aneuploidies?

In contrast to the mutation theory of carcinogenesis, the aneuploidy theory of carcinogenesis claims that aneuploidy is the driving force to develop a malignant phenotype [45-47]. Originally postulated by von Hansemann and Boveri more than a century ago [48,49] the aneuploidy theory has fallen out of favor due to the increasing understanding of the molecular biology of cancer, persuasive literature supporting the idea that gene mutation drives tumorigenesis, and the conceptional influence of the identification of a handful of frequently mutated genes in cancer such as ras, p53, pRb, AKT, PTEN, p16INK4a, and Wnt [17]. By contrast, aneuploidy is found in most, if not all, solid carcinomas [4,5,50] and has been linked with various "hallmarks" of the metastasis cascade, including invasiveness [51-53] and metastasis formation [54-59]. Additionally, gene variability has also been associated with drug and multi-drug resistance [60-62]. In this context, aneuploidy and genetic instability are thus steady sources of karyotypic-phenotypic variations from which, in classical Darwinian terms, selection of cancer-specific aneuploidies and genetic variability encourages the evolution and subsequent malignant progression of cancer cells [45].

Two recent *in-vivo* studies of Sotillo et al. and Weaver et al. on transgenic mice strengthen the aneuploidy theory of carcinogensis [63,64]. Sotillo and colleagues showed that Mad2, which is an essential component of the spindle checkpoint, leads to a wide variety of neoplasias, appearance of broken chromosomes, anaphase bridges, and whole chromosome gains and losses [63]. Interestingly, the authors observed that a continued Mad2 overexpression is not required for tumor maintenance. Thus, once aneuploidy is induced, e.g. by a transient Mad2 overexpression, and a particular level of genetic instability is reached the following steps of tumor progression will run in an autonomous manner, which is in agreement to the autocatalytic karyotypic evolution model of tumor progression proposed by Peter Duesberg [45,46].

An increased aneuploidy and genetic instability was also shown for cells and transgenic mice with reduced levels of the mitosis-specific, centromere-linked motor protein CENP-E [64]. Thereby, elevated levels of spontaneous lymphomas and lung tumors in aged animals were driven by increased rates of aneuploidy. Interestingly, the authors reported that in examples of chemically or genetically induced tumor formation, an increased rate of aneuploidy is a more effective inhibitor than initiator of tumorigenesis [64], thus suggesting a dual role for aneuploidy and chromosomal instability in the initiation and prevention of tumorigenesis.

In addition to these mechanisms explaining that one cause of aneuploidy and genetic instability might be attributed to failures in spindle proteins and motor proteins cell fusion has also been linked with aneuploidy. Like the "aneuploidy theory of carcinogenesis", which was postulated by Boveri and von Hansemann more than a century ago [65,66], the assumption that cell fusion contributes to tumor progression was proposed almost 100 years ago by the German pathologist Otto Aichel [67]. Thereby, Aichel postulated that malignancy could be a consequence of hybridization between somatic cells and leukocytes and that the differences between cells within tumors could be explained by their unequal distribution of chromosomes [67]. Later studies proposed that also metastatic cells as well as the phenotypic diversity of tumors might be the cause of hybridization of leukocytes and tumor cells [68,69]. Moreover, it is assumed that cell fusion can also contribute to drug resistance, increased tumor cell diversity (due to chromosomal aberrations including chromosome nondisjunction, mitotic recombination, translocations, deletions, insertions, and inversions), and an increased malignancy (due to a higher proliferation rate of hybrid cells) [70]. These hallmarks of hybrids [70] are identical to the hallmarks of aneuploidy [45]. In fact, aneuploidy can be induced by various mechanisms, e.g. dysregulation of components of the spindle checkpoint or mitosis-specific proteins [63,64], non-mutagenic carcinogens [47], and cell fusion [71,72].

The Stem Cell Theory of Carcinogenesis

It is widely supposed that the initial step of the carcinogenesis process must involve the "immortilization of a normal somatic cell", so that it can survive long enough to accrue all of the other "hallmarks of cancer" including invasiveness and metastatic behavior [6]. However, till now it remains unclear how a single mutation and/or aneuploidy event converts a mortal cell to be immortal. Although it is well recognized that human cells can be immortalized by transfection with the human telomerase reverse transcriptase (hTERT) gene [73] a causal connection between a genetic change and telomerase activity has not yet been reported. Moreover, the "conventional" models of carcinogenesis fail to explain the many similarities of phenotypic traits of cancer cells and embryonic cells [7,8], such as immortality, telomerase activity, drug resistance, responsiveness to chemokines, migratory activity, and a graduate un-differentiated phenotype. In this context, the stem cell theory of carcinogenesis provides elegant answers to these dilemmas by postulating that tumors derive from adult tissue stem cells [9,10]. The latter have been identified in a variety of tissues including skin [74], breast [75,76], kidney [76], brain [77], liver [78], intestine [79], or skeletal muscle [80]. Tissue stem cells are long-lived, slow dividing, self-renewing cells and therefore highly susceptible for an accumulation of genetic aberrations [10]. Once these cells acquire one or more genetic alterations a patch of genetically altered daughter cells will be formed. Subsequent genetic alterations in single cells will give rise to various subclones exhibiting different properties, e.g. escape from normal growth control. Finally, different clones will emerge at certain time points with respect to the grade of genetic alterations, but do share a common clonal origin [81,82]. The result will be a chimeric network of normal and modified cells, which has been defined as "genetic field" or "field cancerization" [83], and which has been observed in head and neck [81,82], oesphagus [84,85], bladder [86], and breast cancer [87,88]. It is also proposed that field cancerization would account for tumor tissue heterogeneity since i) several different subclones emerge and ii) the tissue itself consists of various cell types including tumor cells, aberrant and normal tissue stem cells as well as partially and fully differentiated aberrant and normal tissue stem cells.

BMDCs and Solid Cancer – The Fatal Plasticity of BMDCs

The so-called stem-cell-plasticity of (human) BMDCs, such as hematopoietic stem and progenitor cells (HSPCs) and mesenchymal stem cells (MSCs) has become of interest for novel clinical therapeutic strategies to regenerate degenerated tissue, thereby restoring organ function. First evidences of the capacity of BMDCs or extrahepatic stem cells to trans-differentiate into hepatocytes were derived from cross-gender and whole transplantation studies in rodents by identifying male-derived cells, as indicated by the Y-chromosome, in the female liver [89,90]. Moreover, male-derived hepatic cells displayed typical liver characteristics such as albumin expression thus suggesting functional trans-differentiation of recruited BMDCs or extrahepatic stem cells [89]. Within the past six to seven years it became evident that BMDCs represent a type of pluripotent cells capable to functionally trans-differentiate into various tissues including the small intestine [12], skeletal muscle [91-93], hepatocytes [71,72,94-100], epithelial cells [101], neurons [102-104], endothelial cells [105,106], and cardiomyocytes [105,106].

One of the most prominent chemokines that regulate BMDC migration is stromal cell-derived factor-1α (SDF-1α) [107]. It provides a gradient toward which BMDCs migrate, and in which they are retained in the bone marrow microenvironment [108]. Within the past years it became evident that SDF-1α not only directs BMDC homing but also directs the recruitment of BMDCs to degenerated tissue [11,109,110]. BMDCs are directed along the hypoxic gradient, which causes SDF-1α up-regulation in different cell types including dendritic cells [111] and endothelial cells [110,111] via a hypoxia-inducible factor-1 (HIF-1) mediated mechanism. Thereby, SDF-1α up-regulation in ischemic tissue is directly proportional to reduced oxygen tension and is correlated with increased adhesion, migration, and homing of circulating BMDCs to ischemic tissue [112].

Although the overall mechanism of trans-differentiation sounds simple – the specific tissue blueprint directs trans-differentiation [113] – the process itself is complex and not yet fully understood. In between the start- and stop-points of trans-differentiation various gene-reprogramming steps have to occur in a sequential step-by-step manner, for which a defined set of instructions is a prerequisite to ensure an accurate trans-differentiation. Thereby, trans-differentiation instructions can be provided by the local tissue milieu [96,99,114], cell fusion [72,115-123] or a combination of both.

Instructions of the local tissue milieu can be delivered by different mechanisms including i) soluble factors secreted by tissue cells [99], ii) intercellular communication via gap junctions [124] and/or nanotubular highways [125-127], and adhesion molecules [128-133]. It can be assumed that all of these mechanisms are participated in trans-differentiation, but it still remains unclear to which extent they are involved in this process. Among the mechanisms how stem cells adopt tissue function the process of cell fusion is the most descriptive one. For instance, mouse progenitor cells of the central nervous system [123] as well as mouse bone marrow cells [121] spontaneously fuse with mouse embryonic stem cells, thereby adopting the phenotype of the recipient cell (embryonic stem cell). The tetraploid hybrid cells exhibit full pluripotent character, including multilineage contribution to chimaeras [123]. The adoption of tissue function by cell fusion appears to be a rapid and efficient process as was demonstrated using a mouse model of tyrosinaemia type I [71,72,115,122]. These mice are dominant negative for the fumarylacetoacetate hydrolase (*FaH*), which is a key enzyme in the tyrosine degradation process. Mutant mice have progressive liver failure and renal tubular damage unless treated with 2-(2-nitro-4-trifluoro-methyl-benzoyl)-1,3-cyclohexadione (NTBC) [134]. In all studies, NTBC application was withdrawn immediately after introduction of bone marrow cells of *FaH* positive mice [71,72,115,122]. Most of the transplanted animals survived indicating the rapidity and efficiency of tissue restoration by cell fusion.

However, it still remains unclear whether cell fusion is a general mechanism for stem cell trans-differentiation or if trans-differentiation directed by cell fusion is solely restricted to certain cell types and tissues. Up to now, multinucleated cells have been predominantly found in those tissues known to harbor polyploid cells such as brain, liver, and heart [71,72,95,135]. Recently, Rizivi et al. reported that BMDCs fuse with both normal and neoplastic intestinal epithelium suggesting that more tissues than brain, liver, and heart may be restored by cell fusion [12].

BMDCs and Solid Cancer – BMDCs: Watch out for your Neighborhood!

Trans-differentiation is a complex mechanism and presupposes the presence of a defined set of instructions regulating the initiation, the maintenance, and the finalization of this process, which is characterized by various gene-reprogramming steps that have to occur in a sequential step-by-step manner [113]. Although the mechanism of stem cell trans-differentiation remains unclear it is generally accepted that stem cells are highly susceptible to factors, agents, or compounds exhibiting differentiating capacities.

By now, various studies have emphasized the fatal susceptibility of stem cells to transforming agents, which include both well-known carcinogens and chronically inflamed tissue, and the possible outcome of a neoplastic phenotype [11,13,136]. Transformed BMDCs (prolonged treatment with 3-methylcholanthrene until malignant transformation was achieved) could form many tumor types, including epithelial tumors, neural tumors, muscular tumors, tumors of fibroblasts, blood vessel endothelial tumors, and tumors of poor differentiation *in vivo* [13]. Thereby, a single transformed BMDC has the ability to self-renew, differentiate spontaneously into various types of tumor cells *in vitro*, express markers associated with multipotency, and form teratoma *in vivo* [13]. These data suggest that not only cancer cells but also CSCs seem to originate from transformed BMDCs. Interestingly, Liu and colleagues used 3-methylcholanthrene for BMDC transformation, which has been shown to be nongenotoxic in Chinese hamster embryonic (CHE) cells unless oxidized by liver enzymes [137,138]. Thus, BMDC transformation was likely caused by aneuploidization rather than by mutations, which is in agreement with the aneuploidy theory of carcinogenesis [45-47].

The correlation between chronic inflammation and cancer has already been postulated by the German physician Rudolph Virchow nearly 150 years ago and has experienced a renaissance in the past decade [113,139,140]. Chronically inflamed tissue is characterized by a chaotic interaction of various cell types including monocytes/macrophages, neutrophil granulocytes, lymphocytes, mast cells, fibroblasts, epithelial cells, and endothelial cells, and the complex interplay between them mediated by chemokines, cytokines, proteases, and reactive oxygen and nitrogen species [139,140]. Reactive oxygen and nitrogen species are secreted by granulocytes and increased concentrations of these radicals can cause DNA damages, can prevent DNA repair, and can lead to the inactivation of tumor suppressor genes [140]. The growth and survival of normal and malignant cells is facilitated by cytokines and chemokines, whereas the tissue is remodeled by proteases. An active role in oxidative damage to DNA and genomic instability has been suggested for matrix-metalloproteinase-3 (MMP-3; also named stromelysin-1), which is frequently found to be upregulated in breast cancer [141]. Exposure of mouse mammary epithelial cells to MMP-3 induces the expression of an alternative splice variant of Rac1, which causes an increase in cellular reactive oxygen species concomitant with an enhanced oxidative DNA damage and genomic instability [141]. Thus chronically inflamed tissue represents an environment exhibiting both proliferatory and DNA damaging conditions, which will be disastrous for dividing cells including stem cells. Theise et al. reported that hepatocellular carcinoma and cholangiocarcinoma, which are predominantly found in livers with features of chronic hepatitis caused by chronic hepatitis B virus, or hepatitis C virus infection, or chronic alcohol liver injury, are derived from hepatic progenitor cells [136]. Furthermore, bronchioalveolar stem cells have been suggested to be the putative cells of origin for lung adenocarcinoma [142]. Using a mouse model of gastric cancer Houghton and colleagues showed recently that chronic infection of C57BL/6 mice with *Helicobacter felis*, a known carcinogen, induces repopulation of the stomach with BMDCs [11]. Due to chronically inflamed conditions recruited BMDCs progress through metaplasia and dysplasia to intraepithelial cancer [11]. Western Blot analysis revealed that SDF-1α was solely found to be up-regulated substantially in chronically inflamed but not in acute inflamed gastric mucosa tissue. Cell fusion has been described as one of the two mechanisms of BMDC trans-differentiation and has further been associated with tumor progression (see above). Although different methods were applied, e.g. analysis of sex chromosome in male mice receiving bone marrow from a female donor, stable BMDC gastric epithelium fusion events were not detected [11].

The Chronically Inflamed Tumor Microenvironment

In his 1986 review Dvorak showed how wound healing and tumor stroma formation share many important properties [139,143]. Therefore, tumors are often described as wounds that do not heal [143]. Both tissues contain inflammatory cells and consists of new blood vessels, connective tissue, and a fribin-gel matrix [139]. Wound healing is usually self-limiting whereas tumors secrete a vascular permeability factor, vascular endothelial growth factor (VEGF), that can lead to persistent extravasation of fibrin and fibronectin and continuous generation of extracellular matrix [139]. A critical role in tumorigenesis has been addressed to platelets and macrophages. Platelet activation is mandatory in wound healing due to secretion of chemotactic factors such as transforming growth factor-β (TGF-β), platelet-derived growth factor (PDGF), and VEGF, which recruit and activate fibroblasts, granulocytes, and monocytes [140]. Additionally, these compounds induce and activate proteolytic enzymes necessary for remodeling of the extracellular matrix, restoration of the capillary system, and re-epithelialization [140], which in turn also promotes neoangiogenesis if platelets are activated by tumor tissue.

Monocyte-derived so-called tumor-associated macrophages (TAMs) are a major component of most, if not all, tumors and play a dual role in tumorigenesis [144]. TAMs are antigen presenting cells and are therefore capable to mediate the elimination of the tumor tissue either by activating the immune system by presenting tumor antigens or by killing themselves neoplastic cells following activation by interleukin-2 (IL-2), interferon, and interleukin-12 (IL-12) [145,146]. On the other hand, evidence has emerged for a symbiotic relationship between tumor cells and TAMs, in which tumor cells attract TAMs and sustain their survival, with TAMs then responding to micro-environmental factors in tumors, such as hypoxia by producing important growth factors, cytokines, chemokines, and enzymes, which stimulates neovascularization and tumor progression [147]. For instance, both neoplastic cells and recruited macrophages secrete factors such as interleukin-6 (IL-6) and colony stimulating factor-1 (CSF-1), which direct monocytes towards a macrophage-like phenotype, thereby impairing dendritic cell differentiation [148]. In fact, tumor-associated dendritic cells are less mature and defective in T-cell stimulatory capacity [140], which of course would be crucial for a T-cell mediated anti-tumor response. Moreover both TAMs and tumor cells produce interleukin-10 (IL-10), which effectively blunts the anti-tumor response by cytotoxic T-cells [140]. Additionally, hypoxic conditions cause the up-regulation of SDF-1α [112] in different cell types including dendritic cells [111] and endothelial cells [110,111], which led to the recruitment of BMDCs into tumor tissue. Several studies revealed that tumor tissue itself secrete chemoattractant factors such as VEGF [149,150] and SDF-1α [151,152] indicating that BMDCs can be recruited directly by tumor tissue.

Tumor tissue is highly heterogeneous [153-159] and resembles chronically inflamed tissue [139,143]. Thus trans-differentiation of tumor recruited BMDCs should run in an incorrect manner due to the presentation of a faulty blueprint if we consider the local tissue milieu as the instructor for trans-differentiation [96,99,114]. Moreover, due to the increasing number of mis-trans-differentiated BMDCs, trans-differentiation of newly recruited BMDCs should also be directed by mis-trans-differentiated BMDCs, which would result in an increased tissue heterogeneity [113].

The recruitment of BMDCs to tumor tissue concomitant with neoplastic BMDC transformation was shown recently by Mengel and co-workers demonstrating that metanephric adenomas revealed microchimerism comprising of both donor- and recipient-derived tumor cells [160]. The authors conclude from their results that except for metanephric adenomas, tumors arising in renal transplants originate completely from graft cells. Thus, the mixed derivation of metanephric adenomas indicates an incorporation of recipient-derived progenitor cells suggesting that adult stem cells can assume neoplastic phenotypes [160]. Moreover, genetic evidence for bone marrow tumor cell hybrids in human cancer was recently obtained in 2 cases where individuals developed renal cell carcinoma following bone marrow transplantation [161,162]. In each case, DNA from bone marrow transplant donors was detected in recipient tumor cells [161,162] suggesting graft BMDC recruitment to donor tumor tissue and subsequent adoption of tissue function by cell fusion. This finding is in agreement with recent data revealing the stable fusion of BMDCs and neoplastic intestinal epithelium [12].

In addition to BMDCs, tissue stem cells have also been described as fusion partners for tumor cells. For instance, Bhatia and colleagues reported recently that normal human prostate epithelial cells, which also express several prostate progenitor/ stem cell markers, fuse spontaneously with 293T and other tumor cells, thereby becoming immortalized and transformed [163]. Similar results were obtained by our group showing that breast stem cells [75] spontaneously fuse with breast cancer cells giving rise to stable hybrid cells (Figure 2). Thereby, breast stem cells stably transfected with the pEGFP-MCS vector, which exhibit a neomycin-resistance [164], were co-cultivated with the HS578T and MDA-MB-435 breast cancer cell line, respectively, both exhibiting a stable hygromycin-resistance, in selection media containing both antibiotics. Short tandem repeat (STR) analysis of hybrid cells revealed an overlap of the genetic background of the parental cells indicating real cell fusion (data not shown).

Cancer Stem Cells

Cancer stem cells (CSCs) have been described as a rare-population of cancer cells with self-renewing and differentiation capacity [14,15]. Moreover, due to their capacity to initiate tumor growth and to be drug-resistant CSCs have now been correlated to metastatic spreading and recurrence of disease after conventional therapy [14,15]. CSCs have first been described in hematopoietic disorders such as acute myeloid leukemia (AML) [165,166] and chronic myeloid leukemia (CML) [166,167], and have now been identified in solid tumor including breast [168], brain [169-171], lung [142], skin [172], and prostate [173].

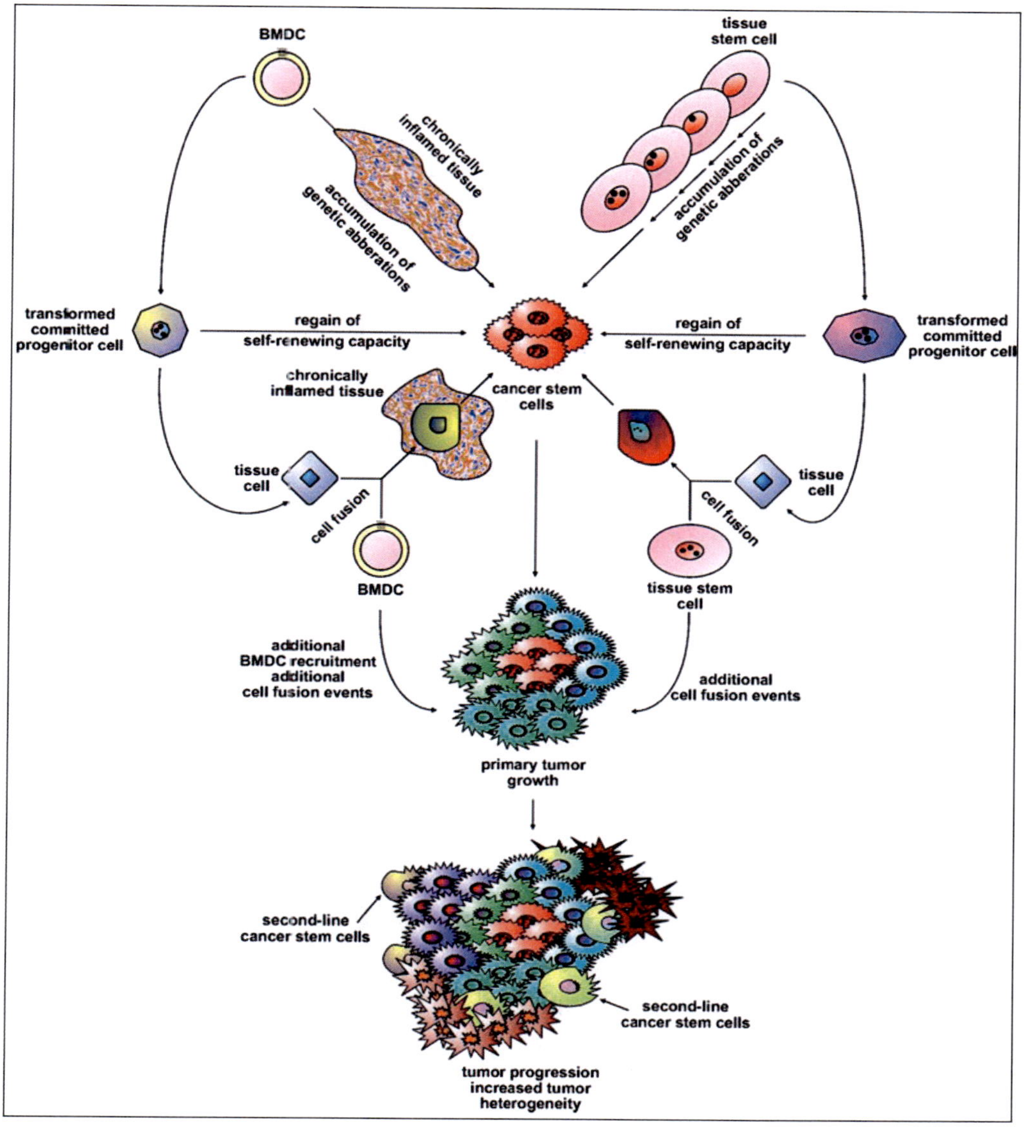

Figure 1. The origin of cancer stem cells. Several models have been proposed to explain the origin of cancer stem cells. Cancer stem cells could originate directly from bone marrow-derived stem cells or tissue stem cells due to accumulation of genetic aberrations either caused by well-known carcinogens and/ or chronically inflamed tissue. It is also conceivable that transformed committed progenitor cells regain self-renewing capacity, thereby giving rise to cancer stem cells. In the third mechanism cancer stem cells originate from a rare fusion event between stem cells and tissue cells. Cancer stem cells initiate primary tumor growth and differentiate, thereby explaining tumor tissue heterogeneity. Due to chronically inflamed conditions within the tumor microenevironment, BMDCs are permanently recruited to tumor tissue. Additionally, so-called second-line/ third-line cancer stem cells might originate from the fusion between tumor cells and stem cells, including tissue stem cells and recruited BMDCs, which further contributes to tumor progression and tissue heterogeneity.

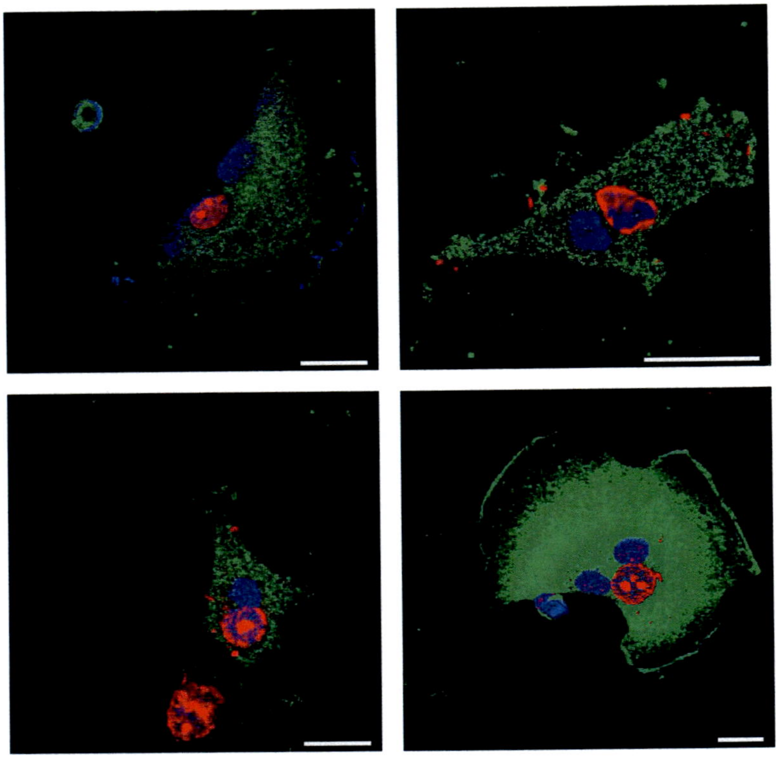

Figure 2. Stable breast stem cell/ breast cancer cell hybrids. The breast stem cell line M13SV1-EGFP-Neo and the breast cancer cell line HS578T-Hyg were co-cultured for 24h and then analyzed for cell fusion events. To distinguish between breast stem cells and breast cancer cells the breast cancer cell line was preincubated with BrdU. Thus fusion cells appear green and contain two or more nuclei (as visualized by Syto60 staining), whereby at least one nucleus exhibit BrdU incorporation. Green: EGFP, red: BrdU, blue: Syto60. The bar represents 20μm.

Do Cancer Stem Cells Originate from (Committed) Tissue Stem Cells?

The origin of CSCs still remains ambiguous. One theory postulates that CSCs may originate from normal tissue stem cells (Figure 1), which e.g. is the case for malignant hematological disorders and which is in agreement with the above-mentioned stem cell theory of carcinogenesis [9,10]. Due to their inherent self-renewing capacity, normal tissue stem cells are ideal candidates for accumulation of genetic aberrations, which ultimately could give rise to a neoplastic phenotype. This theory would be in agreement with certain similarities such as surface markers and signal transduction pathways between CSCs and normal stem cells. For instance, leukemia stem cells have been found to be $CD34^{+}CD38^{-}$, which is virtually identical to normal hematopoietic stem and progenitor cells [15]. Additionally, the universal stem cell marker CD133 is expressed on both neural stem cells and brain tumor stem cells [77,169,174] as well as on prostate stem cells and prostate CSCs [173,175]. Moreover, the Bmi-1 proto-oncogene is crucial in the regulation of self-renewal both for leukemia stem cells and hematopoietic stem cells [176,177] as well as for neural stem cells and brain CSCs [169,178,179]. On the contrary, various studies indicate that CSCs have their own distinguishing molecular profile [15]. For instance, leukemia stem cells can be distinguished from normal hematopoietic stem cells due to a loss of the PTEN tumor suppressor [180]. Breast CSCs have been defined as $CD44^{+}CD24^{-/low}ESA^{+}$, which is opposite to breast stem cells that are $CD44^{-}CD24^{+}$ (unpublished results). Nonetheless, the identification and characterization of unique features of CSCs that discriminate them from normal stem cells will be pivotal for devising specific therapies that would spare normal stem cells [15].

The finding that some CSCs are divergent from normal stem cells may support the theory that CSCs might originate from previously differentiated progenitor cells with oncogenic mutations that regain the ability to self-renew [15]. However, it remains unknown whether the recovery of self-renewing capacity is attributed to the re-activation of normal self-renewal pathways or to the activation of self-renewal pathways in response to signals that do not normally regulate this process [14]. It is also conceivable that genetic aberrations may lead to the recovery of self-renewal capacity by causing the up-regulation of self-renewal specific ligands and/or self-renewal specific receptors, thereby inducing a cell-autonomous stimulatory autocrine loop [14]. These assumptions would be in agreement with several findings suggesting that CSCs may originate from a transformed committed progenitor. Crossing of hMRP8p210BCR/ABL transgenic mice, in which BCR/ABL expression is absent in hematopoietic stem cells (HSCs) and target exclusively to myeloid progenitors and their myelomonocytic progeny, with apoptosis-resistant hMRP8BCL-2 mice gave rise to double-transgenic animals, of which 50% developed acute myeloid leukemias [181]. Thus acute myeloid leukemias can arise in mice without BCR/ABL expression in HSCs [181]. Krivtsov and colleagues reported similar findings by showing that introduction of the MLL-AF9 fusion protein in committed granulocyte macrophage progenitors gave rise to leukemia stem cells [182]. Moreover, analysis of the gene expression pattern of these leukemia stem cells revealed a reactivation of a subset of HSC specific genes, which likely correlates with the regain of self-renewal capacity [182]. In addition to that, activation of the Wnt/β-catenin signal transduction pathway has been shown to sustain the self-renewal capacity of leukemic granulocyte-macrophage progenitors [183]. Studies on brain tumor development indicated that these neoplasias likely derive from more committed neural progenitor cells [184]. These findings suggest that also solid tumors might evolve from more committed progenitor cells.

Do Cancer Stem Cells Originate from BMDCs?

In a recent viewpoint JeanMarie Houghton proposed that the origin of CSCs might be from a different stem cell source – namely the bone marrow [185]. CSCs and BMDCs share several phenotypic similarities, including CD44 surface expression and the ABC transporter Bcrp1/ABCG2, which endows both types of cells with the stem-cell-side population phenotype, as well as similar signal transduction pathways for chemotaxis and metastasis [185]. This assumption is in agreement with the blueprint model of (trans)-differentiation [113], which defines that BMDC trans-differentiation is directed by the blueprint of the surrounding tissue. If BMDC are instructed by a faulty blueprint, e.g. chronically inflamed tissue, BMDC trans-differentiation might deviate from normal organ tissue giving rise to cells still exhibiting stem cell features such as the capacity for self-renewing. In this context, a recent study of Liu et al. is of interest showing that transformed BMDCs could give rise to multiple tumor types including epithelial tumors, neural tumors, muscular tumors, tumors of fibroblasts, blood vessel endothelial tumors, and tumors of poor differentiation *in vivo* [13]. Moreover, a single transformed BMDC has the ability to self-renew, differentiate spontaneously into various types of tumor cells *in vitro*, express markers associated with multipotency, and form teratoma *in vivo* [13]. This in turn suggests that transformed BMDCs could give rise to multipotent CSCs.

Do Cancer Stem Cells Originate from Cell Fusion Events?

Since Liu and colleagues used already *in vitro* transformed BMDCs for their study the role of cell fusion in the particular process of CSCs evolution still remains unclear. Nonetheless, it is speculated that multipotent CSCs may come from a rare fusion event (Figure 1) between stem cells and other cells [15]. As mentioned above, cell fusion is one of the mechanisms how BMDCs, MSCs, or cells of the myelomonocytic lineage adopt tissue function [71,72,95, 115,116,119,121,122,186-188], thereby restoring organ tissue integrity. However, it remains unclear whether tissue stem cells adopt tissue function by cell fusion or whether they differentiate into the appropriate tissue in accordance to the blueprint given by the surrounding tissue.

It is well recognized that cell fusion has been linked with tumor progression. Cell fusion between leukocytes and somatic cells as well as tumor cells, respectively have been correlated with an increased malignancy [67] as well as phenotypic diversity of tumor tissue [68,69]. For instance, artificial and spontaneous macrophage melanoma hybrids showed an enhanced metastatic pattern as demonstrated in *in-vivo* mice studies [189-191]. Moreover, cell fusion has also been linked with drug resistance [70]. Cell fusion is one cause of aneuploidy [71,72], which in turn has been associated with drug and multi-drug resistance [60-62]. Thus it is also conceivable that hybrid cells become drug resistant due to the fact that BMDCs are positive for ATP-binding cassette (ABC) membrane transporter expression, such as ABCG2 (also named breast cancer resistance protein (BCRP)) [192-194], multiple drug resistance-1 (MDR-1; also named ABCB1 or P-glycoprotein), ABCC1, and ABCA2 [195-197]. In this context, a study of Frank and colleagues is of interest showing that progenitor cell fusion is regulated by the membrane transporter ABCB5 [198], whereby it remains unclear if cell fusion is also promoted by other members of the ABC-family. Additionally, BMDCs are positive for CD44 expression [185], which has also been described as a mediator of cell fusion [199]. However, ABCG2 expression is also found on several other stem/ progenitor cells such as hepatic oval cells [200] and neural stem cells [201]. Additionally, CD44 is an adhesion molecule known to be expressed by various normal and neoplastic cell types [202-207]. Moreover, M13SV1 breast stem cells and HS578T breast cancer cells spontaneously fuse with each other although both cell types are negative for CD44 expression (unpublished results).

In fact, considerably less is known about the mechanism of mammalian cell fusion including those proteins that direct the amalgamation of two cells. Recent studies indicate that many proteins involved in cell-cell fusion are also required for seemingly unrelated processes, such as cell migration, axon growth, phagocytosis, and synaptogenesis [208]. In Caenorhabditis elegans, fusogenic factors have been studied in detail and there are a large number of molecules that both promote and inhibit cell fusion [209]. By contrast, nearly a handful of molecules have yet been identified that might be involved in mammalian cell fusion. In addition to the molecules mentioned above, CD47, the macrophage fusion receptor PTPNS1 [199], interleukin-4 (IL-4) and IL-4 receptor [210,211], as well as SDF-1α and CXCR4 [212] have been implicated with cell fusion. For instance, IL-4 and IL-4 receptor promote myoblast fusion with myotubes, which is a crucial step in muscle growth [210,211]. Moreover, various tumors, including gliomas, express high levels of the IL-4 receptor, which raises the possibility that IL-4 might promote cell-cell fusion in gliomas [209]. The interaction between CXCR4 and its' ligand SDF-1α plays a pivotal role in regulating the retention, migration, and mobilization of HSCs during steady state homeostasis and tissue injury [108]. Additionally, the SDF-1α/CXCR4 axis drives the organ-specific metastatic spreading of various tumors, such as breast cancer [213-218], malignant melanoma [219-221], colon and colorectal cancer [219,222-224], prostate cancer [225-229], oral squamous cell carcinoma [230-233], and neuroblastoma [234-236]. *In-vitro* studies of Wright and colleagues suggested a role for SDF-1α/CXCR4 in promoting monocyte cell fusion [212]. Thus it might be speculated that SDF-1α/CXCR4 promote both the recruitment of BMDCs to tumor tissue and the fusion of these two cell populations. Rizivi and colleagues demonstrated recently that BMDCs fuse with normal and neoplastic intestinal epithelium [12]. For their experiments the authors used ROSA26 mice, which received a bone marrow transplant of an EGFP mice strain. Control experiments with non-irradiated mice showed no BMDC incorporation into the intestinal epithelium [12], which points out the necessity of tissue injury and SDF-1α up-regulation to recruit BMDCs. As mentioned above, tumor tissue resembles chronically inflamed tissue, which causes SDF-1α up-regulation in various cell types and subsequent BMDC recruitment. However, it is not yet clear what will be the impact of fusion between BMDCs and neoplastic intestinal tissue on tumor progression since the used mice strain (Min mice) did not survive long enough. BMDC-tumor cell fusions were observed in the majority of intestinal adenomas screened, whereas ubiquitously EGFP expressing tumors were not detected [12]. Additionally, BMDC transplantation did not alter overall polyp number [12]. Thus the authors concluded that BMDCs are not involved in tumor initiation in the small intestine and that fusion between BMDCs and tumor cells may be a common but late event in intestinal tumorigenesis [12]. This assumption is in agreement with our proposed model of "carcinogenesis driven by BMDCs" [113]. Tumor tissue resembles chronically inflamed tissue and thus BMDC recruitment and cell fusion can occur at any tumor stage [113]. Since tumor formation has already occurred Rizivi et al. further rule out that fusion between BMDCs and neoplastic intestinal tissue would give rise to CSCs [12]. However, since Min mice did not survive long enough it is difficult to conclude that BMDC tumor cell fusion might not give rise to CSCs since later stages of intestinal tumorigenesis have not been observed. Moreover, it can not be ruled out that even at later tumor stages second-line or third-line CSCs may originate from fusion events between BMDCs and tumor cells (Figure 1).

Thus the question, which cell type would be the precursor of a cancer stem cell can not be answered sufficiently since several possible mechanisms exists explaining the presence of CSCs. CSCs can originate directly from transformed tissue stem cells, transformed BMDCs, or transformed committed progenitors cells. However, CSCs can also originate from the fusion of tissue stem cells and BMDCs with somatic cells and tumor cells, whereby also transformed tissue stem cells and BMDCs can be the fusion partners.

Cancer Stem Cells and Metastasis Formation

Due to their capacity of tumor initiation CSCs have been associated with secondary tumor growth at distant organ sites [14,15]. This assumption implies that CSCs must fulfill the hallmarks of metastasis formation, which includes detachment from the primary tumor, migration through the connective tissue, accession to the capillary system, extravasation across the endothelial lining, and secondary tumor growth [237]. Till now, considerably less is known about the properties of CSCs with the exception of tumor initiation at low cell numbers. For instance, Al-Hajj and colleagues showed that tumor formation was clearly detectable in mice after injection of solely 100 $CD44^{+}$ $CD24^{-/low}$ ESA^{+} breast cancer stem cells [168].

Nonetheless, it still remains unclear whether CSCs are migratory active and capable to extravasate across the endothelial lining. Tumor cell extravasation, which has been described as a rate-regulating event in metastasis [237-240], resembles leukocyte/ lymphocyte and hematopoietic stem and progenitor cell (HSPC) extravasation. Thus tumor cells use some of the same adhesion molecules, which play a role during physiological processes such as homing of HSPCs and the recruitment of HSPCs and leukocytes/ lymphocytes to sites of inflammation [108,241]. In accordance with the leukocyte adhesion cascade, which has been described as a sequence of adhesion and activation events that ends with extravasation of the leukocyte [241], the "docking and locking hypothesis" has been proposed for the adhesion of tumor cells to the endothelial lining [242]. The initial "docking phase" takes place during rolling of tumor cells on the endothelial cell layer and is facilitated via carbohydrate-selectin interactions [243]. The subsequent "locking phase" on the monolayer is chiefly mediated by α_v or β_1 integrins [244,245], which are thought to induce the transmigratory process by activation of the endothelium, locomotion of tumor cells, and invasion into the underlying tissue [246].

The initial contact of tumor cells to the endothelium is thought to be a critical step in the metastatic cascade [247,248]. Tumor cells, which adhere rapidly to the venular epithelium and extravasate are believed to escape the immune response, as well as shear stress in the blood circulatory system, and thus have a selective advantage. In order to extravasate circulating cells have to be captured by the endothelium first, a process chiefly mediated by selectins [248]. This class of adhesion molecules consists of three, structurally related proteins, which mediate the adhesion between leukocytes, platelets, and endothelial cells. L-selectin and E-selectin are exclusively expressed by leukocytes and activated endothelial cells, respectively, whereas P-selectin expression is detected on both activated platelets and activated endothelial cells [249]. Mucin-like proteins such as CD34, glycosylation dependent cell adhesion molecule-1 (GlyCAM-1), and mucosal addressin cell adhesion molecule-1 (MadCAM-1) are the binding partners for L-selectin [250]. E-selectin binds to the carbohydrates sialyl Lewis^a (sLe^a) and sialyl Lewis^x (sLe^x) [251], whereas P-selectin interacts with P-selectin glycoprotein ligand-1 (PSGL-1) and the mucin CD24 [250].

The P-selectin ligand CD24 has recently become of interest in breast cancer research since breast CSCs are negative or solely express weak levels of CD24 (breast CSCs are CD44$^+$CD24$^{-/low}$ESA$^+$) [168]. By contrast, CD24 overexpression has been associated with breast cancer progression, an enhanced metastasis formation and a poor outcome [252-255]. Aigner and colleagues demonstrated that the interaction between P-selectin and CD24 is crucial for rolling of breast cancer cells on the endothelium [252]. Additionally, highly invasive breast carcinoma cell lines showed elevated CD24 levels suggesting an active role of this P-selectin ligand in the formation of secondary tumors [255]. These findings likely indicate that circulating breast CSCs fail to be captured by activated endothelium due to lack of CD24 expresion. On the other hand, it can not be ruled out that breast CSCs adhere and extravasate passively, while being in contact with CD24 overexpressing breast cancer cells. It is well recognized that not only single tumor cells but also tumor cell spheroids are capable to extravasate across the endothelial lining [237,239,256]. Like a Trojan horse, breast CSCs remaining inside a breast cancer cell spheroid would be able to survive the circulation and to transmigrate across the endothelial barrier. Once this spheroid has entered the desired organ tissue, breast CSCs would initiate secondary tumor growth.

It is further assumed that the inherent plasticity of stem cells would make it easier for CSCs to survive and engraft in a foreign environment where the local organ tissue milieu will be different from that in the primary tumor site [15]. This would be in agreement with the observation that the bone- and lung metastatic gene expression signature of metastatic breast cancer differed markedly from each other and from the overall poor-prognosis signature [257,258]. Depend on the local tissue milieu different set(s) of genes are required to govern the metastatic activity of circulating tumor cells. This would also explain the unability of highly specific organ metastatic tumor cells to grow in a different organ.

In summary, CSCs represent a type of tumor initiating cells that not only could give rise to primary tumor formation but also to secondary tumors at distant organ sites. Thus these cells are excellent candidates for novel anti-cancer therapies. Nonetheless, considerably less is currently known about CSCs and CSC-derived cancer cells in tumor progression and metastasis formation, which of course has to be investigated in further studies.

Conclusions

In the previous paragraphs we briefly discussed the role of stem cells in cancer. It became clear that although big efforts have already been made in cancer research considerably less is still known about this disease. Nevertheless, from day-to-day, from month-to-month we learn more about the complexity of cancer, which will help us to find the targets where cancer cells and cancer stem cells are vulnerable. What was once believed as a disease that has its origin in transformed somatic cells is now considered to be a stem cell-based disease, which has its origin in the generation of cancer stem cells. Cancer stem cells share several similarities with normal stem cells including the capacity to self-renew, to differentiate, to be drug-resistant due to expression of membrane transporters of the ABC family, and to "regenerate tissue integrity" at low cell numbers. The latter properties of CSCs have been associated with two hallmarks of cancer, namely metastasis formation and tumor recurrence.

Currently, various hypotheses are debated to explain the origin of cancer stem cells. It is conceivable that CSCs might derive directly from tissue stem cells and/or BMDCs or indirectly from committed progenitor cells that have regained self-renewal capacity. It is also postulated that cancer stem cells CSCs might originate from rare cell fusion events, whereby several different (stem) cell types have been suggested a fusion partners.

Due to their tumor-initiating capacity as well as their feature to be drug-resistant the future of cancer research will focus on these cells. Till now, the primary cause (>95%) of death in cancer disease is not attributed to primary tumor formation, but to secondary tumor growth at distant organ sites as well as to tumor recurrence after conventional cancer therapy. Thus the challenge of future cancer research will be to identify and to characterize cancer stem cells, to compare them with normal stem cells to find the differences, which can be used as targets for novel anti-cancer strategies.

Acknowledgments

This work was supported by the Westermann-Westdorp-Foundation, Essen, Germany, the Verein zur Förderung der Krebsforschung e.V., Heidelberg, Germany, and the Fritz-Bender-Foundation, Munich, Germany.

References

[1] Heyder C; Gloria-Maercker E; Hatzmann W; Niggemann B; Zanker KS; Dittmar T. Role of the beta1-integrin subunit in the adhesion, extravasation and migration of T24 human bladder carcinoma cells. *Clin Exp Metastasis*, 2005 22, 99-106.

[2] Fearon ER; Vogelstein B. A genetic model for colorectal tumorigenesis. *Cell*, 1990 61, 759-767.

[3] Weinberg RA. Oncogenes and tumor suppressor genes. *CA Cancer J Clin*, 1994 44, 160-170.

[4] Hoglund M; Gisselsson D; Mandahl N; Johansson B; Mertens F; Mitelman F; Sall T. Multivariate analyses of genomic imbalances in solid tumors reveal distinct and converging pathways of karyotypic evolution. *Genes Chromosomes Cancer*, 2001 31, 156-171.

[5] Mertens F; Johansson B; Hoglund M; Mitelman F. Chromosomal imbalance maps of malignant solid tumors: a cytogenetic survey of 3185 neoplasms. *Cancer Res*, 1997 57, 2765-2780.

[6] Trosko JE; Tai MH. Adult stem cell theory of the multi-stage, multi-mechanism theory of carcinogenesis: role of inflammation on the promotion of initiated stem cells. *Contrib Microbiol*, 2006 13, 45-65.

[7] Potter VR. Phenotypic diversity in experimental hepatomas: the concept of partially blocked ontogeny. The 10th Walter Hubert Lecture. *Br J Cancer*, 1978 38, 1-23.

[8] Till JE. Stem cells in differentiation and neoplasia. *J Cell Physiol Suppl*, 1982 1, 3-11.

[9] Trosko JE. The role of stem cells and gap junctional intercellular communication in carcinogenesis. *J Biochem Mol Biol*, 2003 36, 43-48.

[10] Trosko JE; Chang CC. Stem cell theory of carcinogenesis. *Toxicol Lett*, 1989 49, 283-295.

[11] Houghton J; Stoicov C; Nomura S; Rogers AB; Carlson J; Li H; Cai X; Fox JG; Goldenring JR; Wang TC. Gastric cancer originating from bone marrow-derived cells. *Science*, 2004 306, 1568-1571.

[12] Rizvi AZ; Swain JR; Davies PS; Bailey AS; Decker AD; Willenbring H; Grompe M; Fleming WH; Wong MH. Bone marrow-derived cells fuse with normal and transformed intestinal stem cells. *Proc Natl Acad Sci U S A*, 2006 103, 6321-6325.

[13] Liu C; Chen Z; Chen Z; Zhang T; Lu Y. Multiple tumor types may originate from bone marrow-derived cells. *Neoplasia*, 2006 8, 716-724.

[14] Clarke MF; Fuller M. Stem cells and cancer: two faces of eve. *Cell*, 2006 124, 1111-1115.

[15] Li F; Tiede B; Massague J; Kang Y. Beyond tumorigenesis: cancer stem cells in metastasis. *Cell Res*, 2007 17, 3-14.

[16] Hahn WC; Weinberg RA. Rules for making human tumor cells. *N Engl J Med*, 2002 347, 1593-1603.

[17] Pihan G; Doxsey SJ. Mutations and aneuploidy: co-conspirators in cancer? *Cancer Cell*, 2003 4, 89-94.

[18] Muller HJ. Artificial transmutation of the gene. *Science*, 1927 66, 84-87.

[19] Muller HJ. Radiation biology: The nature of the genetic effects produced by radiation. In: Hollander A, Hollander A. *Radiation biology: The nature of the genetic effects produced by radiation*. new York: McGraw Hill; 1954; 351-374.

[20] Muller HJ. Genetic damage produced by radiation. *Science*, 1955 121, 837-840.

[21] Kinzler KW; Vogelstein B. Cancer-susceptibility genes. Gatekeepers and caretakers. *Nature*, 1997 386, 761, 763.

[22] Dannenberg JH; te Riele HP. The retinoblastoma gene family in cell cycle regulation and suppression of tumorigenesis. *Results Probl Cell Differ*, 2006 42, 183-225.

[23] Goodrich DW. The retinoblastoma tumor-suppressor gene, the exception that proves the rule. *Oncogene*, 2006 25, 5233-5243.

[24] Kaelin WG, Jr. The von Hippel-Lindau tumor suppressor protein and clear cell renal carcinoma. *Clin Cancer Res*, 2007 13, 680s-684s.

[25] Kim WY; Kaelin WG, Jr. Molecular pathways in renal cell carcinoma--rationale for targeted treatment. *Semin Oncol*, 2006 33, 588-595.

[26] Herbst A; Kolligs FT. Wnt signaling as a therapeutic target for cancer. *Methods Mol Biol*, 2007 361, 63-91.

[27] Segditsas S; Tomlinson I. Colorectal cancer and genetic alterations in the Wnt pathway. *Oncogene*, 2006 25, 7531-7537.

[28] Levitt NC; Hickson ID. Caretaker tumour suppressor genes that defend genome integrity. *Trends Mol Med*, 2002 8, 179-186.

[29] Kansara M; Thomas DM. Molecular pathogenesis of osteosarcoma. *DNA Cell Biol*, 2007 26, 1-18.

[30] Cheok CF; Bachrati CZ; Chan KL; Ralf C; Wu L; Hickson ID. Roles of the Bloom's syndrome helicase in the maintenance of genome stability. *Biochem Soc Trans*, 2005 33, 1456-1459.

[31] Wolff TA; Wilson JE. Genetic risk assessment and BRCA mutation testing for breast and ovarian cancer susceptibility. *Am Fam Physician*, 2006 74, 1759-1760.

[32] Sogaard M; Kjaer SK; Gayther S. Ovarian cancer and genetic susceptibility in relation to the BRCA1 and BRCA2 genes. Occurrence, clinical importance and intervention. *Acta Obstet Gynecol Scand*, 2006 85, 93-105.

[33] Rodriguez-Viciana P; Tetsu O; Oda K; Okada J; Rauen K; McCormick F. Cancer targets in the Ras pathway. *Cold Spring Harb Symp Quant Biol*, 2005 70, 461-467.

[34] Markovic A; MacKenzie KL; Lock RB. FLT-3: a new focus in the understanding of acute leukemia. *Int J Biochem Cell Biol*, 2005 37, 1168-1172.

[35] Sanders KM; Ward SM. Kit mutants and gastrointestinal physiology. *J Physiol*, 2007 578, 33-42.

[36] Brandt B; Heyder C; Gloria-Maercker E; Hatzmann W; Rotger A; Kemming D; Zanker KS; Entschladen F; Dittmar T. 3D-extravasation model -- selection of highly motile and metastatic cancer cells. *Semin Cancer Biol*, 2005 15, 387-395.

[37] Feldner JC; Brandt BH. Cancer cell motility--on the road from c-erbB-2 receptor steered signaling to actin reorganization. *Exp Cell Res*, 2002 272, 93-108.

[38] Bielas JH; Loeb KR; Rubin BP; True LD; Loeb LA. Human cancers express a mutator phenotype. *Proc Natl Acad Sci U S A*, 2006 103, 18238-18242.

[39] Loeb KR; Loeb LA. Significance of multiple mutations in cancer. *Carcinogenesis*, 2000 21, 379-385.

[40] Loeb LA; Loeb KR; Anderson JP. Multiple mutations and cancer. *Proc Natl Acad Sci USA*, 2003 100, 776-781.

[41] Nowell PC. The clonal evolution of tumor cell populations. *Science*, 1976 194, 23-28.

[42] Royds JA; Iacopetta B. p53 and disease: when the guardian angel fails. *Cell Death Differ*, 2006 13, 1017-1026.

[43] Jass JR. Hereditary Non-Polyposis Colorectal Cancer: the rise and fall of a confusing term. *World J Gastroenterol*, 2006 12, 4943-4950.

[44] Rozen P; Macrae F. Familial adenomatous polyposis: The practical applications of clinical and molecular screening. *Fam Cancer*, 2006 5, 227-235.

[45] Duesberg P; Li R; Fabarius A; Hehlmann R. Aneuploidy and cancer: from correlation to causation. *Contrib Microbiol*, 2006 13, 16-44.

[46] Li R; Sonik A; Stindl R; Rasnick D; Duesberg P. Aneuploidy vs. gene mutation hypothesis of cancer: recent study claims mutation but is found to support aneuploidy. *Proc Natl Acad Sci U S A*, 2000 97, 3236-3241.

[47] Li R; Yerganian G; Duesberg P; Kraemer A; Willer A; Rausch C; Hehlmann R. Aneuploidy correlated 100% with chemical transformation of Chinese hamster cells. *Proc Natl Acad Sci U S A*, 1997 94, 14506-14511.

[48] Boveri T. Foundations of Experimental Embryology: On multipolar mitosis as a means of analysis of the cell nucleus. In: Willier B. H. & Oppenheimer JM, Willier B. H. & Oppenheimer JM. *Foundations of Experimental Embryology: On multipolar mitosis as a means of analysis of the cell nucleus*. Prentice Hall; 1902/1964; 74-97.

[49] Hansemann D. Ueber asymmetrische Zelltheilung in Epithelkrebsen und deren biologische Bedeutung. *Virchows Arch Pathol Anat*, 1890 119, 299-326.

[50] Gebhart E; Liehr T. Patterns of genomic imbalances in human solid tumors. *Int. J. Oncol.*, 2000 16, 383-399.

[51] Hoglund M; Sall T; Heim S; Mitelman F; Mandahl N; Fadl-Elmula I. Identification of cytogenetic subgroups and karyotypic pathways in transitional cell carcinoma. *Cancer Res*, 2001 61, 8241-8246.

[52] Meijer GA; Hermsen MA; Baak JP; van Diest PJ; Meuwissen SG; Belien JA; Hoovers JM; Joenje H; Snijders PJ; Walboomers JM. Progression from colorectal adenoma to carcinoma is associated with non-random chromosomal gains as detected by comparative genomic hybridisation. *J Clin Pathol*, 1998 51, 901-909.

[53] Wilkens L; Flemming P; Gebel M; Bleck J; Terkamp C; Wingen L; Kreipe H; Schlegelberger B. Induction of aneuploidy by increasing chromosomal instability during dedifferentiation of hepatocellular carcinoma. *Proc Natl Acad Sci USA*, 2004 101, 1309-1314.

[54] Al-Mulla F; Keith WN; Pickford IR; Going JJ; Birnie GD. Comparative genomic hybridization analysis of primary colorectal carcinomas and their synchronous metastases. *Genes Chromosomes Cancer*, 1999 24, 306-314.

[55] Aragane H; Sakakura C; Nakanishi M; Yasuoka R; Fujita Y; Taniguchi H; Hagiwara A; Yamaguchi T; Abe T; Inazawa J; Yamagishi H. Chromosomal aberrations in colorectal cancers and liver metastases analyzed by comparative genomic hybridization. *Int J Cancer*, 2001 94, 623-629.

[56] Hermsen M; Postma C; Baak J; Weiss M; Rapallo A; Sciutto A; Roemen G; Arends JW; Williams R; Giaretti W; De Goeij A; Meijer G. Colorectal adenoma to carcinoma progression follows multiple pathways of chromosomal instability. *Gastroenterology*, 2002 123, 1109-1119.

[57] Knosel T; Schluns K; Stein U; Schwabe H; Schlag PM; Dietel M; Petersen I. Chromosomal alterations during lymphatic and liver metastasis formation of colorectal cancer. *Neoplasia*, 2004 6, 23-28.

[58] Nakao K; Shibusawa M; Ishihara A; Yoshizawa H; Tsunoda A; Kusano M; Kurose A; Makita T; Sasaki K. Genetic changes in colorectal carcinoma tumors with liver metastases analyzed by comparative genomic hybridization and DNA ploidy. *Cancer*, 2001 91, 721-726.

[59] Petersen I; Petersen S. Towards a genetic-based classification of human lung cancer. *Anal Cell Pathol*, 2001 22, 111-121.

[60] Duesberg P; Stindl R; Hehlmann R. Explaining the high mutation rates of cancer cells to drug and multidrug resistance by chromosome reassortments that are catalyzed by aneuploidy. *Proc Natl Acad Sci U S A*, 2000 19, 14295-14300.

[61] Duesberg P; Stindl R; Hehlmann R. Origin of multidrug resistance in cells with and without multidrug resistance genes: Chromosome reassortments catalyzed by aneuploidy. *Proc Natl Acad Sci USA*, 2001 98, 11283-11288.

[62] Li R; Hehlmann R; Sachs R; Duesberg P. High rates of drug resistance in cancer cells generated by cancer-specific aneuploidy via frequent chromosomal alterations. *Cancer Genetics Cytogenetics*, 2005 in press.

[63] Sotillo R; Hernando E; Diaz-Rodriguez E; Teruya-Feldstein J; Cordon-Cardo C; Lowe SW; Benezra R. Mad2 overexpression promotes aneuploidy and tumorigenesis in mice. *Cancer Cell*, 2007 11, 9-23.

[64] Weaver BA; Silk AD; Montagna C; Verdier-Pinard P; Cleveland DW. Aneuploidy acts both oncogenically and as a tumor suppressor. *Cancer Cell*, 2007 11, 25-36.

[65] Boveri T. Zur Frage der Entstehung Maligner Tumoren. Gustav Fischer Verlag, Jena, Germany; 1914.

[66] von Hansemann D. *Virchows Arch Pathol Anat*, 1890 119, 299-326.

[67] Aichel O. Vorträge und Aufsätze über Entwicklungsmechanik der Organismen: Über Zellverschmelzung mit quantitativ abnormer Chromosomenverteilung als Ursache der Geschwulstbildung. In: Roux W, Roux W. *Vorträge und Aufsätze über Entwicklungsmechanik der Organismen: Über Zellverschmelzung mit quantitativ abnormer Chromosomenverteilung als Ursache der Geschwulstbildung*. Leipzig, Germany: Wilhelm Engelmann; 1991; 1-115.

[68] Mekler LB. [Hybridization of transformed cells with lymphocytes as 1 of the probable causes of the progression leading to the development of metastatic malignant cells]. *Vestn Akad Med Nauk SSSR*, 1971 26, 80-89.

[69] Mekler LB; Drize OB; Osechinskii IV; Shliankevich MA. [Transformation of a normal differentiated cell of an adult organism, induced by the fusion of this cell with another normal cell of the same organism but with different organ or tissue specificity]. *Vestn Akad Med Nauk SSSR*, 1971 26, 75-80.

[70] Duelli D; Lazebnik Y. Cell fusion: a hidden enemy? *Cancer Cell*, 2003 3, 445-448.

[71] Wang X; Willenbring H; Akkari Y; Torimaru Y; Foster M; Al-Dhalimy M; Lagasse E; Finegold M; Olson S; Grompe M. Cell fusion is the principal source of bone-marrow-derived hepatocytes. *Nature*, 2003 422, 897-901.

[72] Vassilopoulos G; Wang PR; Russell DW. Transplanted bone marrow regenerates liver by cell fusion. *Nature*, 2003 422, 901-904.

[73] Gudjonsson T; Villadsen R; Ronnov-Jessen L; Petersen OW. Immortalization protocols used in cell culture models of human breast morphogenesis. *Cell Mol Life Sci*, 2004 61, 2523-2534.

[74] Clayton E; Doupe DP; Klein AM; Winton DJ; Simons BD; Jones PH. A single type of progenitor cell maintains normal epidermis. *Nature*, 2007 446, 185-189.

[75] Chang CC; Sun W; Cruz A; Saitoh M; Tai MH; Trosko JE. A human breast epithelial cell type with stem cell characteristics as target cells for carcinogenesis. *Radiat Res*, 2001 155, 201-207.

[76] Trosko JE; Chang CC. Isolation and characterization of normal adult human epithelial pluripotent stem cells. *Oncol Res*, 2003 13, 353-357.

[77] Lee A; Kessler JD; Read TA; Kaiser C; Corbeil D; Huttner WB; Johnson JE; Wechsler-Reya RJ. Isolation of neural stem cells from the postnatal cerebellum. *Nat Neurosci*, 2005 8, 723-729.

[78] Alison MR. Liver stem cells: implications for hepatocarcinogenesis. *Stem Cell Rev*, 2005 1, 253-260.

[79] Crosnier C; Stamataki D; Lewis J. Organizing cell renewal in the intestine: stem cells, signals and combinatorial control. *Nat Rev Genet*, 2006 7, 349-359.

[80] Zammit PS; Partridge TA; Yablonka-Reuveni Z. The skeletal muscle satellite cell: the stem cell that came in from the cold. *J Histochem Cytochem*, 2006 54, 1177-1191.

[81] Tabor MP; Brakenhoff RH; Ruijter-Schippers HJ; Van Der Wal JE; Snow GB; Leemans CR; Braakhuis BJ. Multiple head and neck tumors frequently originate from a single preneoplastic lesion. *Am J Pathol*, 2002 161, 1051-1060.

[82] Tabor MP; Brakenhoff RH; van Houten VM; Kummer JA; Snel MH; Snijders PJ; Snow GB; Leemans CR; Braakhuis BJ. Persistence of genetically altered fields in head and neck cancer patients: biological and clinical implications. *Clin Cancer Res*, 2001 7, 1523-1532.

[83] Slaughter DP; Southwick HW; Smejkal W. Field cancerization in oral stratified squamous epithelium; clinical implications of multicentric origin. *Cancer*, 1953 6, 963-968.

[84] Barrett MT; Sanchez CA; Prevo LJ; Wong DJ; Galipeau PC; Paulson TG; Rabinovitch PS; Reid BJ. Evolution of neoplastic cell lineages in Barrett oesophagus. *Nat Genet*, 1999 22, 106-109.

[85] Prevo LJ; Sanchez CA; Galipeau PC; Reid BJ. p53-mutant clones and field effects in Barrett's esophagus. *Cancer Res*, 1999 59, 4784-4787.

[86] Simon R; Eltze E; Schafer KL; Burger H; Semjonow A; Hertle L; Dockhorn-Dworniczak B; Terpe HJ; Bocker W. Cytogenetic analysis of multifocal bladder cancer supports a monoclonal origin and intraepithelial spread of tumor cells. *Cancer Res*, 2001 61, 355-362.

[87] Alvi AJ; Clayton H; Joshi C; Enver T; Ashworth A; Vivanco MM; Dale TC; Smalley MJ. Functional and molecular characterisation of mammary side population cells. *Breast Cancer Res*, 2003 5, R1-8.

[88] Smalley M; Ashworth A. Stem cells and breast cancer: A field in transit. *Nat Rev Cancer*, 2003 3, 832-844.

[89] Theise ND; Badve S; Saxena R; Henegariu O; Sell S; Crawford JM; Krause DS. Derivation of hepatocytes from bone marrow cells in mice after radiation-induced myeloablation. *Hepatology*, 2000 31, 235-240.

[90] Petersen BE; Bowen WC; Patrene KD; Mars WM; Sullivan AK; Murase N; Boggs SS; Greenberger JS; Goff JP. Bone marrow as a potential source of hepatic oval cells. *Science*, 1999 284, 1168-1170.

[91] Camargo FD; Green R; Capetenaki Y; Jackson KA; Goodell MA. Single hematopoietic stem cells generate skeletal muscle through myeloid intermediates. *Nat Med*, 2003 9, 1520-1527.

[92] Fukada S; Miyagoe-Suzuki Y; Tsukihara H; Yuasa K; Higuchi S; Ono S; Tsujikawa K; Takeda S; Yamamoto H. Muscle regeneration by reconstitution with bone marrow or fetal liver cells from green fluorescent protein-gene transgenic mice. *J Cell Sci*, 2002 115, 1285-1293.

[93] Ferrari G; Cusella-De Angelis G; Coletta M; Paolucci E; Stornaiuolo A; Cossu G; Mavilio F. Muscle regeneration by bone marrow-derived myogenic progenitors. *Science*, 1998 279, 1528-1530.

[94] Alison MR; Poulsom R; Jeffery R; Dhillon AP; Quaglia A; Jacob J; Novelli M; Prentice G; Williamson J; Wright NA. Hepatocytes from non-hepatic adult stem cells. *Nature*, 2000 406, 257.

[95] Alvarez-Dolado M; Pardal R; Garcia-Verdugo JM; Fike JR; Lee HO; Pfeffer K; Lois C; Morrison SJ; Alvarez-Buylla A. Fusion of bone-marrow-derived cells with Purkinje neurons, cardiomyocytes and hepatocytes. *Nature*, 2003 425, 968-973.

[96] Jang YY; Collector MI; Baylin SB; Diehl AM; Sharkis SJ. Hematopoietic stem cells convert into liver cells within days without fusion. *Nat Cell Biol*, 2004 6, 532-539.

[97] Korbling M; Katz RL; Khanna A; Ruifrok AC; Rondon G; Albitar M; Champlin RE; Estrov Z. Hepatocytes and epithelial cells of donor origin in recipients of peripheral-blood stem cells. *N Engl J Med*, 2002 346, 738-746.

[98] Lagasse E; Connors H; Al Dhalimy M; Reitsma M; Dohse M; Osborne L; Wang X; Finegold M; Weissman IL; Grompe M. Purified hematopoietic stem cells can differentiate into hepatocytes in vivo. *Nat Med*, 2000 6, 1229-1234.

[99] Newsome PN; Johannessen I; Boyle S; Dalakas E; McAulay KA; Samuel K; Rae F; Forrester L; Turner ML; Hayes PC; Harrison DJ; Bickmore WA; Plevris JN. Human cord blood-derived cells can differentiate into hepatocytes in the mouse liver with no evidence of cellular fusion. *Gastroenterology*, 2003 124, 1891-1900.

[100] Theise ND; Nimmakayalu M; Gardner R; Illei PB; Morgan G; Teperman L; Henegariu O; Krause DS. Liver from bone marrow in humans. *Hepatology*, 2000 32, 11-16.

[101] Krause DS; Theise ND; Collector MI; Henegariu O; Hwang S; Gardner R; Neutzel S; Sharkis SJ. Multi-organ, multi-lineage engraftment by a single bone marrow-derived stem cell. *Cell*, 2001 105, 369-377.

[102] Locatelli F; Corti S; Donadoni C; Guglieri M; Capra F; Strazzer S; Salani S; Del Bo R; Fortunato F; Bordoni A; Comi GP. Neuronal differentiation of murine bone marrow Thy-1- and Sca-1-positive cells. *J Hematother Stem Cell Res*, 2003 12, 727-734.

[103] Mezey E; Chandross KJ. Bone marrow: a possible alternative source of cells in the adult nervous system. *Eur J Pharmacol*, 2000 405, 297-302.

[104] Mezey E; Chandross KJ; Harta G; Maki RA; McKercher SR. Turning blood into brain: cells bearing neuronal antigens generated in vivo from bone marrow. *Science*, 2000 290, 1779-1782.

[105] Jackson KA; Majka SM; Wang H; Pocius J; Hartley CJ; Majesky MW; Entman ML; Michael LH; Hirschi KK; Goodell MA. Regeneration of ischemic cardiac muscle and vascular endothelium by adult stem cells. *J Clin Invest*, 2001 107, 1395-1402.

[106] Orlic D; Kajstura J; Chimenti S; Jakoniuk I; Anderson SM; Li B; Pickel J; McKay R; Nadal-Ginard B; Bodine DM; Leri A; Anversa P. Bone marrow cells regenerate infarcted myocardium. *Nature*, 2001 410, 701-705.

[107] Aiuti A; Webb IJ; Bleul C; Springer T; Gutierrez-Ramos JC. The chemokine SDF-1 is a chemoattractant for human CD34+ hematopoietic progenitor cells and provides a new mechanism to explain the mobilization of CD34+ progenitors to peripheral blood. *J Exp Med*, 1997 185, 111-120.

[108] Weidt C; Niggemann B; Kasenda B; Drell TL; Zänker KS; Dittmar T. Stem cell migration: a quintessential stepping stone to successful therapy. *Curr Stem Cell Res Treat*, 2007 2, 89-103.

[109] Askari AT; Unzek S; Popovic ZB; Goldman CK; Forudi F; Kiedrowski M; Rovner A; Ellis SG; Thomas JD; DiCorleto PE; Topol EJ; Penn MS. Effect of stromal-cell-derived factor 1 on stem-cell homing and tissue regeneration in ischaemic cardiomyopathy. *Lancet*, 2003 362, 697-703.

[110] Kollet O; Shivtiel S; Chen YQ; Suriawinata J; Thung SN; Dabeva MD; Kahn J; Spiegel A; Dar A; Samira S; Goichberg P; Kalinkovich A; Arenzana-Seisdedos F; Nagler A; Hardan I; Revel M; Shafritz DA; Lapidot T. HGF, SDF-1, and MMP-9 are involved in stress-induced human CD34+ stem cell recruitment to the liver. *J Clin Invest*, 2003 112, 160-169.

[111] Pablos JL; Amara A; Bouloc A; Santiago B; Caruz A; Galindo M; Delaunay T; Virelizier JL; Arenzana-Seisdedos F. Stromal-cell derived factor is expressed by dendritic cells and endothelium in human skin. *Am J Pathol*, 1999 155, 1577-1586.

[112] Ceradini DJ; Kulkarni AR; Callaghan MJ; Tepper OM; Bastidas N; Kleinman ME; Capla JM; Galiano RD; Levine JP; Gurtner GC. Progenitor cell trafficking is regulated by hypoxic gradients through HIF-1 induction of SDF-1. *Nat Med*, 2004 10, 858-864.

[113] Dittmar T; Seidel J; Zänker KS; Niggemann B. Infection and Inflammation: Impacts on Oncogenesis: Carcinogenesis Driven by Bone Marrow-Derived Stem Cells. In: Dittmar T; Zänker KS; Schmidt A, Dittmar T; Zänker KS; Schmidt A. *Infection and Inflammation: Impacts on Oncogenesis: Carcinogenesis Driven by Bone Marrow-Derived Stem Cells*. Basel: Karger; 2006; 156-169.

[114] Ianus A; Holz GG; Theise ND; Hussain MA. In vivo derivation of glucose-competent pancreatic endocrine cells from bone marrow without evidence of cell fusion. *J Clin Invest*, 2003 111, 843-850.

[115] Camargo FD; Finegold M; Goodell MA. Hematopoietic myelomonocytic cells are the major source of hepatocyte fusion partners. *J Clin Invest*, 2004 113, 1266-1270.

[116] Kozorovitskiy Y; Gould E. Stem cell fusion in the brain. *Nat Cell Biol*, 2003 5, 952-954.

[117] Oh H; Bradfute SB; Gallardo TD; Nakamura T; Gaussin V; Mishina Y; Pocius J; Michael LH; Behringer RR; Garry DJ; Entman ML; Schneider MD. Cardiac progenitor cells from adult myocardium: homing, differentiation, and fusion after infarction. *Proc Natl Acad Sci U S A*, 2003 100, 12313-12318.

[118] Reinecke H; Minami E; Poppa V; Murry CE. Evidence for fusion between cardiac and skeletal muscle cells. *Circ Res*, 2004 94, e56-60.

[119] Shi D; Reinecke H; Murry CE; Torok-Storb B. Myogenic fusion of human bone marrow stromal cells, but not hematopoietic cells. *Blood*, 2004 104, 290-294.
[120] Spees JL; Olson SD; Ylostalo J; Lynch PJ; Smith J; Perry A; Peister A; Wang MY; Prockop DJ. Differentiation, cell fusion, and nuclear fusion during ex vivo repair of epithelium by human adult stem cells from bone marrow stroma. *Proc Natl Acad Sci U S A*, 2003 100, 2397-2402.
[121] Terada N; Hamazaki T; Oka M; Hoki M; Mastalerz DM; Nakano Y; Meyer EM; Morel L; Petersen BE; Scott EW. Bone marrow cells adopt the phenotype of other cells by spontaneous cell fusion. *Nature*, 2002 416, 542-545.
[122] Willenbring H; Bailey AS; Foster M; Akkari Y; Dorrell C; Olson S; Finegold M; Fleming WH; Grompe M. Myelomonocytic cells are sufficient for therapeutic cell fusion in liver. *Nat Med*, 2004 10, 744-748.
[123] Ying QL; Nichols J; Evans EP; Smith AG. Changing potency by spontaneous fusion. *Nature*, 2002 416, 545-548.
[124] Trosko JE; Chang CC; Wilson MR; Upham B; Hayashi T; Wade M. Gap junctions and the regulation of cellular functions of stem cells during development and differentiation. *Methods*, 2000 20, 245-264.
[125] Francis K; Ramakrishna R; Holloway W; Palsson BO. Two new pseudopod morphologies displayed by the human hematopoietic KG1a progenitor cell line and by primary human CD34(+) cells. *Blood*, 1998 92, 3616-3623.
[126] Fruehauf S; Srbic K; Seggewiss R; Topaly J; Ho AD. Functional characterization of podia formation in normal and malignant hematopoietic cells. *J Leukoc Biol*, 2002 71, 425-432.
[127] Rustom A; Saffrich R; Markovic I; Walther P; Gerdes HH. Nanotubular highways for intercellular organelle transport. *Science*, 2004 303, 1007-1010.
[128] Aplin AE; Howe A; Alahari SK; Juliano RL. Signal transduction and signal modulation by cell adhesion receptors: the role of integrins, cadherins, immunoglobulin-cell adhesion molecules, and selectins. *Pharmacol Rev*, 1998 50, 197-263.
[129] Aplin AE; Juliano RL. Integrin and cytoskeletal regulation of growth factor signaling to the MAP kinase pathway. *J Cell Sci*, 1999 112, 695-706.
[130] Ben-Ze'ev A; Shtutman M; Zhurinsky J. The integration of cell adhesion with gene expression: the role of beta-catenin. *Exp Cell Res*, 2000 261, 75-82.
[131] Reyes-Reyes M; Mora N; Gonzalez G; Rosales C. beta1 and beta2 integrins activate different signalling pathways in monocytes. *Biochem J*, 2002 363, 273-280.
[132] Rosales C; O'Brien V; Kornberg L; Juliano R. Signal transduction by cell adhesion receptors. *Biochim Biophys Acta*, 1995 1242, 77-98.
[133] Yarwood SJ; Woodgett JR. Extracellular matrix composition determines the transcriptional response to epidermal growth factor receptor activation. *Proc Natl Acad Sci U S A*, 2001 98, 4472-4477.
[134] Grompe M; Lindstedt S; al-Dhalimy M; Kennaway NG; Papaconstantinou J; Torres-Ramos CA; Ou CN; Finegold M. Pharmacological correction of neonatal lethal hepatic dysfunction in a murine model of hereditary tyrosinaemia type I. *Nat Genet*, 1995 10, 453-460.

[135] Grompe M. The role of bone marrow stem cells in liver regeneration. *Semin Liver Dis*, 2003 23, 363-372.

[136] Theise ND; Yao JL; Harada K; Hytiroglou P; Portmann B; Thung SN; Tsui W; Ohta H; Nakanuma Y. Hepatic 'stem cell' malignancies in adults: four cases. *Histopathology*, 2003 43, 263-271.

[137] Conney AH. Induction of microsomal enzymes by foreign chemicals and carcinogenesis by polycyclic aromatic hydrocarbons: G. H. A. Clowes Memorial Lecture. *Cancer Res*, 1982 42, 4875-4917.

[138] Oberly TJ; Rexroat MA; Bewsey BJ; Richardson KK; Michaelis KC. An evaluation of the CHO/HGPRT mutation assay involving suspension cultures and soft agar cloning: results for 33 chemicals. *Environ Mol Mutagen*, 1990 16, 260-271.

[139] Balkwill F; Mantovani A. Inflammation and cancer: back to Virchow? *Lancet*, 2001 357, 539-545.

[140] Coussens LM; Werb Z. Inflammation and cancer. *Nature*, 2002 420, 860-867.

[141] Radisky DC; Levy DD; Littlepage LE; Liu H; Nelson CM; Fata JE; Leake D; Godden EL; Albertson DG; Nieto MA; Werb Z; Bissell MJ. Rac1b and reactive oxygen species mediate MMP-3-induced EMT and genomic instability. *Nature*, 2005 436, 123-127.

[142] Kim CF; Jackson EL; Woolfenden AE; Lawrence S; Babar I; Vogel S; Crowley D; Bronson RT; Jacks T. Identification of bronchioalveolar stem cells in normal lung and lung cancer. *Cell*, 2005 121, 823-835.

[143] Dvorak HF. Tumors: wounds that do not heal. Similarities between tumor stroma generation and wound healing. *N Engl J Med*, 1986 315, 1650-1659.

[144] Mantovani A; Bottazzi B; Colotta F; Sozzani S; Ruco L. The origin and function of tumor-associated macrophages. *Immunol Today*, 1992 13, 265-270.

[145] Brigati C; Noonan DM; Albini A; Benelli R. Tumors and inflammatory infiltrates: friends or foes? *Clin Exp Metastasis*, 2002 19, 247-258.

[146] Tsung K; Dolan JP; Tsung YL; Norton JA. Macrophages as effector cells in interleukin 12-induced T cell-dependent tumor rejection. *Cancer Res*, 2002 62, 5069-5075.

[147] Bingle L; Brown NJ; Lewis CE. The role of tumour-associated macrophages in tumour progression: implications for new anticancer therapies. *J Pathol*, 2002 196, 254-265.

[148] Allavena P; Sica A; Vecchi A; Locati M; Sozzani S; Mantovani A. The chemokine receptor switch paradigm and dendritic cell migration: its significance in tumor tissues. *Immunol Rev*, 2000 177, 141-149.

[149] Young MR; Kolesiak K; Wright MA; Gabrilovich DI. Chemoattraction of femoral CD34+ progenitor cells by tumor-derived vascular endothelial cell growth factor. *Clin Exp Metastasis*, 1999 17, 881-888.

[150] Young MR; Petruzzelli GJ; Kolesiak K; Achille N; Lathers DM; Gabrilovich DI. Human squamous cell carcinomas of the head and neck chemoattract immune suppressive CD34(+) progenitor cells. *Hum Immunol*, 2001 62, 332-341.

[151] Sun YX; Wang J; Shelburne CE; Lopatin DE; Chinnaiyan AM; Rubin MA; Pienta KJ; Taichman RS. Expression of CXCR4 and CXCL12 (SDF-1) in human prostate cancers (PCa) in vivo. *J Cell Biochem*, 2003 89, 462-473.

[152] Zou W; Machelon V; Coulomb-L'Hermin A; Borvak J; Nome F; Isaeva T; Wei S; Krzysiek R; Durand-Gasselin I; Gordon A; Pustilnik T; Curiel DT; Galanaud P; Capron

F; Emilie D; Curiel TJ. Stromal-derived factor-1 in human tumors recruits and alters the function of plasmacytoid precursor dendritic cells. *Nat Med*, 2001 7, 1339-1346.

[153] Cahill DP; Kinzler KW; Vogelstein B; Lengauer C. Genetic instability and darwinian selection in tumours. *Trends Cell Biol*, 1999 9, M57-60.

[154] Lengauer C; Kinzler KW; Vogelstein B. Genetic instability in colorectal cancers. *Nature*, 1997 386, 623-627.

[155] Lengauer C; Kinzler KW; Vogelstein B. DNA methylation and genetic instability in colorectal cancer cells. *Proc Natl Acad Sci U S A*, 1997 94, 2545-2550.

[156] Lengauer C; Kinzler KW; Vogelstein B. Genetic instabilities in human cancers. *Nature*, 1998 396, 643-649.

[157] Shih IM; Zhou W; Goodman SN; Lengauer C; Kinzler KW; Vogelstein B. Evidence that genetic instability occurs at an early stage of colorectal tumorigenesis. *Cancer Res*, 2001 61, 818-822.

[158] Duesberg P; Li R; Rasnick D; Rausch C; Willer A; Kraemer A; Yerganian G; Hehlmann R. Aneuploidy precedes and segregates with chemical carcinogenesis. *Cancer Genet Cytogenet*, 2000 119, 83-93.

[159] Duesberg P; Rasnick D; Li R; Winters L; Rausch C; Hehlmann R. How aneuploidy may cause cancer and genetic instability. *Anticancer Res*, 1999 19, 4887-4906.

[160] Mengel M; Jonigk D; Wilkens L; Radermacher J; von Wasielewski R; Lehmann U; Haller H; Mihatsch M; Kreipe H. Chimerism of metanephric adenoma but not of carcinoma in kidney transplants. *Am J Pathol*, 2004 165, 2079-2085.

[161] Chakraborty A; Lazova R; Davies S; Backvall H; Ponten F; Brash D; Pawelek J. Donor DNA in a renal cell carcinoma metastasis from a bone marrow transplant recipient. *Bone Marrow Transplant*, 2004 34, 183-186.

[162] Yilmaz Y; Lazova R; Qumsiyeh M; Cooper D; Pawelek J. Donor Y chromosome in renal carcinoma cells of a female BMT recipient: visualization of putative BMT-tumor hybrids by FISH. *Bone Marrow Transplant*, 2005 35, 1021-1024.

[163] Bhatia B; Schneider-Broussard R; Calhoun T; Tang S; Zhou J; Schroeder L; Shen J; Multani AS; Pathak S; Chang S; Tang DT. Replicative senescence and cell fusion-mediated immortalization and tumorigenic transformation of normal human prostate (NHP) epithelial cells: Implications in prostate tumorigenesis and tumor heterogeneity. *Proc Amer Assoc Cancer Res*, 2005 46, 2060.

[164] Dittmar T; Schafer F; Brandt BH; Zänker KS. Accelerated assessing of antisense RNA efficacy using a chimeric enhanced green fluorescent protein-antisense RNA-producing vector. *Antisense Nucleic Acid Drug Dev*, 2000 10, 401-408.

[165] Bonnet D; Dick JE. Human acute myeloid leukemia is organized as a hierarchy that originates from a primitive hematopoietic cell. *Nat Med*, 1997 3, 730-737.

[166] Sell S. Leukemia: stem cells, maturation arrest, and differentiation therapy. *Stem Cell Rev*, 2005 1, 197-205.

[167] Huntly BJ; Gilliland DG. Leukaemia stem cells and the evolution of cancer-stem-cell research. *Nat Rev Cancer*, 2005 5, 311-321.

[168] Al-Hajj M; Wicha MS; Benito-Hernandez A; Morrison SJ; Clarke MF. Prospective identification of tumorigenic breast cancer cells. *Proc Natl Acad Sci U S A*, 2003 100, 3983-3988.

[169] Hemmati HD; Nakano I; Lazareff JA; Masterman-Smith M; Geschwind DH; Bronner-Fraser M; Kornblum HI. Cancerous stem cells can arise from pediatric brain tumors. *Proc Natl Acad Sci U S A*, 2003 100, 15178-15183.

[170] Singh SK; Clarke ID; Terasaki M; Bonn VE; Hawkins C; Squire J; Dirks PB. Identification of a cancer stem cell in human brain tumors. *Cancer Res*, 2003 63, 5821-5828.

[171] Singh SK; Hawkins C; Clarke ID; Squire JA; Bayani J; Hide T; Henkelman RM; Cusimano MD; Dirks PB. Identification of human brain tumour initiating cells. *Nature*, 2004 432, 396-401.

[172] Fang D; Nguyen TK; Leishear K; Finko R; Kulp AN; Hotz S; Van Belle PA; Xu X; Elder DE; Herlyn M. A tumorigenic subpopulation with stem cell properties in melanomas. *Cancer Res*, 2005 65, 9328-9337.

[173] Collins AT; Berry PA; Hyde C; Stower MJ; Maitland NJ. Prospective identification of tumorigenic prostate cancer stem cells. *Cancer Res*, 2005 65, 10946-10951.

[174] Uchida N; Buck DW; He D; Reitsma MJ; Masek M; Phan TV; Tsukamoto AS; Gage FH; Weissman IL. Direct isolation of human central nervous system stem cells. *Proc Natl Acad Sci U S A*, 2000 97, 14720-14725.

[175] Richardson GD; Robson CN; Lang SH; Neal DE; Maitland NJ; Collins AT. CD133, a novel marker for human prostatic epithelial stem cells. *J Cell Sci*, 2004 117, 3539-3545.

[176] Lessard J; Sauvageau G. Bmi-1 determines the proliferative capacity of normal and leukaemic stem cells. *Nature*, 2003 423, 255-260.

[177] Park IK; Qian D; Kiel M; Becker MW; Pihalja M; Weissman IL; Morrison SJ; Clarke MF. Bmi-1 is required for maintenance of adult self-renewing haematopoietic stem cells. *Nature*, 2003 423, 302-305.

[178] Liu S; Dontu G; Mantle ID; Patel S; Ahn NS; Jackson KW; Suri P; Wicha MS. Hedgehog signaling and Bmi-1 regulate self-renewal of normal and malignant human mammary stem cells. *Cancer Res*, 2006 66, 6063-6071.

[179] Molofsky AV; Pardal R; Iwashita T; Park IK; Clarke MF; Morrison SJ. Bmi-1 dependence distinguishes neural stem cell self-renewal from progenitor proliferation. *Nature*, 2003 425, 962-967.

[180] Yilmaz OH; Valdez R; Theisen BK; Guo W; Ferguson DO; Wu H; Morrison SJ. Pten dependence distinguishes haematopoietic stem cells from leukaemia-initiating cells. *Nature*, 2006 441, 475-482.

[181] Jaiswal S; Traver D; Miyamoto T; Akashi K; Lagasse E; Weissman IL. Expression of BCR/ABL and BCL-2 in myeloid progenitors leads to myeloid leukemias. *Proc Natl Acad Sci U S A*, 2003 100, 10002-10007.

[182] Krivtsov AV; Twomey D; Feng Z; Stubbs MC; Wang Y; Faber J; Levine JE; Wang J; Hahn WC; Gilliland DG; Golub TR; Armstrong SA. Transformation from committed progenitor to leukaemia stem cell initiated by MLL-AF9. *Nature*, 2006 442, 818-822.

[183] Jamieson CH; Ailles LE; Dylla SJ; Muijtjens M; Jones C; Zehnder JL; Gotlib J; Li K; Manz MG; Keating A; Sawyers CL; Weissman IL. Granulocyte-macrophage progenitors as candidate leukemic stem cells in blast-crisis CML. *N Engl J Med*, 2004 351, 657-667.

[184] Vescovi AL; Galli R; Reynolds BA. Brain tumour stem cells. *Nat Rev Cancer*, 2006 6, 425-436.

[185] Houghton J. Bone-marrow-derived cells and cancer--an opportunity for improved therapy. *Nat Clin Pract Oncol*, 2007 4, 2-3.

[186] Alison MR; Poulsom R; Otto WR; Vig P; Brittan M; Direkze NC; Lovell M; Fang TC; Preston SL; Wright NA. Recipes for adult stem cell plasticity: fusion cuisine or readymade? *J Clin Pathol*, 2004 57, 113-120.

[187] Eisenberg LM; Eisenberg CA. Stem cell plasticity, cell fusion, and transdifferentiation. *Birth Defects Res Part C Embryo Today*, 2003 69, 209-218.

[188] Vassilopoulos G; Russell DW. Cell fusion: an alternative to stem cell plasticity and its therapeutic implications. *Curr Opin Genet Dev*, 2003 13, 480-485.

[189] Chakraborty AK; Sodi S; Rachkovsky M; Kolesnikova N; Platt JT; Bolognia JL; Pawelek JM. A spontaneous murine melanoma lung metastasis comprised of host x tumor hybrids. *Cancer Res*, 2000 60, 2512-2519.

[190] Pawelek JM. Tumour cell hybridization and metastasis revisited. *Melanoma Res*, 2000 10, 507-514.

[191] Rachkovsky M; Sodi S; Chakraborty A; Avissar Y; Bolognia J; McNiff JM; Platt J; Bermudes D; Pawelek J. Melanoma x macrophage hybrids with enhanced metastatic potential. *Clin Exp Metastasis*, 1998 16, 299-312.

[192] Kim M; Turnquist H; Jackson J; Sgagias M; Yan Y; Gong M; Dean M; Sharp JG; Cowan K. The multidrug resistance transporter ABCG2 (breast cancer resistance protein 1) effluxes Hoechst 33342 and is overexpressed in hematopoietic stem cells. *Clin Cancer Res*, 2002 8, 22-28.

[193] Scharenberg CW; Harkey MA; Torok-Storb B. The ABCG2 transporter is an efficient Hoechst 33342 efflux pump and is preferentially expressed by immature human hematopoietic progenitors. *Blood*, 2002 99, 507-512.

[194] Zhou S; Schuetz JD; Bunting KD; Colapietro AM; Sampath J; Morris JJ; Lagutina I; Grosveld GC; Osawa M; Nakauchi H; Sorrentino BP. The ABC transporter Bcrp1/ABCG2 is expressed in a wide variety of stem cells and is a molecular determinant of the side-population phenotype. *Nat Med*, 2001 7, 1028-1034.

[195] Benchaouir R; Rameau P; Decraene C; Dreyfus P; Israeli D; Pietu G; Danos O; Garcia L. Evidence for a resident subset of cells with SP phenotype in the C2C12 myogenic line: a tool to explore muscle stem cell biology. *Exp Cell Res*, 2004 294, 254-268.

[196] Hirschmann-Jax C; Foster AE; Wulf GG; Nuchtern JG; Jax TW; Gobel U; Goodell MA; Brenner MK. A distinct "side population" of cells with high drug efflux capacity in human tumor cells. *Proc Natl Acad Sci U S A*, 2004 101, 14228-14233.

[197] Lechner A; Leech CA; Abraham EJ; Nolan AL; Habener JF. Nestin-positive progenitor cells derived from adult human pancreatic islets of Langerhans contain side population (SP) cells defined by expression of the ABCG2 (BCRP1) ATP-binding cassette transporter. *Biochem Biophys Res Commun*, 2002 293, 670-674.

[198] Frank NY; Pendse SS; Lapchak PH; Margaryan A; Shlain D; Doeing C; Sayegh MH; Frank MH. Regulation of progenitor cell fusion by ABCB5 P-glycoprotein, a novel human ATP-binding cassette transporter. *J Biol Chem*, 2003 278, 47156-47165.

[199] Vignery A. Osteoclasts and giant cells: macrophage-macrophage fusion mechanism. *Int J Exp Pathol*, 2000 81, 291-304.

[200] Shimano K; Satake M; Okaya A; Kitanaka J; Kitanaka N; Takemura M; Sakagami M; Terada N; Tsujimura T. Hepatic oval cells have the side population phenotype defined by expression of ATP-binding cassette transporter ABCG2/BCRP1. *Am J Pathol*, 2003 163, 3-9.

[201] Cai J; Cheng A; Luo Y; Lu C; Mattson MP; Rao MS; Furukawa K. Membrane properties of rat embryonic multipotent neural stem cells. *J Neurochem*, 2004 88, 212-226.

[202] Screaton GR; Bell MV; Jackson DG; Cornelis FB; Gerth U; Bell JI. Genomic structure of DNA encoding the lymphocyte homing receptor CD44 reveals at least 12 alternatively spliced exons. *Proc Natl Acad Sci U S A*, 1992 89, 12160-12164.

[203] Zetter BR. Adhesion molecules in tumor metastasis. *Semin Cancer Biol*, 1993 4, 219-229.

[204] Naor D; Sionov RV; Ish-Shalom D. CD44: Structure, function, and association with the malignant process. *Adv Cancer Res*, 1997 71, 241-319.

[205] Ladeda V; Aguirre Ghiso JA; Bal de Kier Joffe E. Function and expression of CD44 during spreading, migration, and invasion of murine carcinoma cells. *Exp Cell Res*, 1998 242, 515-527.

[206] Schneider J; Pollan M; Ruibal A; Jimenez E; Lucas AR; Nunez MI; Sanchez J; Tejerina A. Histologic grade and CD44 are independent predictors of axillary lymph node invasion in early (TI) breast cancer. *Tumour Biol*, 1999 20, 319-330.

[207] Avigdor A; Goichberg P; Shivtiel S; Dar A; Peled A; Samira S; Kollet O; Hershkoviz R; Alon R; Hardan I; Ben-Hur H; Naor D; Nagler A; Lapidot T. CD44 and hyaluronic acid cooperate with SDF-1 in the trafficking of human CD34+ stem/progenitor cells to bone marrow. *Blood*, 2004 103, 2981-2989.

[208] Chen EH; Olson EN. Unveiling the mechanisms of cell-cell fusion. *Science*, 2005 308, 369-373.

[209] Bjerkvig R; Tysnes BB; Aboody KS; Najbauer J; Terzis AJ. Opinion: the origin of the cancer stem cell: current controversies and new insights. *Nat Rev Cancer*, 2005 5, 899-904.

[210] Horsley V; Jansen KM; Mills ST; Pavlath GK. IL-4 acts as a myoblast recruitment factor during mammalian muscle growth. *Cell*, 2003 113, 483-494.

[211] Horsley V; Pavlath GK. Forming a multinucleated cell: molecules that regulate myoblast fusion. *Cells Tissues Organs*, 2004 176, 67-78.

[212] Wright LM; Maloney W; Yu X; Kindle L; Collin-Osdoby P; Osdoby P. Stromal cell-derived factor-1 binding to its chemokine receptor CXCR4 on precursor cells promotes the chemotactic recruitment, development and survival of human osteoclasts. *Bone*, 2005 36, 840-853.

[213] Burger JA; Kipps TJ. CXCR4: a key receptor in the crosstalk between tumor cells and their microenvironment. *Blood*, 2006 107, 1761-1767.

[214] Cabioglu N; Yazici MS; Arun B; Broglio KR; Hortobagyi GN; Price JE; Sahin A. CCR7 and CXCR4 as novel biomarkers predicting axillary lymph node metastasis in T1 breast cancer. *Clin Cancer Res*, 2005 11, 5686-5693.

[215] Kang H; Watkins G; Parr C; Douglas-Jones A; Mansel RE; Jiang WG. Stromal cell derived factor-1: its influence on invasiveness and migration of breast cancer cells in vitro, and its association with prognosis and survival in human breast cancer. *Breast Cancer Res*, 2005 7, R402-410.

[216] Kato M; Kitayama J; Kazama S; Nagawa H. Expression pattern of CXC chemokine receptor-4 is correlated with lymph node metastasis in human invasive ductal carcinoma. *Breast Cancer Res*, 2003 5, R144-150.

[217] Muller A; Homey B; Soto H; Ge N; Catron D; Buchanan ME; McClanahan T; Murphy E; Yuan W; Wagner SN; Barrera JL; Mohar A; Verastegui E; Zlotnik A. Involvement of chemokine receptors in breast cancer metastasis. *Nature*, 2001 410, 50-56.

[218] Zlotnik A. Involvement of chemokine receptors in organ-specific metastasis. *Contrib Microbiol*, 2006 13, 191-199.

[219] Kim J; Mori T; Chen SL; Amersi FF; Martinez SR; Kuo C; Turner RR; Ye X; Bilchik AJ; Morton DL; Hoon DS. Chemokine receptor CXCR4 expression in patients with melanoma and colorectal cancer liver metastases and the association with disease outcome. *Ann Surg*, 2006 244, 113-120.

[220] Murakami T; Cardones AR; Hwang ST. Chemokine receptors and melanoma metastasis. *J Dermatol Sci*, 2004 36, 71-78.

[221] Murakami T; Maki W; Cardones AR; Fang H; Tun Kyi A; Nestle FO; Hwang ST. Expression of CXC chemokine receptor-4 enhances the pulmonary metastatic potential of murine B16 melanoma cells. *Cancer Res*, 2002 62, 7328-7334.

[222] Kim J; Takeuchi H; Lam ST; Turner RR; Wang HJ; Kuo C; Foshag L; Bilchik AJ; Hoon DS. Chemokine receptor CXCR4 expression in colorectal cancer patients increases the risk for recurrence and for poor survival. *J Clin Oncol*, 2005 23, 2744-2753.

[223] Ottaiano A; di Palma A; Napolitano M; Pisano C; Pignata S; Tatangelo F; Botti G; Acquaviva AM; Castello G; Ascierto PA; Iaffaioli RV; Scala S. Inhibitory effects of anti-CXCR4 antibodies on human colon cancer cells. *Cancer Immunol Immunother*, 2005 54, 781-791.

[224] Schimanski CC; Schwald S; Simiantonaki N; Jayasinghe C; Gonner U; Wilsberg V; Junginger T; Berger MR; Galle PR; Moehler M. Effect of chemokine receptors CXCR4 and CCR7 on the metastatic behavior of human colorectal cancer. *Clin Cancer Res*, 2005 11, 1743-1750.

[225] Arya M; Patel HR; McGurk C; Tatoud R; Klocker H; Masters J; Williamson M. The importance of the CXCL12-CXCR4 chemokine ligand-receptor interaction in prostate cancer metastasis. *J Exp Ther Oncol*, 2004 4, 291-303.

[226] Mochizuki H; Matsubara A; Teishima J; Mutaguchi K; Yasumoto H; Dahiya R; Usui T; Kamiya K. Interaction of ligand-receptor system between stromal-cell-derived factor-1 and CXC chemokine receptor 4 in human prostate cancer: a possible predictor of metastasis. *Biochem Biophys Res Commun*, 2004 320, 656-663.

[227] Sun YX; Schneider A; Jung Y; Wang J; Dai J; Wang J; Cook K; Osman NI; Koh-Paige AJ; Shim H; Pienta KJ; Keller ET; McCauley LK; Taichman RS. Skeletal localization and neutralization of the SDF-1(CXCL12)/CXCR4 axis blocks prostate cancer metastasis and growth in osseous sites in vivo. *J Bone Miner Res*, 2005 20, 318-329.

[228] Vaday GG; Hua SB; Peehl DM; Pauling MH; Lin YH; Zhu L; Lawrence DM; Foda HD; Zucker S. CXCR4 and CXCL12 (SDF-1) in Prostate Cancer: Inhibitory Effects of Human Single Chain Fv Antibodies. *Clin Cancer Res*, 2004 10, 5630-5639.

[229] Wang J; Wang J; Sun Y; Song W; Nor JE; Wang CY; Taichman RS. Diverse signaling pathways through the SDF-1/CXCR4 chemokine axis in prostate cancer cell lines leads to altered patterns of cytokine secretion and angiogenesis. *Cell Signal*, 2005 17, 1578-1592.

[230] Almofti A; Uchida D; Begum NM; Tomizuka Y; Iga H; Yoshida H; Sato M. The clinicopathological significance of the expression of CXCR4 protein in oral squamous cell carcinoma. *Int J Oncol*, 2004 25, 65-71.

[231] Ishikawa T; Nakashiro K; Hara S; Klosek SK; Li C; Shintani S; Hamakawa H. CXCR4 expression is associated with lymph-node metastasis of oral squamous cell carcinoma. *Int J Oncol*, 2006 28, 61-66.

[232] Uchida D; Begum NM; Almofti A; Nakashiro K; Kawamata H; Tateishi Y; Hamakawa H; Yoshida H; Sato M. Possible role of stromal-cell-derived factor-1/CXCR4 signaling on lymph node metastasis of oral squamous cell carcinoma. *Exp Cell Res*, 2003 290, 289-302.

[233] Uchida D; Begum NM; Tomizuka Y; Bando T; Almofti A; Yoshida H; Sato M. Acquisition of lymph node, but not distant metastatic potentials, by the overexpression of CXCR4 in human oral squamous cell carcinoma. *Lab Invest*, 2004 84, 1538-1546.

[234] Geminder H; Sagi-Assif O; Goldberg L; Meshel T; Rechavi G; Witz IP; Ben Baruch A. A possible role for CXCR4 and its ligand, the CXC chemokine stromal cell-derived factor-1, in the development of bone marrow metastases in neuroblastoma. *J Immunol*, 2001 167, 4747-4757.

[235] Russell HV; Hicks J; Okcu MF; Nuchtern JG. CXCR4 expression in neuroblastoma primary tumors is associated with clinical presentation of bone and bone marrow metastases. *J Pediatr Surg*, 2004 39, 1506-1511.

[236] Zhang L; Yeger H; Das B; Irwin MS; Baruchel S. Tissue microenvironment modulates CXCR4 expression and tumor metastasis in neuroblastoma. *Neoplasia*, 2007 9, 36-46.

[237] Heyder C; Gloria-Maercker E; Hatzmann W; Zaenker KS; Dittmar T. Visualization of tumor cell extravasation. *Contrib Microbiol*, 2006 13, 200-208.

[238] Haier J; Korb T; Hotz B; Spiegel HU; Senninger N. An intravital model to monitor steps of metastatic tumor cell adhesion within the hepatic microcirculation. *J Gastrointest Surg*, 2003 7, 507-514; discussion 514-505.

[239] Heyder C; Gloria-Maercker E; Entschladen F; Hatzmann W; Niggemann B; Zänker KS; Dittmar T. Realtime visualization of tumor cell/endothelial cell interactions during transmigration across the endothelial barrier. *J Cancer Res Clin Oncol*, 2002 128, 533-538.

[240] Orr FW; Wang HH; Lafrenie RM; Scherbarth S; Nance DM. Interactions between cancer cells and the endothelium in metastasis. *J Pathol*, 2000 190, 310-329.

[241] Ley K. Molecular mechanisms of leukocyte recruitment in the inflammatory process. *Cardiovasc Res*, 1996 32, 733-742.

[242] Honn KV; Tang DG. Adhesion molecules and tumor cell interaction with endothelium and subendothelial matrix. *Cancer Metastasis Rev*, 1992 11, 353-375.

[243] Tang DG; Chen YQ; Newman PJ; Shi L; Gao X; Diglio CA; Honn KV. Identification of PECAM-1 in solid tumor cells and its potential involvement in tumor cell adhesion to endothelium. *J Biol Chem*, 1993 268, 22883-22894.

[244] Kannagi R. Carbohydrate-mediated cell adhesion involved in hematogenous metastasis of cancer. *Glycoconj J*, 1997 14, 577-584.

[245] Pilch J; Habermann R; Felding-Habermann B. Unique ability of integrin alpha(v)beta 3 to support tumor cell arrest under dynamic flow conditions. *J Biol Chem*, 2002 277, 21930-21938.

[246] Tang DG; Honn KV. Adhesion molecules and tumor metastasis: an update. *Invasion Metastasis*, 1994 14, 109-122.

[247] Witz IP. Tumor-microenvironment interactions: the selectin-selectin ligand axis in tumor-endothelium cross talk. *Cancer Treat Res*, 2006 130, 125-140.

[248] Witz IP. The involvement of selectins and their ligands in tumor-progression. *Immunol Lett*, 2006 104, 89-93.

[249] Chapman PT; Haskard DO. Leukocyte adhesion molecules. *Br Med Bull*, 1995 51, 296-311.

[250] Elangbam CS; Qualls CW, Jr.; Dahlgren RR. Cell adhesion molecules--update. *Vet Pathol*, 1997 34, 61-73.

[251] Ugorski M; Laskowska A. Sialyl Lewis(a): a tumor-associated carbohydrate antigen involved in adhesion and metastatic potential of cancer cells. *Acta Biochim Pol*, 2002 49, 303-311.

[252] Aigner S; Ramos CL; Hafezi-Moghadam A; Lawrence MB; Friederichs J; Altevogt P; Ley K. CD24 mediates rolling of breast carcinoma cells on P-selectin. *Faseb J*, 1998 12, 1241-1251.

[253] Baumann P; Cremers N; Kroese F; Orend G; Chiquet-Ehrismann R; Uede T; Yagita H; Sleeman JP. CD24 expression causes the acquisition of multiple cellular properties associated with tumor growth and metastasis. *Cancer Res*, 2005 65, 10783-10793.

[254] Kristiansen G; Sammar M; Altevogt P. Tumour biological aspects of CD24, a mucin-like adhesion molecule. *J Mol Histol*, 2004 35, 255-262.

[255] Schindelmann S; Windisch J; Grundmann R; Kreienberg R; Zeillinger R; Deissler H. Expression profiling of mammary carcinoma cell lines: correlation of in vitro invasiveness with expression of CD24. *Tumour Biol*, 2002 23, 139-145.

[256] Heyder C; Gloria-Maercker E; Hatzmann W; Zänker KS; Dittmar T. Circulating cancer cells: Flow cytometry, video microscopy and confocal laser scanning microscopy. In: Hayat MA, Hayat MA. *Handbook of Immunohistochemistry and in situ hybridization of human carcinomas, Volume 4: Molecular genetics, gastrointestinal carcinoma, and ovarian carcinoma: Circulating cancer cells: Flow cytometry, video microscopy and confocal laser scanning microscopy*. Burlington: Elsevier Academic Press; 2006; 77-88.

[257] Minn AJ; Gupta GP; Siegel PM; Bos PD; Shu W; Giri DD; Viale A; Olshen AB; Gerald WL; Massague J. Genes that mediate breast cancer metastasis to lung. *Nature*, 2005 436, 518-524.

[258] Minn AJ; Kang Y; Serganova I; Gupta GP; Giri DD; Doubrovin M; Ponomarev V; Gerald WL; Blasberg R; Massague J. Distinct organ-specific metastatic potential of individual breast cancer cells and primary tumors. *J Clin Invest*, 2005 115, 44-55.

In: New Cell Differentiation Research Topics
Editor: Hitoshi Saitama, pp. 153-181
ISBN: 978-1-60021-937-5

Chapter IV

Trends in the Neural Transdifferentiation of Bone Marrow-Derived Stem Cells

***Paul Lu*[1, 2,*] *and Tonya Mead*[1]**

[1]Department of Neurosciences, University of California-San Diego, La Jolla, CA 92093, USA;

[2]Veterans Affairs San Diego Healthcare System, San Diego, CA 92165, USA.

Abstract

Studies exploring the potential neural transdifferentiation of bone marrow-derived stem cells (BMSCs) are of increasing value in stem cell research. New research demonstrates that in addition to providing a good model for studying adult stem cell plasticity, BMSCs have considerable potential for promoting neural repair. Use of these cells as a therapeutic model is facilitated by their easy accessibility for autologous transplantation. Methods for inducing neural differentiation include adding simple chemicals, cytokines, growth factors, and brain tissue extracts to cultured BMSCs, as well as co-culturing BMSCs with organotypic brain slices or neural cells. Induced marrow-derived stem cells resemble mature neurons in morphology, expression of neural marker genes, and electrophysiological properties, but whether these changes represent true neural differentiation merits further investigation. Notably, naïve BMSCs express several neural marker genes and exhibit small voltage-dependent membrane currents even before neural induction. This suggests a couple of possibilities. First, these properties may be related to the multidifferentiation potential of BMSCs and facilitate neural differentiation. Alternatively, the shared neural marker genes may not be as specific to neural cells as previously thought.

* Correspondence concerning this article should be addressed to: Paul Lu, Ph.D, Department of Neurosciences-0626, University of California, San Diego, La Jolla, CA 92093-0626. (858) 534-8857; (858) 534-5220 (FAX); plu@ucsd.edu (electronic mail).

INTRODUCTION

Bone marrow stem cells (BMSCs) form a heterogeneous population of hematopoietic stem cells (HSCs), mesenchymal stem cells (MSCs, also called marrow stromal cells), and a small number of multipotent adult progenitor cells (MAPCs) (Waller et al., 1995; Jiang et al., 2002a, 2002b; Kucia et al., 2006). As tissue-specific adult stem cells, HSCs and MSCs self-renew and differentiate into cells of their normal lineage. HSC populations produce mature blood cells, whereas MSCs normally differentiate into the adipocytes, chondrocytes, and osteoblasts that compose stromal and skeletal tissue (Prockop 1997; Pittenger et al., 1999; Kondo et al., 2003). Although germ layer commitment is thought to limit the differentiation potential of adult stem cells, recent studies indicate that bone marrow-derived stem cells may be coerced to differentiate into non-mesodermal cells, such as neural cells typically of ectoderm origin. This phenomenon is called transdifferentiation.

Neural cells (neurons or glia) produced by the transdifferentiation of cultured BMSCs may have many therapeutic applications from replacing the neuron loss caused by Parkinson's and Alzheimer's diseases to promoting recovery in cases of spinal cord injury. Marrow stem cells are easily obtained through bone marrow biopsy, and can be amplified to desired quantities. Furthermore, since these cells can be used for autologous transplantation, the risk of immunogenic response is decreased. Because of these features, BMSCs may represent a readily available and less ethically challenging alternative to embryonic stem cells, and could ultimately lead to practical, effective therapies for the treatment of neurological diseases and CNS trauma.

Studies pursuing the therapeutic potential of BMSCs in the brain utilize various neural cell induction protocols. The most common methods for neural induction include adding simple chemicals, cocktails of cytokines and growth factors, or brain tissue extracts to cultured BMSCs; introducing a particular gene to the cultured cells; or by co-culturing BMSCs with organotypic brain slices or neural cells. This article will review recent studies that used these techniques, and evaluate their relative success.

New research in this field has revealed the importance of using multiple measures when confirming the validity of a transdifferentiation protocol. In many early studies, an assertion of transdifferentiation was based on the demonstration of relevant changes in morphology and the presence of select neural markers. Because these minimal guidelines have proven unreliable for confirmation of neural transdifferentiation, some studies now include further characterization of the treated cells by performing a more complete analysis of gene and protein expression, and by investigating changes in cellular electrophysiology. For confirmation of neural cell transdifferentiation, relevant changes in these measures should differ from baseline levels in uninduced BMSCs, and should replicate known characteristics of neural cells. In this chapter, we will focus on in vitro studies of this transdifferentiation phenomenon, with particular concentration on research published within the last three years.

NEURAL TRANSDIFFERENTIATION: SIMPLE CHEMICAL INDUCTION

The first indication that non-neural stem cells could be used to generate neurons came from an in vitro study by Woodbury and colleagues (2000). The authors discovered that they could induce MSCs to rapidly differentiate into neuron-like cells by adding β-mercaptoethanol (BME) alone, or together with dimethylsulfoxide (DMSO) and butylated hydroxyanisol (BHA) to the culture medium. Treatment with BME (1-10 mM) in serum-free medium initiated changes in cell morphology within 60 minutes. Three hours later, the flat and elongated perikarya of the MSCs became more spherical, and the responsive cells had long, elaborate processes resembling neurites. In addition to the changes in morphology, BME treatment increased expression of a neuronal protein, neuron-specific enolase (NSE), as determined by Western blot analysis. Similar results were observed with DMSO/BHA induction. When given a combination of 2% DMSO and 200 μM BHA, many MSCs took on a neuron-like appearance and appeared to express the neuronal markers NSE, tau, and neurofilament M (NF-M). NeuN, a protein expressed by postmitotic neurons, was only stained in a subset of the cells with a neuron-like appearance (Table 1). The induced cells did not express glial fibrillary acidic protein (GFAP), an astrocytic protein, indicating the simple chemical protocol induced neuron-specific transdifferentiation.

The results from this initial study were promising, but additional experimentation raised questions about the nature of naïve (untreated) MSCs and whether the induction technique could promote terminal differentiation of the cells. In a follow-up study, Woodbury and colleagues (2002) explored the potential mechanisms behind their chemically induced transdifferentiation. They compared gene expression profiles before and after MSC induction for a number of germ layer markers. They were surprised to find that even before neuronal induction, MSC populations and clonal lines expressed a mixture of mesodermal; germinal; endodermal; and ectodermal genes, including several whose expression was thought to be restricted to neural cells. The authors confirmed the expression of six neural genes by reverse-transcriptase polymerase chain reaction (RT-PCR). These genes coded for the proteins: GFAP, aldolase C, APP, NMDA glutamate binding subunit, syntaxin, and neuroD, a transcription factor regulating neuronal fate during development (Table 1). Low-level NF-M mRNA expression was also detected in the naïve MSCs maintained in serum-free medium; this expression increased after the MSCs were cultured in neuronal induction medium. Based on these findings, the authors proposed that although MSCs are multipotent, capable of differentiating in to a limited set of cell types, they are also multidifferentiated, expressing selective genes normally restricted to non-mesodermal lineages. A later study from this group provided further support for this hypothesis by demonstrating the expression of several genes related to dopamine neurons in naïve MSCs (Table 1) (Kramer et al., 2006). Woodbury and colleagues predicted that the multidifferentiated state of MSCs allows them to be converted to cells of a non-mesenchymal lineage through modulation of their gene expression (2002). Importantly, the authors observed another unexpected phenomenon: morphologic reversion. The neuron-like cells reverted to their original undifferentiated morphology after the serum-free neuronal induction medium was replaced with serum-containing MSC culture medium 24 hours after induction. The authors propose that this may result from residual stem cell plasticity, but they do not investigate whether the transdifferentiation of these cells eventually becomes irreversible.

Table 1. Neural Gene Expression in Naïve and Chemically Induced MSCs (RT-PCR; *Immunolabeling and Western Blot*)

Naïve MSCs	Chemically Induced MSC	References
NSE	*NSE, NF-M, tau, NeuN, nestin, trkA*	Woodbury et al. 2000
GFAP, ALDO, APP, GLUT, SYN, NeuroD, NF-M	GFAP, ALDO, APP, GLUT, SYN, NeuroD, NF-M, Tau; *Tau, Tuj1, synaptophysin, Chat, TH*	Woodbury et al. 2002
N/A	*Tau, NSE, NeuN, TUC-4*	Munoz-Elias et al. 2003
N/A	*NeuN, GFAP*	Rismanchi et al. 2003
NSE; *NSE, nestin*	NSE; *NSE, NeuN, nestin*	Lu et al. 2004
Vimetin, Tuj1, NF-M; *vimentin, NF-M, nestin, tau*	Vimetin, Tuj1, NF-M; *vimentin, Tuj1 (rare), NF-M, nestin, tau, GFAP, no synaptophysin*	Neuhuber et a. 2004
Tuj1, NF, NSE, GFAP; *Tuj1, NF, NSE, GFAP*	Tuj1, NF, NSE, GFAP; *Tuj1, NF, NSE, GFAP*	Suon et al. 2004
OCT-4, SOX1, SOX2, SOX10, NTRK1, NeuroD1, Neurog2, OTX1, MSI1, NES, CALB, TUBB4/III, SNCA, GFAP, MBP, TH; *nestin, GFAP, GalC, Tuj1*	OCT-4, SOX1, SOX2, SOX10, NTRK1, NeuroD1, Neurog2, OTX1, MSI1, NES, CALB, TUBB4/III, SNCA, GFAP, MBP, TH; *nestin, GFAP, GalC, Tuj1, MAP2*	Hermann et al. 2004; 2006
NeuN, tau, Tuj1, NSE	*NeuN, tau, Tuj1, NSE, NF-H*	Bertani et al. 2005
Nestin, Tuj1, NF-M; *Nestin,*	Nestin, Tuj1, NF-M; *Nestin, Tuj1*	Yamaguchi et al. 2006
NSE, NF, nestin, trkA	*NSE, NF, nestin, trkA*	Choi et al. 2006
Patched, Smoothened, Nurr1, GFRα1, Otx1, GLI, FGFR1, FGFR4; *Nurr1,*	*Nurr1*	Kramer et al. 2006

Following their initial discoveries, Woodbury and colleagues continued to explore the mechanisms underlying their method of neuronal transdifferentiation (Munoz-Elias et al., 2003). In doing so, they revealed additional characteristics that set this paradigm apart from normal cellular differentiation. The formation of neuronal and glial cells from neural stem cells and progenitor cells is connected to the cell cycle control system, with neural stem cells or progenitors undergoing asymmetric division before terminal differentiation into neurons or glia (for review see Temple, 2001; Bally-Cuif and Hammerschmidt, 2003; Cremisi et al., 2003). In this study, the suppression of mitosis through the application of the DNA synthesis inhibitor cytosine arabinoside (Ara-C) had no apparent effect on the neuronal transdifferentiation of MSCs. The same proportion of induced MSCs displayed neuron-like structure and expressed the neuronal marker tau. Even more surprising was the finding that the neuron-like cells were still capable of mitosis after induction with Woodbury's protocol. The authors also discovered that although the proportion of induced cells that take on neuronal morphology is surprisingly high (>70%), it is not the result of a cell-selective mechanism. Apoptosis occurred in less than 1% of induced cells at 72 hours. This result is incongruous with neural differentiation in which the programmed cell death of a moderate proportion of cells is expected (Esdar et al., 2001).

The transdifferentiation of MSCs using Woodbury's protocol demonstrates properties that are remarkably inconsistent with normal neuronal differentiation. The majority of induced MSCs become strikingly neuronal in appearance, but this transition is both rapid and non-permanent. Normally, stem cells must undergo mitosis prior to differentiation. With Woodbury's protocol, the structural transition occurs too rapidly for mitosis to take place, and inhibition of mitosis does not affect the cells' physical transformation. Further, a neural progenitor must exit the cell cycle before it becomes a neuron or glial cell through terminal differentiation. After Woodbury's neuronal transdifferentiation, the neuron-shaped cells continue to undergo cytokinesis. These studies also highlight the importance of characterizing naïve MSCs. The expression of neural and other non-mesodermal lineage markers in uninduced MSCs complicates the interpretation of studies in which neural transdifferentiation is based solely on morphology and the expression of a few neural markers. When assessing the validity of a transdifferentiation protocol, it is important to keep in mind that the presence of neural markers in induced MSCs may not represent de novo expression. Together, these studies demonstrate the value of examining the treated cells from many different perspectives before making determinations about their nature.

Fact or Artifact?

Woodbury and colleagues' hypothesis that naïve MSCs exist in a multidifferentiated state has found some support in later studies (Woodbury et al., 2002), but whether their simple chemical transdifferentiation technique produces neurons from MSCs has met with skepticism. A number of different groups have investigated the effects of Woodbury's protocol on MSCs, and provide a better understanding of the underlying mechanisms (Hofstetter et al. 2002; Hung et al., 2002; Rismanchi et al., 2003; Lu et al., 2004; Neuhuber et al., 2004; Suon et al. 2004; Bertani et al., 2005; Egusa et al., 2005; Tao et al., 2005; Choi et al., 2006; Hermann et al., 2006; Mareschi et al., 2006; Yamaguchi et al., 2006). The first three studies described in this section provide a foundation for the argument that the post-induction cells are not neuron-like despite their superficial appearance, and produce a variety of evidence in support of this claim (Lu et al., 2004; Neuhuber et al., 2004; Suon et al. 2004).

The other studies reviewed in this section provide additional information about chemically induced MSCs by investigating the specificity and reversibility of the induction effects, quantifying post-induction cell death, measuring changes in the expression of relevant neural genes and proteins, and testing whether the cells display electrophysiological characteristics. These studies clarify problems with the simple chemical induction method and suggest that the "neuron-like" qualities produced using this technique may be simply artifactual.

Our lab became interested in neural transdifferentiation for its therapeutic potential in cases of spinal cord injury. Following an unsuccessful in vivo study (Lu et al., 2005b), we carefully studied the in vitro properties of the simple chemical induction paradigm to gain a better understanding of the mechanisms involved in the transformation (Lu et al., 2004). First, we examined whether the induction effects are unique to MSCs by testing the protocol on other cell types, including primary rat fibroblasts, primary human keratinocytes, HEK293 cells, and rat PC-12 cells. Rat MSCs were used as a positive control. We were surprised to find that keratinocytes were the only cells not responsive to BME or DMSO/BHA stimulation. All other cell types developed the neuron-like morphological characteristics of a spherical, refractile cell body and long, thin extensions that approximate neurites. Time-lapse microscopy revealed that the neuron-like structure of induced MSCs was not a result of neurite extension. Instead, the ends of the neurite-like branches formed the remaining points of adhesion to the culture plate after exposure to the induction chemicals resulted in plasma membrane retraction. We also observed cell death in a small percentage of cells after chemical induction, this percentage increased with a longer exposure to the induction chemicals. These observations led us to question whether the structural response to the chemical induction protocol was simply a reflection of a toxic reaction. To test this possibility, we replaced the induction chemicals with various stressors including detergents, high molarity sodium chloride, and extremes of pH. A portion of MSCs took on pseudo-neuronal morphology in response to each manipulation. A cornerstone of the simple chemical induction paradigm is its ability to increase neural marker expression in the cells that take on neuronal morphology. In this study, induced cells showed a greater intensity of labeling for the neural markers NSE and NeuN, but treatment with the protein synthesis inhibitor cyclohexamide did not block this effect, and RT-PCR analysis showed no change in NSE mRNA levels after induction (Table 1). Our results call into question whether the morphological and gene/protein expression changes associated with chemical induction represent manifestations of cellular toxicity rather than the regulated modifications of a complicated cellular differentiation process.

As our group carried out the aforementioned study, another group (Neuhuber et al., 2004) was performing complementary experiments and reached similar conclusions. Neuhuber and colleagues independently confirmed that cell types other than MSCs, in this case rat dermal fibroblasts, respond to neural induction by appearing to form neurite-like processes. The authors were also able to induce a neuron-like morphology in both MSCs and fibroblasts with DMSO alone. As in our study, time-lapse imaging revealed that the "neurites" were not caused by the extension of new processes; rather they formed the remaining attachment points after retraction of the cytoplasm toward the nucleus. The authors linked these changes in morphology to disruption of the actin cytoskeleton through the use of actin filament-depolymerizing agents. Consistent with the MSC multidifferentiation hypothesis, Neuhuber and colleagues found that uninduced MSCs expressed the neuronal markers NF-M and tau. After induction, the cultured cells either began expressing or appeared to increase their expression of β-III-tubulin, NF-M, and tau (Table 1). Synaptophysin, another neuronal marker, was not expressed either before or after induction. The absence of synaptophysin in the induced cell population drew the authors' caveat that rather than promoting the systematic gene expression necessary for neuronal differentiation, simple chemical induction may be causing anomalous changes in gene expression. Further support came from the finding that pre-induction and post-induction fibroblasts showed the same pattern of neural gene expression changes. The authors concluded that verification of neuronal transdifferentiation from MSCs might require a more complete dissection of the molecular signaling and commitment events involved.

In a related study, Suon and colleagues (2004) tested a number of neural induction methods on human MSCs (hMSCs). The chemicals used in Woodbury's protocol had limited success in hMSCs, with only 25% of the cells transitioning to a neuronal morphology. By combining the PKA activators forskolin and IBMX with the PKC activator TPA, the authors were able to induce nearly all hMSCs. As in the previous studies, Suon's cocktail led to the rapid and dramatic conversion of MSC structure to neuron-like morphology. This change was transient with cells either spontaneously reverting to their original shape or undergoing cell death. Additionally, Western blot analysis showed increased expression of a number of neural cell markers including β-III-tubulin, neurofilament (NF), NSE, and GFAP, as well as decreased expression of the mesenchymal marker fibronectin (Table 1). A major concern of the study was to determine whether this simple chemical transdifferentiation process requires transcriptional, translational, or purely mechanical modification. Neither the protein synthesis inhibitor cyclohexamide nor the mRNA inhibitor actinomycin D prevented the transformation of the cell structure to a neuronal appearance, indicating the biochemical and structural changes were regulated by independent mechanisms. Incubation with colchicine, a microtubule disruption agent, prevented hMSCs from developing a neuron-like structure demonstrating that the change in morphology was caused by simple mechanical modification. Together, these three studies (Lu et al., 2004; Neuhuber et al., 2004; Suon et al., 2004) provide evidence that what was previously thought to be the neural transdifferentiation of MSCs by chemical induction may simply be the result of cytoskeletal rearrangement and non-specific changes in gene expression. Our study (Lu et al., 2004), in particular, suggests that these changes may not be simply non-specific. We provide evidence that the structural and epigenetic changes may be artifacts of cell shrinkage and cytoskeletal collapse caused by the bath application of toxic chemicals.

Other studies bolster this argument with additional evidence of post-induction cell death and the reversibility of the induction effects. Rismanchi and colleagues (2003) found that the Woodbury differentiation protocol caused significant cell death immediately after bath application of the induction medium, and continuing toxicity was evident at the 72-hour post-induction time point. The percent of adherent cells that were dead at the 12-hour, 24-hour, and 72-hour time points did not differ between cultures kept in maintenance or growth media, and did not show a time-related increase. When floating cells were considered, the proportion of dead cells at 72 hours increased to 53% for cells kept in maintenance medium. Those kept in growth medium for the same time frame appear to have better viability, but the authors note that cell death in this group is likely an underestimation due to the higher levels of cellular debris. This study also examined the reversibility of NeuN expression in cells induced using Woodbury's protocol. No immunoreactivity was found in pre-induction MSC cultures. Immediately after induction, the percent of cells expressing NeuN jumped to nearly 50%. Only about 15% of the induced cells transferred to maintenance medium continued to express NeuN at the three time points, and nearly all of the cells transferred to growth medium reverted to a pre-induction NeuN-negative state. These results suggest the Woodbury technique is toxic to many MSCs, and that the surviving population were not terminally differentiated 72 hours after induction.

In a later paper, Tao and colleagues (2005) confirm the findings of Rismanchi's group. Although their study focuses on the use of a cocktail of growth factors to induce the neuronal transdifferentiation of human MSCs, they also investigate the effects of Woodbury's chemical induction protocol (DMSO/BHA). As in previous studies, the cellular response to DMSO/BHA application was rapid, with the cytoskeletal changes taking place within the 30-minutes to 6-hours post-induction time frame. Forty-eight hours after induction, the hMSCs began to lose their branches and detach. An assay of the detached cells confirmed they were dead. This is in stark contrast with the growth factor-induced differentiation, which was stable for at least 3 months.

Other studies compared the expression of relevant neural genes and proteins in naïve MSCs, induced MSCs, and neural cells. Hermann and colleagues (2004, 2006) provided information about gene expression in naïve human MSCs. They found that many early (*NES, MSI1, OTX1, SOX1, SOX2, SOX10, Neurog2, CALB1, NeuroD1*) and late neuroectodermal (*GFAP, MBP, TUBB4/III, NTRK1, TH*) genes are already expressed in these cells (Table 1). The authors also confirmed the finding in Suon et al., 2004 that direct transdifferentiation using Woodbury's simple chemical protocol does not generate significant changes of morphology in human MSCs (Hermann et al., 2006). They found no change in the proportion of cells that stained for the proteins β-III-tubulin and GFAP, but a decrease in fibronectin-positive cells. The only gene expression changes were decreases in the early neuroectodermal genes *MSI1, SOX1,* and *Neurog2*. The ten other genes tested showed no significant change in expression.

Bertani et al. (2005) provided further insight into the changes in protein and gene expression in human MSCs after DMSO/BHA induction. Consistent with earlier work in the field, they found apparent upregulation of a number of neural markers (NSE, NF-200, NeuN, and Tau) by immunocytochemistry (Table 1). However, they were unable to produce similar results using Western blot analysis. By treating uninduced hMSCs with EDTA or cytochalasin D, which have identical effects on the cellular morphology, they found that these cells showed the same staining pattern for the neural markers. Even more conspicuous was their finding of de novo expression of β-galactosidase and GFP in induced hMSCs, neither of which could have been produced by the cells. These results suggest that chemically induced MSCs have a strong, non-specific affinity for antibody binding, and call into question the immunocytochemistry results from other studies. In addition to their protein analyses, Bertani and colleagues also performed a DNA microarray study using 21,329 oligonucleotides to more comprehensively evaluate gene expression in chemically treated hMSCs. They established that of the 100 genes with the greatest expression changes after 6 and 48 hours of DMSO/BHA induction, only 5 overlapped with the genes differentially expressed between uninduced hMSCs and fetal brain cells. A similar overlap of 7 genes was seen when the induced hMSC expression changes were compared to the differences between uninduced hMSCs and adult liver cells. In total, only 55 and 104 genes showed significant modulation after 6 and 48 hours of hMSC induction, respectively, whereas 854 genes differed between uninduced hMSCs and fetal brain cells. These results combine to show that the effects of Woodbury's induction protocol are inconsistent with the protein and gene expression changes necessary for neural transdifferentiation. The gene and protein profiles of the induced hMSCs bear no more resemblance to a neural phenotype than to a hepatic phenotype.

A similar study, conducted by Yamaguchi et al. (2006), compared gene expression before and after chemical induction using microarray and RT-PCR techniques. They employed a new chemical induction protocol using DMSO, bFGF, and retinoic acid (RA), combining elements from other established protocols (Woodbury et al., 2000, 2002; Jiang et al., 2002a; Kondo et al., 2005). Consistent with earlier studies, they identified 115 multi-lineage genes expressed in naïve mouse MSCs, including 35 genes associated with neural cells, neuroendocrine cells, and neurologic disorders. Only 9 of the 35 genes related to neural cells were significantly altered after chemical induction, and only 5 of the 9 genes showed significantly increased expression. These genes produce adrenomedullin, ceroid-lipofuscinosis, enolase, prostaglandin D2 synthase 2, and GABA-A receptor-associated protein like 2. The authors also determined that the gene expression levels of the popular neural cell markers nestin, Tuj-1 (β-III-tubulin), and NF-M do not change after chemical induction (Table 1). Of particular interest is the finding that chemical induction caused a dramatic reduction in actin gene expression (normalized signal intensity from 527.1 to 7.5). This supports the earlier finding that chemical induction disrupted the actin cytoskeleton (Neuhuber et al., 2004). These data, in combination with the two previously described studies, indicate that there is no molecular basis for genuine neuronal transdifferentiation of MSCs using these simple chemical induction methods.

Protein expression changes after simple chemical induction mirror the results described in the gene expression studies. Choi et al. (2006) conducted a proteomics profile analysis comparing naïve MSCs to MSCs after 3 hours of induction using Woodbury's protocol. Although 80 proteins were significantly modulated by induction, the authors identified only 20 of these by mass spectrometry. The identified proteins with altered expression were unrelated to neuronal differentiation. Those with increased expression included cytoskeletal proteins (vimentin and γ-actin), metabolic enzymes (glyceraldehyde-3-phosphate dehydrogenase), an RNA-binding protein (annexin A2), a ribosomal protein, an inflammatory/stress-responsive protein (BiP), and a hypothetical protein. The authors did show increased expression of the neural markers NSE, NF, nestin, and trkA by Western blot analysis (Table 1). It is possible that these proteins were among the 60 modulated by induction, but not identified through mass spectrometry. Suon et al. (2004) also found increased NSE and NF expression using the same technique, but Bertani et al. (2005) saw no change in these markers. The reason for this variance between studies remains unclear. Nevertheless, increased expression of NSE and NF alone is not sufficient for neuronal transdifferentiation.

The electrophysiological properties of neurons make them unique among cells. A few studies utilizing the simple chemical induction technique have investigated whether induced BMSCs display any neuron-like electrical properties. One early study used patch clamp electrophysiology on size-sieved bone marrow stem cells (SS-BMSC) induced with β-mercaptoethanol and RA for 24 hours and then transferred to serum-free medium (Hung et al., 2002). This medium is similar to Woodbury's pre-induction medium, but does not include the DMSO/BHA combination used most frequently in simple chemical transdifferentiation studies. By varying the voltage from a holding potential of –70 mV, the authors found changes in the whole-cell current suggesting the presence of voltage-sensitive sodium and potassium channels. This possibility was not further tested by the use of antagonists. It is important to note that the presence of voltage-sensitive sodium and potassium channels does not on its own indicate neuron-like characteristics as these channels can also be found in non-neuronal cells, including myocytes (Le Treut et al., 1990; Adda et al., 1996). The authors also demonstrated that both glutamate and concentrated potassium, when released near an induced SS-BMSC, caused a transient increase in intracellular calcium. The response in the induced cells was significantly greater than in uninduced SS-BMSCs. Unfortunately, the authors did not investigate the source of this calcium or whether the effect could be blocked by antagonists.

In addition to investigating changes in protein expression with MSC induction, Choi and colleagues (2006) also examined whether MSCs, induced by Woodbury's method, share any electrophysiological characteristics with neurons. As in the previous study, the treated cells were tested with whole-cell patch clamp. When the cells were depolarized in steps from a holding potential of –70 mV, they showed small step-wise increases in their outward currents, but did not display a voltage-dependent inward current.

This is consistent with an early study showing a small outward current, but no inward sodium current using the voltage step protocol on chemically-induced MSCs (Hofstetter et al., 2002). Hofstetter and colleagues also argued that the outward current does not represent a voltage-gated, outward potassium current because of its low amplitude and complications from recording leaks and leak currents. The resting membrane potential for induced MSCs in this study was –11.4 ± 8.7 mV, which differs greatly from the resting potential of a typical neuron (-60 to –70 mV).

Another recent electrophysiology study supports our hypothesis that the simple chemical induction protocol contributes to cellular toxicity. In this study, Mareschi and colleagues (2006) compared several methods of neural transdifferentiation from hMSC. They found that cells induced with β-mercaptoethanol or DMSO/BHA were so stressed that it was technically impossible to record from them using patch-clamp electrophysiology.

A few studies have found evidence that even naïve MSCs may express voltage-sensitive ion channels. In human MSCs, Mareschi and colleagues (2006) found voltage-activated outward currents partially sensitive to tetraethylammonium (TEA), a compound that can reversibly block potassium currents. These unusual currents could only be elicited at unphysiological potentials above +80 mV. A separate study using murine MSCs found voltage-activated inward currents with small amplitudes and high between-cell variability (Egusa et al., 2006). The sources of these currents were not investigated. Neither the outward currents of Mareschi's study nor the inward currents of Egusa's are comparable to neuronal currents, but they do demonstrate the importance of using uninduced MSCs as controls for neural transdifferentiation protocols, and further, highlight the need for a greater understanding of the molecular and electrophysiological characteristics of naïve MSCs.

The studies reviewed in this section have amassed overwhelming evidence that the simple chemical induction methods for neural transdifferentiation do not convert BMSCs into neurons. A number of the characteristics of cellular differentiation are missing from the pseudo-transdifferentiation process. Whereas asymmetric division precedes stem cell differentiation, mitosis is not necessary for the production of neuron-like cells by chemical induction. Additionally, the pseudo-neuronal cells do not exit the cell cycle, continuing to demonstrate cellular division after development of their neuron-like structure. The neuron-shaped cells can also revert to their pre-induction structural and epigenetic state. Chemical induction can generate neuron-like morphology in wide-ranging cell types. Although the final shape of the cells resembles that of neurons, it develops through disruption of the actin cytoskeleton and retraction of the plasma membrane rather than through the normal process of neurite extension. The structural changes in the induced MSCs may signify a toxic response to the induction medium; chemicals known to cause cellular stress can produce similar physical changes in these cells. The alterations in gene and protein expression do not appear to change in a coordinated or systematic way in these studies and do not give the MSCs a neuron-like expression profile. Finally, induced MSCs do not display the electrophysiological properties characteristic of neurons. Despite the flaws in this transdifferentiation method, the studies it inspired led to a better understanding of the criteria necessary for confirming transdifferentiation.

NEURAL TRANSDIFFERENTIATION: COCKTAILS OF GROWTH FACTORS AND CYTOKINES

The first study to utilize growth factors for neural transdifferentiation was conducted by Sanchez-Ramos et al. (2000). They investigated the potential generation of neural cells from human and mouse MSCs given supplements of epidermal growth factor (EGF) followed by a differentiation medium containing brain-derived neurotrophic factor (BDNF). Both growth factors are commonly used to induce the differentiation of embryonic stem cells into neural stem cells (NSCs) (Johnson et al., 2007), or to culture and maintain NSCs (Moe et al., 2005). After the MSCs were cultured for 1 to 2 weeks in low-serum medium containing retinoic acid (RA) and BDNF, the typically large and flat MSCs became either spindle-shaped cells with long processes or small, oval-shaped cells with short processes. Immunocytochemical analysis revealed that small proportions of these cells were nestin-positive (0.5%), NeuN-positive (0.5%), and GFAP-positive (0.8%). Notably, Western blot analysis revealed that nestin, NeuN and GFAP were also expressed in the presence of medium alone. Later studies verified that naïve MSCs express these neural cell markers (Woodbury et al., 2002; Rismanchi et al., 2003; Blondheim et al., 2006; Deng et al., 2006). Sanchez-Ramos and colleagues (2000) caution against placing too much emphasis on the presence of nestin, as this protein has been found in cells from non-neural lineages. It is also important to note that only a very small proportion of the MSCs in this study expressed neural markers and became morphologically similar to neural precursors. The small number of multipotent adult progenitor cells (MAPCs) found in bone marrow may represent the receptive population in this and other studies. It will be important in future studies to determine whether a technique is successful in differentiating only the progenitor population, or whether it is truly capable of inducing transdifferentiation in the larger MSC population.

Following this initial study, many researchers used a modified version of the growth factor neural induction medium by reducing or removing serum and using variety of growth factors or cytokines (Kabos et al., 2002; Kim et al, 2002; Jiang et al., 2002a, 2002b; Jin et al., 2003; Wislet-Gendebien et al., 2003; Croft and Przyborski, 2004; Hermann et al., 2004; 2006; Kang et al., 2004; Suzuki et al., 2004; Bossolasco et al., 2005; Kondo et al., 2005; Long et al., 2005; Magaki et al., 2005; Tao et al., 2005; Chen et al., 2006; Tatard et al., 2007). The exclusion of serum is believed to maintain the expansion of stem cells, especially nestin-positive cells, and to prevent the proliferation of differentiated cells.

It now appears that serum removal might be a critical factor in the neural transdifferentiation effects described in the studies mentioned above. Croft and Przyborski (2006) systematically studied the role of serum in the process of neural cell induction. They found that only a small proportion of naïve MSCs expressed low levels of nestin when cultured in serum-containing medium. The removal of serum, on the other hand, resulted in a high level of nestin expression in 24% of MSCs. This is consistent with an earlier MSC study in which the presence of serum corresponded with decreased nestin expression (Wislet-Gendebien et al., 2003). Croft and Przyborski also revealed that the addition of N2 supplement or FGF to the culture medium did not significantly alter the effect. Serum-free medium caused a similar increase in the expression of other neuronal proteins including β-III-tubulin, NeuroD1, GAP-43, NeuN, and synaptophysin. Interestingly, all of the nestin-positive cells developed a neuron-like morphology after 24 hours in serum-free medium. The authors investigated these structural changes using time-lapse imaging, and discovered that they were strikingly similar to the effects of Woodbury's induction protocol (Neuhuber et al., 2004; Lu et al., 2004). Culturing MSCs in serum-free medium led to cytoskeletal collapse in about 24% of the cells as a result of actin disruption. This response was enhanced to 91% with the addition of 2% DMSO to the culture medium. As with Woodbury's protocol, the response to serum deprivation was reversible and not specific to MSCs. Fibroblasts also adopted neuron-like cell morphology and expressed nestin and β-III-tubulin in response to serum withdrawal. These results suggest that in the absence of serum, cellular changes leading to neuron-like characteristics in MSCs are likely a response to cellular stress. Caution should be used in the interpretation of transdifferentiation studies using serum-free medium.

One particular study systematically compares the effects of different concentrations of serum, growth factors, and supplements, used alone or in various combinations, on neural cell differentiation from human MSCs (Bossolasco et al., 2005). Consistent with the study from Sanchez-Ramos et al. (2000), serum-containing medium is less successful in producing the morphological and gene expression changes associated with neural cell differentiation. The most effective media tested by Bossolasco and colleagues was serum-free basal medium, NS-A, designed to optimize the growth and expansion of CNS precursor cells, supplemented with the growth factors bFGF, EGF, and LIF. However, the morphology of induced hMSCs appeared more astrocytic than neuronal. Conversion of hMSCs to a neuron-like morphology was only observed with NS-A supplemented with astroblast conditioning medium. When bFGF and EGF were added to this cocktail, hMSCs acquired astrocyte-like morphology. Based on these results, growth factors appear to be less effective than astroblast conditioning medium for direct neuronal transdifferentiation (see Lu and Tuszynski, 2005b for comment on this study).

Using neurotrophins in a two-step process has proven more successful for neural transdifferentiation than the direct method used in the previous study. The first step involves culturing unfractionated bone marrow or MSCs in serum-free media supplemented with EGF and bFGF. This generates cellular spheres that bear a strong resemblance to neurospheres (Kabos et al., 2002; Croft and Przyborski, 2004; Hermann et al., 2004, 2006). The preferred method for differentiation of these cells, the second step, varies between studies. After withdrawal of EGF and bFGF, the cellular spheres are cultured on a surface coated with laminin or poly-L-lysine. Croft and Przyborski (2004) used only serum-free medium for their neural differentiation step. Although the cellular spheres had a high level of nestin expression, it decreased after differentiation. Additionally, the more mature neuronal and astrocytic markers, β-III-tubulin and GFAP, increased to detectable levels in the newly differentiated adherent cells. The cellular spheres in the study by Kabos and colleagues (2002) also expressed nestin. Their differentiation step included serum-free medium supplemented with either RA and cyclic AMP or bFGF. After differentiation, 5 to 10% of the adherent cells expressed neuronal or glial markers, including GFAP, neurofilament 200, NSE, and MAP2. The differentiation step developed by Hermann and colleagues has proven to be the most successful in producing neuron-like cells (2004; 2006). They used a low-serum medium supplemented with RA and recombinant human BDNF (rh-BDNF). Compared to pre-differentiated hMSCs, the differentiated cells showed decreased expression of *FN1* (fibronectin) and the pro-neural genes *SOX1, OTX1, NeuroD1, and Neurog2*. Differentiated cells also increased their expression of the mature neural cell genes *GFAP, MBP, TUBB4/III, SNCA,* and *TH.* A proportion of the cells expressed β-III-tubulin, MAP2ab, and GFAP, demonstrating no co-labeling of the neuronal marker MAP2ab and the astrocytic protein GFAP. Of particular interest was the finding that about 11% of differentiated cells expressed tyrosine hydroxylase (TH), the rate-limiting enzyme for dopamine synthesis, and a marker for catecholaminergic neurons. The differentiated cells not only produced dopamine, but also increased their dopamine release in response to potassium-induced membrane depolarization. Finally, the hMSCs differentiated with Hermann's two-step process were no longer multipotent; they could no longer be differentiated into osteoblasts.

The recent studies using growth factors and cytokines for the neural transdifferentiation of BMSCs suggest that a two-step process, similar to that used in the differentiation of NSCs, is more effective than direct transdifferentiation. Although the results from these studies are promising, it is important for future research to provide information about the electrophysiology of cells produced by these methods. It will also be important to determine whether specific subpopulations of BMSCs respond to a particular transdifferentiation protocol. In the two-step growth factor transdifferentiation method, only a small proportion of the initial bone marrow cell population survives and reacts to the medium by forming cellular aggregates. It is possible that this may represent the selective expansion and differentiation of the multipotent adult progenitor cells described by Jiang et al. (2002a, 2002b) rather than the transdifferentiation of adult stem cells.

Neuronal Transdifferentiation: Co-Culture with Neural Cells, Tissue, or Conditioned Medium

In a recent study, autoreactive T cells were used to induce the transdifferentiation of MSCs to neural stem cells (Moviglia et al., 2006). The anti-nervous-tissue T cells were collected from the circulated lymphocytes of patients with spinal cord injury or amyotrophic lateral sclerosis. The T cells were then activated and cultured with MSCs from the same patients. Under these conditions, hMSCs developed a spindle-shaped, polarized morphology resembling that of bipolar neurons. A large percentage of the hMSCs labeled for either GFAP, nestin, or β-III-tubulin using immunocytochemistry, but this was not confirmed by Western blot analysis.

The co-culture of BMSCs with glial cells or their conditioned medium has been tested in a number of studies as a method of neural transdifferentiation (Zurita et al., 2005; Bossolasco et al., 2005; Reali et al., 2006). Schwann cells secrete neurotrophic factors, extracellular matrix proteins, and cell adhesion molecules, and have been shown to facilitate axonal regeneration in vivo. Zurita and colleagues (2005) utilized these properties by culturing rat MSCs with Schwann cells. Nestin expression increased and peaked over the course of the first 24 hours, and was expressed in only about 5% of the MSCs at 72 hours. By this same time point, a portion of the cells began to show positive labeling for NF-200, NSE, and β-III-tubulin. After one week, when the majority of co-cultured MSCs had taken on neuron-like morphology, about 60% were positive for NSE, 52% for NF-200, and 34% β-III-tubulin. None of the cells labeled for GFAP. It is also important to note that the morphology of the Schwann cells and their expression of S-100 remained unaltered throughout the study. The authors proposed that trophic factors released by Schwann cells and the intercellular contacts between the cells may facilitate BMSC transdifferentiation.

In vivo, astrocytes support the proliferation, survival, and maturation of neurons during development. Because of these properties, they seem particularly suited for encouraging neural transdifferentiation. Bossolasco and colleagues (2005), in addition to testing various other conditions, cultured hMSCs with astroblasts, astrocyte precursor cells. Co-culture with astroblasts caused the hMSCs to take on astrocyte-like morphology. The authors also cultured MSCs with astroblast-conditioned medium, which caused the MSCs to develop both neuron-like and astrocyte-like morphologies. A portion of MSCs in both culture conditions labeled for GFAP expression, and RT-PCR analysis demonstrated that these cells expressed the genes for GFAP, GAP43, NSE, β-III-tubulin, MAP2, and NF-M. Conditioned medium from human fetal astrocytes was used by another group to induce adult human hematopoietic stem cells (hHSCs) (Reali et al., 2006). When hHSCs were plated in DMEM with 10% serum, they did not attach and died several days later. Interestingly, the hHSCs cultured in astrocyte-conditioned medium were able to survive and attach to the culture flask. The attached cells had spindle-shaped morphology with flat, short processes, and did not resemble neural cells. However, these cells labeled for several neural markers, including nestin, GFAP, NF-M, and

NSE. RT-PCR and Western blot analyses confirmed the expression of these markers. Interestingly, the presence of poly-L-lysine on the surface of the culture plate was required for NF-M protein expression, and the absence of the poly-L-lysine substrate was essential for GFAP expression. On the other hand, the authors did not observe a difference in cellular morphology based on the presence or absence of a plating substrate.

While some studies co-cultured BMSCs with a particular glial cell population, others used populations containing NSCs (Sanchez-Ramos et al., 2000; Wislet-Gendebien et al., 2003; Alexanian, 2005). Sanchez-Ramos and colleagues cultured mouse MSCs with an embryonic midbrain cell suspension to determine whether a neural environment would augment neural cell transdifferentiation (2000). They used β-galactosidase-expressing bone marrow cells prepared from transgenic lacZ mice or labeled MSCs with fluorescent cell trackers to differentiate between the co-cultured cells. The cells were also given a medium supplemented with RA and BDNF. After two weeks, approximately 2–5% of β-galactosidase-labeled or cell-tracker-labeled MSCs were positive for NeuN, a much greater proportion than the authors observed with BDNF and RA alone. However, the cells were still spindle or ovoid shaped with short processes, and did not resemble mature neurons. The MSCs did not express the mature neuronal marker MAP2 despite their proximity to MAP2-positive neurons of fetal midbrain origin. Although 5-8% of the MSCs were positive for GFAP, the authors noted that the majority of these cells resembled the NeuN-positive cells in morphology rather than demonstrating astrocyte-like morphology.

Wislet-Gendebien et al. (2003) tested the effect of culturing nestin-positive rat MSCs with GFP-expressing mouse embryonic stratum NSCs. The cells first formed heterogeneous spheres when they were cultured together in EGF-containing medium. After the spheres were plated on a polyornithine-coated surface for 5 days, about 40% of the nestin-positive MSCs labeled for GFAP expression and adopted astrocytic morphology. These cells did not differentiate into a neuron-like morphology under the same conditions, nor did they express neuronal (NeuN, β-III-tubulin, NSE, and MAP2B) or oligodendrocytic (O4 and A2B5) markers.

Another study used proliferating and fixed adult NSCs and conditioned medium (NSCcm) to determine if either the juxtacrine or paracrine influences of NSCs are sufficient for the neural transdifferentiation of mouse MSCs (Alexanian, 2005). The author found that culturing MSCs in NSCcm supplemented with bFGF increased their cell size and the proportion of nestin-positive and weakly labeled Sox2-positive cells, but did not produce cellular spheres. Further, neurosphere-like structures did not develop when MSCs were cultured on paraformaldehyde- or methanol-fixed NSCs unless they were supplemented with NSCcm. When MSCs were cultured together with NSCs in conditioned medium supplemented with bFGF, they produced dense clusters and neurosphere-like structures. Many MSCs in close contact with NSCs showed strong labeling for both nestin and Sox-2. When the MSC-containing cellular spheres were dissociated and grown for an additional week in medium without bFGF, the cells developed into β-III-tubulin-positive cells with a neuronal morphology and GFAP-positive cells with a glial morphology. This study demonstrated that both the juxtacrine and paracrine effects of NSCs played a part in producing neural characteristics in mouse MSCs.

Neural tissue explants have been used in a number of studies to promote the neuronal transdifferentiation of BMSCs (Abouelfetouh et al., 2004; Kondo et al., 2005; Mizobe et al., 2006). For example, Abouelfetouh et al. (2004) cultured mouse MSCs with hippocampal slices from 10-day-old rats. They observed the development of neuron-like cells only within the boundaries of the slice preparation. The neuron-like MSCs formed axon-like projections by day 3 that continued to increase in length over time. By day 14, these cells had formed network-like connections. The number of neuron-like cells that survived to day 14 averaged only about 7 cells per hippocampal slice. The addition of RA to the culture medium appeared to initially facilitate the survival and differentiation of the neuron-like cells, but the effect was negligible after 14 days. Only about 10% of the neuron-like cells stained for NeuN, and paradoxically, they were those with shorter processes and a less mature appearance. Neither the neuron-like nor the non-neuron-like cells remaining in culture on day 5 labeled for GFAP or Iba1, a microglial marker. In a later study, the same group used postnatal rat hippocampal slices in culture with mouse MSC populations selected for their expression of the cell surface markers Sca-1 or Thy-1 (Mizobe et al., 2006). Both of these selected subgroups produced a greater proportion of neuron-like cells than either Sca-1- or Thy-1-negative MSCs. Although the neuron-like cells did not express NeuN or GFAP, they were positive for nestin, NF-M, and NSE. Whether these neuron-like cells could become functional neurons or not remains unknown since no electrophysiological analyses were conducted in these studies.

Kondo et al. (2005) used embryonic inner ear explants to induce mouse MSCs to transdifferentiate into PNS sensory neurons. The MSCs were first co-cultured with the organ of Corti explants. In co-culture alone, the MSCs survived and proliferated, but with the addition of serum-free neuronal induction medium containing forskolin, IBMX, bFGF, BDNF, RA, and Sonic hedgehog (Shh), the MSCs withdrew from the cell cycle. The cells developed neuron-like morphology, became positive for the sensory neuron markers Brn3a, GluR4, and calretinin, and extended processes toward the sensory hair cells within the organ of Corti explants.

Although the next study did not use hippocampal slices in culture with MSCs, it demonstrated that adult hippocampus-derived soluble factors induce a neuron-like phenotype in MSCs (Rivera et al., 2006). In addition to converting MSCs to a neuronal morphology, hippocampus-conditioned medium (HCcm) induced expression of neurofilament-H (NF-H) in about 40% of MSCs and GAP43 in 5%. The induced MSCs did not express GFAP. Rivera and colleagues also determined that the effect of the HCcm on NF-H expression was independent of either NGF or BDNF signaling. The percentage of NF-H-positive cells did not differ from controls with the addition of NGF, BDNF, or both to the culture medium. Further, the addition of BDNF and NGF antibodies to induced MSCs did not decrease the effect of the HCcm on NF-H expression.

Although the previously described co-culture studies demonstrated promising methods for the neural transdifferentiation of BMSCs, they all limited their characterization to changes in BMSC morphology and the expression of a few neural markers. As discussed earlier, although these changes are important, they are not sufficient for classifying the altered cells as neurons or glia. For neurons, in particular, it will be important to demonstrate neuron-like electrophysiological properties, response to neurotrophins, the presence of synapses, and network activity. One recent study examined several of these properties and attempted to clarify the mechanisms underlying their successful co-culture method. Wislet-Gendebien and colleagues (2005) cultured nestin-positive rat MSCs (npMSCs) with postnatal mouse cerebellar granule neurons (mCGs) (2005). Building on previous work, they demonstrated the importance of using nestin-positive MSCs for neural transdifferentiation. Whereas nestin-positive cells expressed neural markers after 5 days in co-culture (40% GFAP-positive, 19% β-III-tubulin-positive, and 19% NeuN-positive), these markers could not be detected in nestin-negative MSCs. The expression of several genes related to neural cell development (*Sox2, Sox10, Pax6, FZD, ErbB2,* and *ErbB4*) was significantly higher in npMSCs compared to nestin-negative MSCs (nnMSCs). However, other similar genes were not detected in npMSCs, such as *Sox1, Sox3, Sox4, Sox11, Pax3, Notch3, ErbB3*, and *NF-L*. Of the cells that expressed GFAP or NeuN on day 5, none were positive for both markers, indicating the presence of a combination of astrocyte-like and neuron-like cells. By day 8, several additional neuron markers were expressed, including MAP2ab (an axonal protein), SMI31 (a dendritic protein), and synaptophysin (a synaptic protein). To determine whether this co-culture method can generate functional neurons from npMSCs, the authors performed electrophysiological analyses. Using whole-cell patch-clamp recording, they determined that the neuron-like cells maintain a negative resting membrane potential (about –56 mV at 15 days) close to that of a typical neurons (–60 to –70 mV). By applying hyperpolarization and depolarization steps from a holding potential of –80 mV, the authors also discovered the presence of outward currents by day 5 that were sensitive to TEA, a potassium channel blocker, and inward currents by day 8 that were sensitive to TTX, a sodium channel blocker. The neuron-like cells also responded to a current injection with a single action potential. On the other hand, the authors were unable to induce trains of action potentials, and did not observe spontaneous firing. Importantly, the authors also demonstrated responses to GABA, glycine, and glutamate, and determined that this activity was mediated by their ionotropic receptors. These results reveal that the nestin-positive MSCs, when cultured with mCGs, can acquire many of the structural and functional characteristics of neurons, but do not become fully mature neurons.

Neural Transdifferentiation: Transduction of Regulatory Genes

Fewer studies have utilized gene transduction as a method for converting BMSCs to neural cells.

In one of these studies, mouse MSCs were transduced with Noggin, a bone morphogenic protein (BMP) inhibitor involved in neural induction during early embryogenesis (Kohyama et al., 2001). The authors did not characterize the amount, duration, or bioactivity of the transferred gene in the MSCs. Nevertheless, after Noggin transduction, the cells formed floating neurosphere-like clusters. They then adopted a neuron-like morphology after being cultured on ornithine/fibronectin-coated dishes in serum-free medium containing NGF, BDNF, and NT-3. About 50% of the treated MSCs were MAP2-positive and 5% were GFAP-positive after 2 weeks in culture. However, serum deprivation alone can upregulate neural protein expression and induce a neuron-like morphology in MSCs (Croft and Przyborski

2006). The authors did find evidence that the Noggin-transduced MSCs had some neuron-like electrophysiological properties. These cells showed a strong increase in intracellular calcium in response to elevated potassium and a mild increase with glutamate stimulation.

Dezawa and colleagues (2004) took a slightly different approach and transduced both rat and human MSCs with Notch intracellular domain (NICD). NICD was chosen for its ability to alter the expression of transcription factors that influence neuronal and glial differentiation in CNS development (reviewed in Lundkvist and Lendahl, 2001). NICD transduction increased the proportion of MSCs expressing the neural stem- and progenitor-cell proteins GLAST, 3-PGDH, and nestin, however, NICD transduction alone was not able induce neuronal transdifferentiation. When transduced MSCs were cultured in the trophic factors forskolin, bFGF, and CNTF, they began to display neuron-like morphology and express MAP-2ab, NF-M, and β-III-tubulin. The de novo expression of the neuronal marker MAP-2ab was confirmed by Western blot analysis. Importantly, the authors demonstrated that naïve MSCs did not express MAP-2ab in response to the trophic factor cocktail. The transduced and trophic factor-treated cells also appeared to exit the cell cycle, with very few MAP-2ab-positive cells incorporating BrdU after trypsin treatment. The authors examined the electrophysiological characteristics of these cells, and found that they had resting potentials (–50 to –60 mV) close to those of functional neurons (–60 to –70 mV) and voltage-sensitive potassium currents. After 14 days, inward currents could not be elicited. The additional administration of NGF and BDNF to the induced cells enabled about 40% of the cells to produce tetrodotoxin-sensitive inward currents and action potentials. The authors also tested the possibility of producing dopaminergic neurons for the treatment of Parkinson's disease. For this, the induced MSCs were treated with GDNF. The proportion of cells positively labeled for tyrosine hydroxylase increased by about 37%. The GDNF-treated cells also increased their expression of transcription factors involved in the differentiation of dopaminergic neurons, including *Nurr1*, *Lmx1b*, *En1*, and *Ptx3*. Surprisingly, the cells released dopamine in response to a high potassium depolarizing stimulus. Finally, after transplanting the cells into the striatum of rat models of Parkinson's disease, the animals demonstrated significant behavioral recovery.

Mechanisms of Neural Transdifferentiation

Recent studies from the neural transdifferentiation literature have not only clarified the criteria necessary for determining the success of a particular transdifferentiation method, but have also provided the framework for future studies investigating the mechanisms behind various transdifferentiation techniques. One of the most intriguing mechanistic questions was posited by Woodbury et al. (2002). Are BMSCs multidifferentiated? Because of their initial work, we now know that adult BMSCs express a variety of genes from all germ layer lineages. Many of these shared genes were, until recently, thought to be expressed only by neural stem cells and their descendants. One possibility is that the expression of these genes is simply not as specific to neural cells as was previously thought. Alternatively, the concept of multidifferentiation suggests that these genes remain specific to neuroectodermal cells, but that BMSCs express them because the cells are partially differentiated toward a neuroectodermal lineage. Another consideration for future studies is whether a particular transdifferentiation method successfully reprograms BMSCs into neural cells, or whether the technique artificially creates a neuron-like appearance by introducing cellular stress. A related question is whether a particular transdifferentiation method can produce a fully functional neuron. A few of the studies reviewed here have produced cells with several neuron-like electrophysiological characteristics, but none of the altered cells have possessed a neuron's complete repertoire. Finally, recent studies have determined that bone marrow contains a very small population of pluripotent stem cells (MAPCs) that resemble embryonic stem cells and retain the ability differentiate into neural cells (Jiang et al., 2002a, 2002b, 2003; Kucia et al., 2006). Because most transdifferentiation methods utilize multiple expansions of bone marrow cells, it is possible that this subpopulation is selectively enhanced in culture. If a method successfully converts only MAPCs to a neural cell phenotype, by definition the technique would induce differentiation, not transdifferentiation.

Multidifferentiation or Shared Genes?

The concept of BMSC multidifferentiation was initially proposed in Woodbury et al. (2002). The authors discovered that untreated rat MSCs expressed genes associated with differentiated cells of non-mesodermal lineages, including some that encoded proteins thought to be exclusive to neural cells of ectodermal origin, such as *Gfap* (GFAP, glial fibrillary acidic protein) and *Neurod1* (NeuroD1, neurogenic differentiation factor 1). Other neural genes expressed by naïve rat MSCs were *Aldoc* (ALDO, aldolase C), *App* (APP, amyloid precursor protein), *Grina* (GLUT, NMDA glutamate binding subunit), and *Stx12* (SYN, syntaxin) (Table 1). Following Woodbury's study, Dezawa and colleagues (2004) investigated the expression of genes for transcription factors involved in neural development. They found that untreated rat MSCs expressed *Mash1*, *Math1*, *Neurog1*, *Hes1*, *Hes5*, *Stat1*, and *Stat3*. Hermann and colleagues (2004, 2006) established the expression of additional neural-related genes in naïve human MSCs, including the pluripotent stem cell marker *OCT-4*; the pro-neural genes *SOX1, SOX2, SOX10, OTX1, Neurog2,* and *CALB1*; the neural stem cell marker genes *NES* (nestin) and *MSI1*; the glial marker gene *MBP*; and the neuronal genes *TUBB4/III* (β-III-tubulin) and *NTRK1* (neurotrophic tyrosine kinase, receptor, type 1) (Table 1). Furthermore, Yamaguchi et al. (2006) determined the presence and expression levels of 35 neuronal, neuroendocrine-related, or neurologic disorder-associated genes in mouse MSCs.

Although untreated MSCs express a remarkable number of ectodermal genes, the expression of these genes is typically low. Also, the MSCs do not express a complete complement of the genes found in neural cells. The human MSCs in the studies from Hermann and colleagues (2004, 2006) did not express the pro-neural gene *OTX2*, and had barely detectable levels of *NeuroD1* and the neuronal genes *SNCA* (α-synuclein) and *TH* (tyrosine hydroxylase). In Woodbury et al. (2002), the rat MSCs did not express *Nfm* (NF-M) or *Tau* (microtubule-associated protein tau). Blondheim and colleagues (2006) also reported that human MSCs expressed a select assortment of neural genes. They detected the following neural genes in human MSCs: *GPC4*, *NES*, *NEGF2* (neurite growth-promoting factor 2),

NFH (NF-H), *NFM* (NF-M), *NSE*, *NTRK2* (neurotrophic tyrosine kinase, receptor, type 2), *PTCH*, *RARA*, and *SMO*. These cells also expressed dopaminergic genes, including: *En1*, *Nurr1*, *PITX3*, *Aldh1* (aldehyde dehydrogenase 1), *AADC*, *COMT*, and *GTPCH1*. On the other hand, the authors determined that naïve human MSCs did not express the following neural and dopaminergic genes: *RET* (c-Ret), *NDN* (necdin), *NFL* (NF-L), *TPH* (tryptophan hydroxylase), *DAT* (dopamine transporter), *DRD2* (dopamine receptor D2), *MAOB* (monoamine oxidase B), and *VMAT2* (vesicular monoamine transporter 2).

These studies reveal that naïve BMSC, particularly MSCs, express a much broader spectrum of neural genes than previously supposed. One school of thought is that MSCs are multidifferentiated, suggesting that because they are already partially differentiated toward several different lineages, they possess an innate ability to differentiate across germ layer boundaries. An alternative possibility is that BMSCs simply share the expression of "neural genes" that are not, in fact, neural cell specific. If the expression of the "neural" genes represents shared expression rather than multidifferentiation, the shared genes should be only a small proportion of neural genes. Systematic comparisons between MSC and NSC gene expression profiles may begin to address this question of mechanism by revealing the identities of the communally expressed genes and what proportion of neural genes they represent.

Reprogramming or Stress Response?

The reprogramming of BMSCs through the induction of epigenetic changes is thought to underlie the neural transdifferentiation of these cells. Selective alterations in gene expression could, in concept, convert one cell type to another since the somatic cells of an individual typically possess the same genome. Unfortunately, many studies limit their investigation to changes in cell morphology and the expression of a small number of selected genes. We have established that cellular stress and toxicity can lead to a neuronal appearance and alter the expression patterns of many genes, including some neural genes. It is now clear that a more comprehensive understanding of global gene expression changes is necessary to determine the success of a neural transduction technique.

The gene expression studies using Woodbury's simple chemical induction technique do not portray a transition from MSCs to neural cells. In Woodbury et al. (2002), the ectodermal genes measured by RT-PCR did not show consistently increased expression. The expression of *App* (APP, amyloid precursor protein) was unchanged after induction; *Aldoc* (ALDO, aldolase C), *Grina* (GLUT, NMDA glutamate binding subunit), *Stx12* (SYN, syntaxin), *Gfap* (GFAP), and *Neurod1* (NeuroD1) decreased; and only *Nfm* (NF-M), *Tau* (microtubule-associated protein tau), and *Syp* (synaptophysin) increased. The same induction technique did not change *Eno2* (NSE, neuron-specific enolase) expression in Lu et al. (2004). Hermann and colleagues (2006) reported no change in the pluripotency marker *OCT4*, and a slight, but not significant decrease in the expression of the pro-neural genes *OTX1* and *NeuroD1*; the NSC marker gene *NES* (nestin); the glial genes *GFAP* and *MBP* (myelin basic protein); and the neuronal genes *SNCA* (α-synuclein) and *TH* (tyrosine hydroxylase). The neural genes *SOX1*, *Neurog2* (neurogenin 2), and *MSI1* (musashi 1) had decreased expression. Only *TUBB4/III* (β-III-tubulin) showed a non-significant increase in expression. Bertani and colleagues (2005) determined that only 5 genes (from 100) altered by simple chemical induction in hMSCs brought them closer to the gene expression patterns of fetal brain cells. They further determined that this induction technique was no more successful in converting the hMSCs to a neural lineage than to a hepatic one. A similar study using a slightly different transdifferentiation protocol found that only 14% of the neural cell genes expressed by naïve mouse MSCs showed increased expression after induction (Yamaguchi et al., 2006). They also determined that this induction protocol did not affect the expression of *Nes* (nestin), *Tubb3* (β-III-tubulin), or *Nfm* (NF-M). These studies demonstrate the lack of a consistent pattern in neural gene expression changes with simple chemical induction. The incoherency of the gene expression changes may reflect either a coordinated change that is non-neural in nature or an uncoordinated response to cellular stress.

The gene expression changes caused by growth factor induction of BMSCs formed a more coherent pattern. In the two studies by Hermann and colleagues (2004, 2006), the formation of cellular spheres in the first step of the transdifferentiation process was accompanied by increased expression of most pro-neural and neural stem cell genes measured (*OTX1*, *NeuroD1*, *Neurog2*, *MSI1*, and *NES*), whereas most of the glial and neuronal genes typical of more advanced stages of neural cell differentiation had unchanged expression (*GFAP*, *MBP*, *TUBB4/III*, *SNCA*, and *TH*). In the second step, the migration and differentiation of the cells was characterized by a reversal of the pattern of gene expression observed in the first step. The pro-neural genes returned to baseline expression levels (*OTX1*, *NeuroD1*, and *Neurog2*), the NSC genes remained elevated (*NES* and *MSI1*), and the glial and neuronal genes showed increased expression (*GFAP*, *MBP*, *TUBB4/III*, *SNCA*, and *TH*). A related study using a single-step transdifferentiation process obtained similar results (Tao et al. 2005). In this study the growth-factor-induced transition to neuronal morphology was accompanied by increased expression of mature neuronal and glial genes (*NFM*, *MAP2*, *GAD*, and *MBP*).

Another study has been relatively successful in producing neuron-like cells from rat and human MSCs by having a more direct effect on gene expression through the transduction of the Notch intracellular domain (NICD) (Dezawa et al., 2004). The NICD is a proteolytically cleaved portion of the Notch receptor that binds with and activates a number of transcription factors. Although NICD transduction alone was not sufficient for the neuronal transdifferentiation of MSCs, treatment of the transduced cells with forskolin, bFGF, and CNTF followed by subsequent treatment with GDNF caused increased expression of *Nurr1*, *Lmx1b*, *En1*, and *Ptx3* genes. These genes encode transcription factors with important roles in the differentiation of dopaminergic neurons.

The studies reviewed here suggest that although some neural transdifferentiation techniques are unsuccessful and produce random gene expression changes in response to cellular stress, other techniques produce a more logical sequence of changes in the expression of neural genes in induced BMSCs. The demonstration of coordinated epigenetic changes will clarify the mechanisms of action in neural transdifferentiation studies, and may prove critical for confirming the success of a transdifferentiation method.

Fully Functional Neurons or Partially Functional Neuron-like Cells?

Until recently, studies investigating adult neural stem cell differentiation relied exclusively on changes in morphology or protein expression to determine the formation of mature neurons (Reh, 2002). Then, in 2002, two studies demonstrated for the first time that adult NSCs can differentiate into functional neurons in vitro, and can become part of a functional network in vivo (Song et al., 2002; Van Praag et al., 2002). These studies led to a reassessment of the criteria for determining whether a cell has become a functional neuron. Specifically, neurons are post-mitotic; polarized, having both dendrites and a single axon; able to produce an action potential upon depolarization; and able to communicate with other neurons through neurotransmitter binding on dendrites and release from the axon (Reh et al., 2002; Song et al., 2002). As described in this review, most investigations of adult BMSC transdifferentiation to neuron-like cells are limited to the early criteria used for NSCs; they focus primarily on changes in morphology and gene or protein expression. The pitfalls of relying on such limited criteria are clearly demonstrated by the simple chemical induction experiments, and argue for placing a greater emphasis on functional studies.

Whether the transdifferentiation methods described in this review can transform a BMSC into a fully functional neuron remains unclear. In three of the simple chemical induction studies, patch clamp electrophysiology revealed the presence of a voltage-dependent outward current, but antagonists were not used to confirm the presence of voltage-sensitive potassium channels (Hofstetter et al., 2002; Hung et al., 2002; Choi et al., 2006). Only one of the studies described a voltage-dependent inward current (Hung et al., 2002). The resting membrane potential for the chemically induced MSCs was substantially different from what is typically found in mature neurons (Hofstetter et al., 2002). In the growth factor induction studies, only Hermann and colleagues (2004) demonstrated neuron-like functional qualities. They were able to elicit dopamine release from their treated cells through potassium-induced membrane depolarization. Similarly, Dezawa and colleagues (2004) could elicit dopamine release from their neuron-like cells in response to high levels of potassium. Their neuron-like cells were produced using a complex cocktail of growth factors in combination with NICD transduction. Surprisingly, the only trophic factor common to both experiments was bFGF. Finally, in the co-culture studies, only Wislet-Gendebien and colleagues (2005) performed functional analyses. In this case, the induced MSCs developed a neuron-like resting membrane potential; demonstrated voltage-sensitive sodium and potassium currents; produced a single action potential upon current injection; and responded to the neurotransmitters GABA, glycine, and glutamate. Although this technique produced cells with many of the characteristics of fully functional neurons, the treated cells did not generate action potential volleys or spontaneous action potentials. Further, the cells displayed the functional characteristics of isolated neurons, but their ability to function as part of a network was not determined. Confirmation that the MSCs induced by Wislet-Gendebien's co-culture method have become fully functional neurons will hinge on whether the cells can show depolarization-induced neurotransmitter release, the formation of synapses, and the presence of coordinated activity with other neurons.

As others have suggested, the transdifferentiation of induced MSCs into functional neurons will probably mirror the stages of functional development found in NSC differentiation (Wislet-Gendebien et al., 2005). Carleton and colleagues (2003) described a progression of functional changes in neural progenitors migrating from the subventricular zone to the olfactory bulb where they eventually formed active connections with the local neurons. This sequence began with the expression of GABA-A receptors, followed by AMPA receptors, and then, finally, NMDA receptors. Spontaneous synaptic activity developed later, followed by spiking activity. The resting membrane potential of the cells became more negative throughout this transition until it reached –61.2 mV. A complementary study described the in vitro functional development of neurons from adult neural progenitors (Moe et al., 2005). Over the course of 4 weeks, the resting membrane potential became more negative. A voltage-gated potassium current developed first on day 7, followed on day 10 by the appearance of voltage-sensitive calcium channels. Voltage-gated sodium currents developed by day 15, and mature action potentials were observed near the end of the fourth week. GABAergic and glutamatergic synapses were also present at this time.

So far, no study investigating the transdifferentiation of BMSC has demonstrated the development of fully functional neurons. It will be important for future studies to do more comprehensive electrophysiology. It will also be important to test the functional changes in these cells over time. A particular induction method may not be able to produce fully functional neurons, but it may initiate a functional progression similar to those discussed above, indicating the method has been at least partially successful in converting BMSCs to a neuronal phenotype.

Differentiation or Transdifferentiation?

Finally, it is important to determine whether the methods used to convert BMSCs to neural cells induce differentiation or transdifferentiation. The mechanism of action in this case depends on the nature of the affected cells. Recent studies have demonstrated the presence of a small population of multipotent adult progenitor cells (MAPCs) in bone marrow that co-purify with MSCs (Jiang et al., 2002a, 2002b, 2003). MAPCs appear to retain a high level of epigenetic plasticity similar to that of embryonic stem cells. They may be able to proliferate indefinitely and differentiate into cells from all three germ layers (Jiang et al., 2002a, 2002b). These characteristics suggest that MAPCs have not undergone even the relatively minor cell fate changes that normally restrict the differentiation potential of adult stem cells.

Whether MAPCs resemble embryonic stem cells remains unclear. Jiang and colleagues (2002b) have demonstrated that endothelial, neural, and endodermal differentiation methods developed for embryonic stem cells can convert bone marrow MAPCs to cells with characteristics of endothelium, neural cells, and hepatocytes, respectively. In this study, however, the characterization of the induced cells was limited to cellular morphology and the expression of a few proteins. The same group later produced a more comprehensive analysis of neuronal differentiation from MAPCs (Jiang et al., 2003). When co-cultured with fetal astrocytes, MAPCs developed neuronal morphology and showed increased or de novo expression of 16 neural genes. Patch-clamp electrophysiology demonstrated the development of an increasingly negative resting membrane potential, a large voltage-dependent inward current sensitive to TTX, repetitive spiking in response to depolarization, and fluctuations under current clamp conditions that resembled spontaneous synaptic potentials. Although these findings provide strong support for the ability of MAPCs to differentiate into neurons, further evidence is needed to confirm their ability to differentiate into cells of other germ layer lineages.

Assuming that MAPCs resemble embryonic stem cells in their epigenetic plasticity, conversion of these cells to neurons would be a process of differentiation rather than transdifferentiation. Observations from recent studies suggest that a small population within MSC cultures may be responsible for the development of neuron-like cells. Often, only a small percentage of cells develop neural cell structure and neural marker expression (Sanchez-Ramos et al., 2000; Kabos et al., 2002). Furthermore, many of the studies use protocols that could selectively enhance a subset of the cultured cells (Kabos et al., 2002; Croft and Przyborski, 2004; Hermann et al., 2004, 2006). Distinguishing between differentiation and transdifferentiation is important not only for clarifying the mechanism of action but also for determining whether a particular subset of bone marrow cells should be selectively enhanced for therapeutic testing.

In conclusion, many studies in the last few years have attempted to transdifferentiate BMSCs into functional neurons in vitro using induction protocols that vary from the use of simple chemicals or growth factors in the culture medium to co-culture with neural cells or the transduction of a particular gene. None, so far, has demonstrated the development of a cell with all of the characteristics of a fully functional neuron, but a select few have been able to show several neuronal traits, including the ability to produce an action potential. Although this review does not address all of the recent studies on the neural transdifferentiation of BMSCs, those discussed here provide a strong overview of the current techniques being used, the relative success of each method, and the issues that need to be addressed in this rapidly expanding field.

Acknowledgments

The authors' research related to this commentary is supported by the NIH (R01 NS09881), Veterans Administration, Canadian Spinal Research Organization, Swiss Institute for Research into Paraplegia.

References

Abouelfetouh A, Kondoh T, Ehara K, Kohmura E (2004) Morphological differentiation of bone marrow stromal cells into neuron-like cells after co-culture with hippocampal slice. *Brain Res 1029*:114-119.

Adda S, Fleischmann BK, Freedman BD, Yu M, Hay DW, Kotlikoff MI (1996) Expression and function of voltage-dependent potassium channel genes in human airway smooth muscle. *J Biol Chem 271*:13239-13243.

Alexanian AR (2005) Neural stem cells induce bone-marrow-derived mesenchymal stem cells to generate neural stem-like cells via juxtacrine and paracrine interactions. *Exp Cell Res 310*:383-391.

Bally-Cuif L, Hammerschmidt M (2003) Induction and patterning of neuronal development, and its connection to cell cycle control. *Curr Opin Neurobiol 13*:16-25.

Bertani N, Malatesta P, Volpi G, Sonego P, Perris R (2005) Neurogenic potential of human mesenchymal stem cells revisited: analysis by immunostaining, time-lapse video and microarray. *J Cell Sci 118*:3925-3936.

Blondheim NR, Levy YS, Ben-Zur T, Burshtein A, Cherlow T, Kan I, Barzilai R, Bahat-Stromza M, Barhum Y, Bulvik S, Melamed E, Offen D (2006) Human mesenchymal stem cells express neural genes, suggesting a neural predisposition. *Stem Cells Dev 15*:141-164.

Bossolasco P, Cova L, Calzarossa C, Rimoldi SG, Borsotti C, Deliliers GL, Silani V, Soligo D, Polli E (2005) Neuro-glial differentiation of human bone marrow stem cells in vitro. *Exp Neurol 193*:312-325.

Carleton A, Petreanu LT, Lansford R, Alvarez-Buylla A, Lledo PM (2003) Becoming a new neuron in the adult olfactory bulb. *Nat Neurosci 6*:507-518.

Chen CW, Boiteau RM, Lai WF, Barger SW, Cataldo AM (2006) sAPPalpha enhances the transdifferentiation of adult bone marrow progenitor cells to neuronal phenotypes. *Curr Alzheimer Res 3*:63-70.

Choi CB, Cho YK, Prakash KV, Jee BK, Han CW, Paik YK, Kim HY, Lee KH, Chung N, Rha HK (2006) Analysis of neuron-like differentiation of human bone marrow mesenchymal stem cells. *Biochem Biophys Res Commun 350*:138-146.

Cremisi F, Philpott A, Ohnuma S (2003) Cell cycle and cell fate interactions in neural development. *Curr Opin Neurobiol 13*:26-33.

Croft AP, Przyborski SA (2004) Generation of neuroprogenitor-like cells from adult mammalian bone marrow stromal cells in vitro. *Stem Cells Dev 13*:409-420.

Croft AP, Przyborski SA (2006) Formation of neurons by non-neural adult stem cells: potential mechanism implicates an artifact of growth in culture. *Stem Cells 24*:1841-1851.

Deng J, Petersen BE, Steindler DA, Jorgensen ML, Laywell ED (2006) Mesenchymal stem cells spontaneously express neural proteins in culture and are neurogenic after transplantation. *Stem Cells 24*:1054-1064.

Dezawa M, Kanno H, Hoshino M, Cho H, Matsumoto N, Itokazu Y, Tajima N, Yamada H, Sawada H, Ishikawa H, Mimura T, Kitada M, Suzuki Y, Ide C (2004) Specific induction of neuronal cells from bone marrow stromal cells and application for autologous transplantation. *J Clin Invest 113*:1701-1710.

Egusa H, Schweizer FE, Wang CC, Matsuka Y, Nishimura I (2005) Neuronal differentiation of bone marrow-derived stromal stem cells involves suppression of discordant phenotypes through gene silencing. *J Biol Chem 280*:23691-23697.

Esdar C, Milasta S, Maelicke A, Herget T (2001) Differentiation-associated apoptosis of neural stem cells is effected by Bcl-2 overexpression: impact on cell lineage determination. *Eur J Cell Biol 80*:539-553.

Hermann A, Liebau S, Gastl R, Fickert S, Habisch HJ, Fiedler J, Schwarz J, Brenner R, Storch A (2006) Comparative analysis of neuroectodermal differentiation capacity of human bone marrow stromal cells using various conversion protocols. *J Neurosci Res 83*:1502-1514.

Hermann A, Gastl R, Liebau S, Popa MO, Fiedler J, Boehm BO, Maisel M, Lerche H, Schwarz J, Brenner R, Storch A (2004) Efficient generation of neural stem cell-like cells from adult human bone marrow stromal cells. *J Cell Sci 117*:4411-4422.

Hofstetter CP, Schwarz EJ, Hess D, Widenfalk J, El Manira A, Prockop DJ, Olson L (2002) Marrow stromal cells form guiding strands in the injured spinal cord and promote recovery. *Proc Natl Acad Sci U S A 99*:2199-2204.

Hung SC, Cheng H, Pan CY, Tsai MJ, Kao LS, Ma HL (2002) In vitro differentiation of size-sieved stem cells into electrically active neural cells. *Stem Cells 20*:522-529.

Jiang Y, Vaessen B, Lenvik T, Blackstad M, Reyes M, Verfaillie CM (2002b) Multipotent progenitor cells can be isolated from postnatal murine bone marrow, muscle, and brain. *Exp Hematol 30*:896-904.

Jiang Y, Henderson D, Blackstad M, Chen A, Miller RF, Verfaillie CM (2003) Neuroectodermal differentiation from mouse multipotent adult progenitor cells. *Proc Natl Acad Sci U S A 100* Suppl 1:11854-11860.

Jiang Y, Jahagirdar BN, Reinhardt RL, Schwartz RE, Keene CD, Ortiz-Gonzalez XR, Reyes M, Lenvik T, Lund T, Blackstad M, Du J, Aldrich S, Lisberg A, Low WC, Largaespada DA, Verfaillie CM (2002a) Pluripotency of mesenchymal stem cells derived from adult marrow. *Nature 418*:41-49.

Jin K, Mao XO, Batteur S, Sun Y, Greenberg DA (2003) Induction of neuronal markers in bone marrow cells: differential effects of growth factors and patterns of intracellular expression. *Exp Neurol 184*:78-89.

Johnson MA, Weick JP, Pearce RA, Zhang SC (2007) Functional neural development from human embryonic stem cells: accelerated synaptic activity via astrocyte coculture. *J Neurosci 27*:3069-3077.

Kabos P, Ehtesham M, Kabosova A, Black KL, Yu JS (2002) Generation of neural progenitor cells from whole adult bone marrow. *Exp Neurol 178*:288-293.

Kang SK, Putnam LA, Ylostalo J, Popescu IR, Dufour J, Belousov A, Bunnell BA (2004) Neurogenesis of Rhesus adipose stromal cells. *J Cell Sci 117*:4289-4299.

Kim BJ, Seo JH, Bubien JK, Oh YS (2002) Differentiation of adult bone marrow stem cells into neuroprogenitor cells in vitro. *Neuroreport 13*:1185-1188.

Kohyama J, Abe H, Shimazaki T, Koizumi A, Nakashima K, Gojo S, Taga T, Okano H, Hata J, Umezawa A (2001) Brain from bone: efficient "meta-differentiation" of marrow stroma-derived mature osteoblasts to neurons with Noggin or a demethylating agent. *Differentiation 68*:235-244.

Kondo M, Wagers AJ, Manz MG, Prohaska SS, Scherer DC, Beilhack GF, Shizuru JA, Weissman IL (2003) Biology of hematopoietic stem cells and progenitors: implications for clinical application. *Annu Rev Immunol 21*:759-806.

Kondo T, Johnson SA, Yoder MC, Romand R, Hashino E (2005) Sonic hedgehog and retinoic acid synergistically promote sensory fate specification from bone marrow-derived pluripotent stem cells. *Proc Natl Acad Sci U S A 102*:4789-4794.

Kramer BC, Woodbury D, Black IB (2006) Adult rat bone marrow stromal cells express genes associated with dopamine neurons. *Biochem Biophys Res Commun 343*:1045-1052.

Kucia M, Reca R, Campbell FR, Zuba-Surma E, Majka M, Ratajczak J, Ratajczak MZ (2006) A population of very small embryonic-like (VSEL) CXCR4(+)SSEA-1(+)Oct-4+ stem cells identified in adult bone marrow. *Leukemia 20*:857-869.

Le Treut T, Boudier JL, Jover E, Cau P (1990) Localization of voltage-sensitive sodium channels on the extrasynaptic membrane surface of mouse skeletal muscle by autoradiography of scorpion toxin binding sites. *J Neurocytol 19*:408-420.

Long X, Olszewski M, Huang W, Kletzel M (2005) Neural cell differentiation in vitro from adult human bone marrow mesenchymal stem cells. *Stem Cells Dev 14*:65-69.

Lu P, Tuszynski MH (2005a) Can bone marrow-derived stem cells differentiate into functional neurons? *Exp Neurol 193*:273-278.

Lu P, Blesch A, Tuszynski MH (2004) Induction of bone marrow stromal cells to neurons: differentiation, transdifferentiation, or artifact? *J Neurosci Res 77*:174-191.

Lu P, Jones LL, Tuszynski MH (2005b) BDNF-expressing marrow stromal cells support extensive axonal growth at sites of spinal cord injury. *Exp Neurol 191*:344-360.

Lundkvist J, Lendahl U (2001) Notch and the birth of glial cells. *Trends Neurosci 24*:492-494.

Magaki T, Kurisu K, Okazaki T (2005) Generation of bone marrow-derived neural cells in serum-free monolayer culture. *Neurosci Lett 384*:282-287.

Mareschi K, Novara M, Rustichelli D, Ferrero I, Guido D, Carbone E, Medico E, Madon E, Vercelli A, Fagioli F (2006) Neural differentiation of human mesenchymal stem cells: Evidence for expression of neural markers and eag K+ channel types. *Exp Hematol 34*:1563-1572.

Mizobe T, Kidoguchi K, Tamaki M, Sasayama T, Kondoh T, Kohmura E (2006) Sca-1 and Thy-1 accelerate neuron-like differentiation in bone marrow stromal cells. *Kobe J Med Sci 52*:119-125.

Moe MC, Varghese M, Danilov AI, Westerlund U, Ramm-Pettersen J, Brundin L, Svensson M, Berg-Johnsen J, Langmoen IA (2005) Multipotent progenitor cells from the adult human brain: neurophysiological differentiation to mature neurons. *Brain 128*:2189-2199.

Moviglia GA, Varela G, Gaeta CA, Brizuela JA, Bastos F, Saslavsky J (2006) Autoreactive T cells induce in vitro BM mesenchymal stem cell transdifferentiation to neural stem cells. *Cytotherapy 8*:196-201.

Munoz-Elias G, Woodbury D, Black IB (2003) Marrow stromal cells, mitosis, and neuronal differentiation: stem cell and precursor functions. *Stem Cells 21*:437-448.

Neuhuber B, Gallo G, Howard L, Kostura L, Mackay A, Fischer I (2004) Reevaluation of in vitro differentiation protocols for bone marrow stromal cells: disruption of actin cytoskeleton induces rapid morphological changes and mimics neuronal phenotype. *J Neurosci Res 77*:192-204.

Pittenger MF, Mackay AM, Beck SC, Jaiswal RK, Douglas R, Mosca JD, Moorman MA, Simonetti DW, Craig S, Marshak DR (1999) Multilineage potential of adult human mesenchymal stem cells. *Science 284*:143-147.

Prockop DJ (1997) Marrow stromal cells as stem cells for nonhematopoietic tissues. *Science 276*:71-74.

Reali C, Scintu F, Pillai R, Cabras S, Argiolu F, Ristaldi MS, Sanna MA, Badiali M, Sogos V (2006) Differentiation of human adult CD34+ stem cells into cells with a neural phenotype: role of astrocytes. *Exp Neurol 197*:399-406.

Reh TA (2002) Neural stem cells: form and function. *Nat Neurosci 5*:392-394.

Rismanchi N, Floyd CL, Berman RF, Lyeth BG (2003) Cell death and long-term maintenance of neuron-like state after differentiation of rat bone marrow stromal cells: a comparison of protocols. *Brain Res 991*:46-55.

Rivera FJ, Sierralta WD, Minguell JJ, Aigner L (2006) Adult hippocampus derived soluble factors induce a neuronal-like phenotype in mesenchymal stem cells. Neurosci Lett 406:49-54.

Sanchez-Ramos J, Song S, Cardozo-Pelaez F, Hazzi C, Stedeford T, Willing A, Freeman TB, Saporta S, Janssen W, Patel N, Cooper DR, Sanberg PR (2000) Adult bone marrow stromal cells differentiate into neural cells in vitro. *Exp Neurol 164*:247-256.

Song HJ, Stevens CF, Gage FH (2002) Neural stem cells from adult hippocampus develop essential properties of functional CNS neurons. *Nat Neurosci* 5:438-445.

Suon S, Jin H, Donaldson AE, Caterson EJ, Tuan RS, Deschennes G, Marshall C, Iacovitti L (2004) Transient differentiation of adult human bone marrow cells into neuron-like cells in culture: development of morphological and biochemical traits is mediated by different molecular mechanisms. *Stem Cells Dev 13*:625-635.

Suzuki H, Taguchi T, Tanaka H, Kataoka H, Li Z, Muramatsu K, Gondo T, Kawai S (2004) Neurospheres induced from bone marrow stromal cells are multipotent for differentiation into neuron, astrocyte, and oligodendrocyte phenotypes. *Biochem Biophys Res Commun 322*:918-922.

Tao H, Rao R, Ma DD (2005) Cytokine-induced stable neuronal differentiation of human bone marrow mesenchymal stem cells in a serum/feeder cell-free condition. *Dev Growth Differ 47*:423-433.

Tatard VM, D'Ippolito G, Diabira S, Valeyev A, Hackman J, McCarthy M, Bouckenooghe T, Menei P, Montero-Menei CN, Schiller PC (2007) Neurotrophin-directed differentiation of human adult marrow stromal cells to dopaminergic-like neurons. *Bone 40*:360-373.

Temple S (2001) The development of neural stem cells. *Nature 414*:112-117.

van Praag H, Schinder AF, Christie BR, Toni N, Palmer TD, Gage FH (2002) Functional neurogenesis in the adult hippocampus. *Nature 415*:1030-1034.

Waller EK, Olweus J, Lund-Johansen F, Huang S, Nguyen M, Guo GR, Terstappen L (1995) The "common stem cell" hypothesis reevaluated: human fetal bone marrow contains separate populations of hematopoietic and stromal progenitors. *Blood 85*:2422-2435.

Wislet-Gendebien S, Leprince P, Moonen G, Rogister B (2003) Regulation of neural markers nestin and GFAP expression by cultivated bone marrow stromal cells. *J Cell Sci 116*:3295-3302.

Wislet-Gendebien S, Hans G, Leprince P, Rigo JM, Moonen G, Rogister B (2005) Plasticity of cultured mesenchymal stem cells: switch from nestin-positive to excitable neuron-like phenotype. *Stem Cells 23*:392-402.

Woodbury D, Reynolds K, Black IB (2002) Adult bone marrow stromal stem cells express germline, ectodermal, endodermal, and mesodermal genes prior to neurogenesis. *J Neurosci Res 69*:908-917.

Woodbury D, Schwarz EJ, Prockop DJ, Black IB (2000) Adult rat and human bone marrow stromal cells differentiate into neurons. *J Neurosci Res 61*:364-370.

Yamaguchi S, Kuroda S, Kobayashi H, Shichinohe H, Yano S, Hida K, Shinpo K, Kikuchi S, Iwasaki Y (2006) The effects of neuronal induction on gene expression profile in bone marrow stromal cells (BMSC)--a preliminary study using microarray analysis. *Brain Res 1087*:15-27.

Zurita M, Vaquero J, Oya S, Miguel M (2005) Schwann cells induce neuronal differentiation of bone marrow stromal cells. *Neuroreport 16*:505-508.

In: New Cell Differentiation Research Topics
Editor Hitoshi Saitama, pp. 183-211

ISBN: 978-1-60021-937-5

Chapter V

3D Modelling of Epithelial Glandular Differentiation and the Stem Cell Niche

J.F. Pearson and S.H. Lang*

YCR Cancer Research Unit, Department of Biology, University of York,
York YO10 5DD, United Kingdom.

Abstract

This review focuses on the development and technique of three dimensional (3D) cultures to study epithelial cell differentiation and tissue architecture. 3D cultures employ human basal epithelial cells, stem cells and heterotypic cultures to faithfully reproduce tissue complexity. These models have arisen from a requirement to model human cells in a tissue-like structure. They are more accessible than animal based models and have a more relevant complexity than 2D cell culture. A 3D environment is important to properly integrate chemical and biomechanical signals in the establishment of epithelial differentiation and architecture. Epithelial cell function is engineered by combining the influences of matrix, neighbouring cells and soluble growth factors.

In the last 10 years there has been a great expansion in the use of 3D model systems for cell research. These can be divided into two broad categories: 1) 3D spheres generated by cell aggregation using techniques such as rotary wall vessel culture, agar, pellet culture and liquid media and 2) 3D models which produce tissue-like architecture through culturing within gel or synthetic scaffolds. More recently 3D models are being employed to define the stem cell niche, allowing the control of self renewal and commitment to differentiation to be explored.

* Funded by the BBSRC (BB/E021409/1) and The Wellcome Trust (GR076612MA).

INTRODUCTION

At present research on three dimensional (3D) cell systems is more vital and productive than ever. There are many ongoing studies that use human tissue to reconstruct artificial organs for the replacement or support of natural organs after failure. These include the liver and kidney [Mueller-Klieser, 1997], and osteochrondral grafts composed of the patients own cells within a 3D matrix which, when grafted back into the damaged joint, provides a template for tissue development [Martin et al., 2007]. Whilst the reconstruction of glandular tissue for grafting is not yet possible, 3D models of glandular epithelium provide an important cell based assay to establish gene function and provide a tool for researchers to study normal and abnormal differentiation in a structurally appropriate context. They also provide new techniques for screening the efficacy and toxicity of novel drugs in the pharmaceutical industry. For example, 3D reconstruction of the skin provides a means of testing detergents and cosmetics without using animal models [Garlick, 2007; Poumay and Coquette, 2007] and models of human cancers provide an invaluable approach for the development and testing of new anti-cancer drugs as both the tumour cell and its environment can be targeted [Debnath and Brugge, 2005; Kim, 2005; Smalley et al., 2006]. Such systems are becoming increasingly popular since they have the potential to reduce the need for animal experimentation and have the ability to bridge the gap between results generated from animal models and those from human clinical studies.

Traditionally, tissue culture has been dominated by the growth of cell lines as monolayers. While these 2D systems are well documented and easy to set up with good viability of cells, they lack the ability to reproduce the morphology, 3D architecture and some biochemical features of cells in the original tissue. 2D cultures are often faced with significant transport limitations thus affecting how they respond to their cellular environment and integrate multiple signals such as soluble growth factors and signalling pathways. In addition, not all epithelial cell types are amenable to adherence onto artificial substrata, thus limiting the diversity of cell types available to study. Those cell types that are suitable for 2D study, primarily immortalised cell lines, provide researchers with a reproducible, well characterised system that is universally recognised. However, the process of immortalisation, be it virally mediated or spontaneous, and the effect of long-term culture, varying culture conditions and multiple passages, can irreversibly alter the biological properties of cells [Peehl, 2004; Peehl, 2005]. As a result these cell lines seldom represent the *in vivo* tissue from which they were originally derived. Today new insights are being gleaned from the ability to culture both established cell lines and primary epithelial cultures as 3D spheroids.

A 3D environment has many advantages over conventional monolayer culture. Primarily, it offers researchers the possibility to explore the potential mechanisms that underlie the process of tissue formation, such as migration, proliferation, adhesion, differentiation and apoptosis, under conditions that emulate an *in vivo* environment. In addition, the 3D environment enables cells to form cell-cell and cell-matrix interactions that may otherwise be severely constrained or precluded entirely in monolayer culture. 3D cultures can also be homotypic or heterotypic, thereby creating a cellular heterogeneity similar to that within intact organs. 3D culture of multiple cell types enables the interaction between epithelial and endothelial cells and the surrounding stromal cells to be studied. In the prostate, of which stroma accounts for more then 80% of the total volume, co-culture of primary epithelial cells with stroma has proved essential to the formation of polarised 3D spheroids that resemble *in vivo* acini [Lang et al., 2001b]. The relative simplicity and tractability of these heterologous co-cultures allows cellular analysis in a controlled and reproducible format while providing a realistic interpretation of an *in vivo* response to an experimental treatment.

This can all be accomplished without the cost, complexity and ethical approval inherent in the use of animal models and *in vivo* studies, which are currently the best way to investigate complex cell interactions in tissues. 3D modelling *in vitro* therefore provides researchers with an important adjunct to the use of such models, particularly when whole tissue analysis is required. The following chapter shall focus on the use of different 3D culture systems, to faithfully reproduce glandular tissue-like architecture *in vitro* to study epithelial cell differentiation.

3D Cell Culture Methods

A variety of methods are employed to grow cells in 3D. These can be divided into two broad categories: 1) 3D aggregates and 2) spheroids in gel matrix systems. For 3D aggregates, conditions are provided in which the adhesive forces between the cells is greater than that for the substrate on which they are plated. This has been achieved either through spontaneous aggregation of cells in suspension, or by preventing cell adhesion to the tissue culture flask by coating the tissue culture plastic with a non-adhesive substrate, such as agar or poly HEMA, or by perpetual motion using rotary culture. In contrast, gel matrix based techniques involve embedding the cells within a substrate that contains a scaffolding of collagen, or other organic or synthetic fibre, which provides an inert or biological 3D support.

The following section summarizes existing methods for culturing glandular epithelial cells in 3D, focussing primarily on the glandular tissues of the prostate and breast. In particular it will discuss how these structures relate to the *in vivo* tissue, with regards to architecture and function, and how these multi-component tissue-like structures can be used to advance our understanding of normal glandular differentiation.

3D Aggregates

Spontaneous Aggregates

The concept of 3D modelling arose from the observation that some cell lines spontaneously grow as 3D aggregates in culture. Three decades ago, Langlois and colleagues [1979] developed a new human cell line, DU4475, derived from a cutaneous metastatic nodule. The cells grew as a suspension culture displaying clusters and free floating aggregates with serpentine–like outgrowths. Whilst DU4475 aggregates retain features of epithelial tissue, such as duct-like structures and the presence of desmosomal tight junctions, the cells were highly homogeneous and did not express α-lactalbumin or estrogen hormone receptor activity, both of which are often accepted as a prerequisite for any bona fide human mammary cell line.

Perhaps the most notable example of a true acinus-like spheroid cell line is the colon carcinoma cell line LIM1863. First characterised by Whitehead et al. [1987], the LIM1863 cell line is unique in that the epithelial cells grow as floating clumps of cells that are morphologically and functionally organised. The organoids consist of columnar luminal cells, complete with a well-defined brush border, and goblet cells all arranged around a central lumen into which mucus is secreted. The lumen also contains dead cells or cellular debris indicating that a process of differentiation, maturation and cell shedding has occurred. However, it must be noted that no endocrine cells have been found in these cultures and so the organoids are not totally representative of the intestinal crypts.

Although these models provide invaluable insights into 3D modelling, they are both derived from a tumour and so cannot be used to study the processes involved in the growth and differentiation of normal glandular epithelial cells. Moreover, whilst DU4475 and LIM 1863 cell lines form spontaneous aggregates, very few do and so in order to study and compare different cell lines intervention is required to generate 3D structures.

Non-adherent Cultures

Traditionally non-adherent culture methods have been used to discriminate between tumour and non-tumour cells as only tumour cells can initiate the growth of non-adherent cultures. However, recent research has used these methods to culture stem cells [Dontu et al., 2003; Dontu et al., 2004; Reynolds and Weiss, 1996], see later section, and this has been considered early evidence for the stem cell basis of certain tumours [Foster et al., 2002; Hudson, 2004; Lam and Reiter, 2006]. The simplest method of inducing spheroid formation in cells that do not spontaneously aggregate or are difficult to aggregate is by liquid overlay. This method involves either coating the plastic tissue culture surfaces with a thin layer of agar or poly HEMA (polyhydroxyethyl methacrylate), or by using ultra-low attachment plastic ware, which prevents cell adhesion by inhibiting electro-static attraction and preventing the deposition of a matrix. Although the underlying molecular basis is not well understood, it encourages many cell types, in particular tumour derived lines, to undergo spontaneous homotypic aggregation. The inability of the cells to adhere to the tissue culture flask encourages them to migrate towards each other and form small aggregates; these aggregates then increase in size as a result of cell division.

The colon cancer cell line SNU 61, when plated in suspension on poly HEMA coated plates, have formed 3D structures within 72 hours of culture. The variably sized aggregates formed with an irregular outer surface with a small proportion of cells floating as single cells. These aggregates rarely formed glands with distinct lumens and the luminal surface of the glands showed no brush border, moreover, the cells exhibited reversely directed microvilli growing on their basal side as opposed to their luminal side [Kim et al., 1998]. Similarly, when the prostate tumour cell lines; PC-3, LNCaP, DU145 and RWPE-2, were cultured on agar only a small fraction (0.09-2.6 %) survived and formed spherical colonies [Bello et al., 1997; Mitrofanova et al., 2003; Tang et al., 2007; Webber et al., 2001]. Unsurprisingly, since this is a tumourigenic assay, the non-tumour cell line RWPE-1, did not grow when plated on agar [Bello et al., 1997; Webber et al., 2001].

Rotary Cultures

Another approach to prevent adherence of cells to the culture vessel is to continuously stir the cell suspension in specially designed spinner flasks. This method, developed by Sutherland [1970], has been used most widely for the culturing of large numbers of multicellular tumour spheroids. The spheroids formed tend to be composed of an outer shell of viable proliferating cells surrounding an inner core of dead or dying cells, characteristics often associated with solid tumours. This method can expose the cells to high fluid turbulence and shear forces that can damage fragile animal cells affecting membrane integrity, growth and differentiation [Goodwin et al., 1997; Goodwin et al., 1993]. A more sophisticated method was therefore designed by NASA engineers and called the rotary cell culture system (RCCS).

The RCCS works by simulating microgravity through the mechanism of perpetual fluid rotation. The tissue culture flask is rotated on its axis thus providing end over end mixing of the cell suspension. Fluid and shear forces are kept to minimum by ensuring that the tissue culture flask is filled completely with media with aeration provided by a semi-permeable membrane thereby eliminating bubbles that cause hydrodynamic forces. As a result, the spheroids produced are larger in size and significantly more differentiated morphologically and phenotypically compared to those produced in spinner flasks [Kim, 2005]. One of the main advantages of RCCS is the ability to co-culture multiple cell types to form complex heterologous 3D *in vitro* structures. This has been demonstrated with the prostate cancer cell line, LNCaP, which forms 3D solid organoids that over time lose the ability to produce PSA, a marker of terminal differentiation, even in the presence of androgens. However, if these 3D organoids are cultured in the presence of prostate stromal cells, androgen responsiveness and PSA production are maintained [Wang et al., 2005; Zhau et al., 1997]. DU145 prostate cancer cells have also been grown in the RCCS; the resulting solid aggregates were found to be less aggressive, less proliferative and more differentiated than other cell cultivation methods. The authors suggested that these characterises are representative of *in vivo* tissue growth, as the prostate is a slow growing organ [Clejan et al., 2001].

RCCS is a relatively easy method of generating large numbers of solid spheroids that are representative of *in vivo* tumours. It is also amenable to co-culture with multiple cell types providing researches with a valuable method of investigating tumour-stromal interactions. However, the use of this method for generating glandular organs from normal tissues needs further exploration.

Gel-Matrix Systems

The mammary and prostate glands are both secretory tissues with an extensively branched system of epithelial ducts terminating in multiple alveoli or acinar structures. These acini are composed of a bilayer of epithelial cells surrounding a lumen. Terminally differentiated columnar luminal cells line the lumen and secrete mammary (Lactin) or prostate specific (PSA) proteins into the lumen. These cells sit on top of undifferentiated cuboidal basal cells in the prostate and differentiated myoepithelial cells in the mammary gland. In the prostate, the basal and luminal cells can be identified on the basis of their cytokeratin profile; basal cells expressed cytokeratins 1, 5, 10 and 15, whereas luminal cells express cytokeratins 8 and 18. Cells that are morphologically and phenotypically intermediate between basal and luminal cells have also been identified within the normal prostatic epithelium, demonstrating that basal and luminal cells are probably linked in a hierarchical pathway [Hudson et al., 2000; Lang et al., 2001a; Robinson et al., 1998; van Leenders et al., 2000]. A basement membrane, rich in laminin and type I and IV collagen, then separates these epithelial cells from the surrounding stroma. The stroma is composed of mesenchymally-derived adipocytes, fibroblasts, neural cells and a proteinaceous network of stromal matrix proteins, primarily collagens I and III, glycosaminoglycans, and non-collagenous glycoproteins [Cunha et al., 2004; Marker et al., 2003; Weaver and Bissell, 1999]. In this chapter, the term extracellular matrix (ECM) will denote all the acellular, proteinaceous components that exist within the tissue, including the basement membrane.

The ECM is essential in morphogenesis and organogenesis. Epithelial cells bind to the ECM proteins via cell adhesion molecules such as integrin and cadherin receptors, and this enables migration into the mesenchyme and subsequent formation of acini and ducts [Watt, 2002; Weaver et al., 1997]. In particular, laminin I is involved in epithelial cell adhesion and polarisation, while collagen IV forms a scaffold for the attachment of other ECM components [Bello-DeOcampo et al., 2001; Gudjonsson et al., 2002; Weaver et al., 2002]. *In vivo*, glandular epithelial cells exist in an environment that is governed by cell-cell [Hagios et al., 1998; Liu et al., 1997; O'Brien et al., 2002; Zahir and Weaver, 2004], cell-matrix [Barcellos-Hoff et al., 1989; Freeman et al., 1994; Hagios et al., 1998; Streuli and Bissell, 1990; Zahir and Weaver, 2004], and cell-soluble stromal derived factors, such as growth factors and hormones [Darcy et al., 2000], all of which influence cell growth and differentiation. Animal and human studies have shown that stroma and extracellular matrix are essential for functional and morphological differentiation of prostate epithelia [Bayne et al., 1998; Condon, 2005; Cunha et al., 2002; Cunha et al., 2003; Fong et al., 1991; Lang et al., 2000; Liu et al., 1997], and the epithelia of other organs including colon [Kim et al., 1998], endometrium [Yang et al., 2002], urothelium [Howlett et al., 1986], skin [Schoop et al., 1999; Watt, 1991] and breast [Shekhar et al., 2001; Weaver et al., 1996; Weaver et al., 2002]. Provision of such interactions is therefore important to ensuring normal epithelial morphogenesis and function in a 3D *in vitro* model.

To date, the gel-matrix system remains the most popular method of culturing epithelial cells in 3D. ECM is provided in the form of either type I collagen gel or Matrigel®, an artificial basement membrane preparation, in concert with specific media, leading to the formation of spheroids. The matrix appears to provide both a physical contribution and a signalling contribution through collagen binding, simulating the characteristics of the ECM. 3D growth in culture using gel-matrix systems has been demonstrated in diverse tissue types, including osteoblasts, condrocytes, hepatocytes, endothelial cells, astrocytes and neuronal progenitor cells, to name just a few [Edelman and Keefer, 2005]. This review will focus predominantly on the contribution that type I collagen and Matrigel® have made to the 3D modelling of glandular differentiation in the breast and prostate.

Type I Collagen Gels

In tissue, type I collagen is a major component of the ECM regulating the mechanical properties of connective tissue. It has therefore been used extensively to provide a scaffold within which 3D spheroids can develop. On the whole, research has focussed on the use of collagen gels to study the developmental progression of neuronal cells [Edelman and Keefer, 2005], and in keratinocyte raft cultures in which keratinocytes are cultured on a lattice including fibroblasts and collagen [Poumay and Coquette, 2007]. Nevertheless, there has also been considerable use of collagen gels in the modelling of glandular organs, with mixed success. One of the first groups, Emerman and colleagues [1977; 1977], showed that cultures of mammary epithelial cells could be induced to secrete multiple casein milk proteins in response to culturing on collagen gels, indicating the importance of the substratum in influencing milk production. This has since been confirmed by other groups [Lee et al., 1984; Parry et al., 1985]. In 1990, Streuli and Bissell [1990] observed that normal mouse mammary cells could be induced to develop into multicellular 3D structures which are able to organise an endogenously synthesised basement membrane rich in laminin I and type IV collagen. Production of the basement membrane also correlated with increased expression of milk proteins, transferrin and β-casein, indicating that the cells had undergone functional differentiation. In contrast, cells cultured on plastic produced low levels of transferrin and β-casein mRNA and milk protein secretion was barely detectable. However, these structures were not reminiscent of the ductal acinar system of the *in vivo* tissue. More recently, Gudjonsson et al. [2002] found that plating luminal and myoepithelial mammary cells within collagen gels resulted in the formation of double-layered acini with a central lumen. The cells were polarised with basal expression of β4-integrin and deposition of a basement membrane, as indicated by collagen IV expression, reminiscent of *in vivo* breast tissue. Interestingly, formation of the breast-like acini correlated with up-regulation of the laminins 1, 5, 10 and 11. In 2002, Weaver and colleagues [2002] showed that non-malignant S-1 mammary epithelial cells formed solid spheroids in collagen gels which were unpolarised with respect to the expression of β-catenin, β4 integrin, α6 integrin and laminin 5. However, the addition of laminin I, a major component of the basement membrane, restored polarity suggesting that it is the interaction between the epithelial cells and basement membrane laminins, but not other ECM molecules such as collagen, that is important in driving the formation of polarised acini.

It has also been possible to generate ductal structures using collagen. However, instead of plating cells on top of collagen that is attached to the cell culture flask, the collagen is suspended within media. The resulting matrix is less restrained and so the cells are able to move more freely within the matrix. As a result, single T47D mammary carcinoma cells were able to proliferate to form a branched structure, similar to the mammary gland, and stain positive for the presence of human milk fat globule [Keely et al., 1995; Wozniak and Keely, 2005].

Similar work has also been conducted with prostate cells. Prostate epithelial cells isolated from the rat have also been shown to form 3D spherical and branching acinus-like structures containing a lumen enclosed in a single layer of cuboidal cells. The acini were polarised with microvilli on the luminal surface and a distinct basal lamina surrounding the acini. In addition, PSA and PAP proteins were often secreted into the lumen. Interestingly, the acini developed from single cells that proliferated to form small aggregates. After 2-3 days, small secretory vacuoles appeared in their cytoplasm and these either enlarged or fused together to form larger vacuoles. A lumen therefore forms by the coalescing of the vacuoles of two or more neighbouring cells [Ma et al., 1997].

An interesting investigation was also performed by Hall et al. [2002] to look at the effect of stromal cells (PS) on the migration and morphogenesis of prostatic epithelial cells (PEC), derived from both tumour (tPS and tPEC) and benign (bPS and bPEC) tissues. Based on the keratinocyte raft culture method [Atula et al., 1997; Hull et al., 1983; Stark et al., 1999], collagen gels were seeded with bPS and tPS, allowed to proliferate and, after 5 days, formed a dense three-dimensional matrix within the gel. The primary epithelial cells were then seeded on top of the collagen. When bPEC and tPEC were grown on tPS they grew as an epithelial monolayer on the surface of the gel, whereas when grown on bPS individual cells invaded into the collagen and, in the case of bPEC, formed acinus-like structures not seen in the absence of stromal cells. Similarly, the normal cell line, PNT1A, only invaded into bPS, however, the tumourigenic PC-3 cell line invaded into both bPS and tPS but no acinus-like structures were formed. The acini formed from bPEC grown on bPS stained more strongly for the basal cytokeratins (1, 5, 10 and 14) then for luminal cytokeratin 18, nevertheless, prostate specific proteins (PSA and PSMA) were expressed. The presence of fibroblasts within collagen also induces contraction and folding of the collagen/stromal cell substratum [Zhu et al., 2001], encompassing the epithelial cells which then form 3D structures within this envelope [Hall et al., 2002].

Collagen matrices have provided important insights into the mechanism of lumen formation and have enabled researchers to generate acini and ductal structures that are reminiscent of *in vivo* glandular tissue. In addition, the ability to co-culture epithelial and fibroblast cells has emphasised the importance of normal stromal/epithelial interactions in the maintenance and formation of differentiated acini and suggests that the generation of an acinus-like structure is dependant on motility and invasion capabilities in addition to cell-matrix and cell-cell adhesion.

Matrigel®

Matrigel® is a reconstituted basement membrane derived from Engelbreth-Holm Swarm (EHS) tumour. It is primarily composed of the matrix components laminin I, collagen IV and entactin. As alluded to previously, it is the laminin I in the Matrigel which mimics the basement membrane and is responsible for driving the morphogenic process of both mammary [Gudjonsson et al., 2002; Weaver et al., 2002] and prostate cells [Bello-DeOcampo et al., 2001].

Matrigel® has long been used in establishing 3D models of the mammary gland *in vitro*. One of the first groups to demonstrate the ability of primary epithelial cells to recapitulate the growth, as well as the structure and functional differentiation of the *in vivo* mammary gland, was Petersen and colleagues [1992]. They showed that normal breast epithelial cells formed multi-cellular spheroid colonies composed of a central lumen surrounded by polarised luminal epithelial cells expressing cytokeratin 18 and 19 at the apical membrane. In addition, two thirds of the acini deposited an endogenous basement membrane that was rich in type IV collagen. These observations are supported by many other groups showing that the normal breast epithelial cell lines, MCF-10A and S1, and primary epithelial cells cultured from both human, mouse and rat mammary tissue, all form polarised acini with one or two layers of cells surrounding a central lumen. In addition, milk proteins, primarily caesins, were secreted vectorially into the luminal space [Barcellos-Hoff et al., 1989; Blatchford et al., 1999; Darcy et al., 2000; Debnath et al., 2002; Debnath et al., 2003; Gudjonsson et al., 2002; Shekhar et al., 2001; Shekhar et al., 2000; Weaver et al., 1997].

A) KSFM **B) KSFM with 2% Serum**

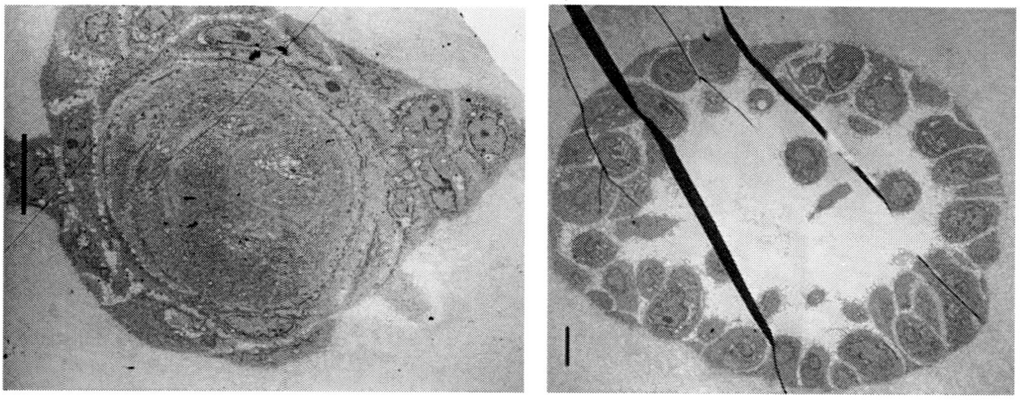

Figure 1. TEM of a whole prostate spheroid grown from primary epithelial cells in Matrigel® with either A) KSFM or B) KSFM supplemented with 2% serum. Bar represents 10 μM.

Recapitulation of *in vivo* tissue architecture has also been possible using normal prostate cell lines and primary epithelial cultures. The RWPE-1 cell line formed acini of polarised epithelium with a distinct lumen, a laminin basement membrane, basal expression of $\alpha6\beta1$ integrin [Bello-DeOcampo et al., 2001; Webber et al., 1997] and luminal expression of PSA [Webber et al., 1997] and cytokeratin 18 [Bello-DeOcampo et al., 2001]. However, the most comprehensive study of prostate epithelial morphogenesis has been conducted by Lang and colleagues. PNT2-C2, Shmac 2 and ShMac 6, all highly differentiated normal epithelial cell lines expressing an intermediate phenotype in monolayer (luminal and basal cytokeratin, CD44 and PSA expression), formed solid spheroids in Matrigel® with no apparent lumen [Lang et al., 2001a; Lang et al., 2006]. Overall, PNT2-C2 showed a similar phenotype in Matrigel® to that in monolayer culture, however, $\beta1$ integrin and CD44 expression was predominantly expressed at the basal membrane [Lang et al., 2001a]. It has also been possible to grow acini from epithelial cells isolated from human prostate tissue. In serum free media, the same media used to generate polarised RWPE-1 acini, solid masses of cells formed which resemble hyperplastic growth (see figure 1A). The centre of the spheroids was composed of tightly packed stratified cells expressing luminal cytokeratin 18. In contrast, the outer cells were more loosely packed and cuboidal in shape and expressed basal cytokeratins (1, 5, 10 and 14). In addition, the prostate specific proteins, PSA and PSMA, markers of differentiation, were expressed throughout the spheroid while the basal cells expressed CD44, E-Cadherin and Desmoglein indicating strong cell-cell adhesion. However, there was no androgen receptor expression. Although these spheroids are phenotypically similar to prostatic tissue, they lack a lumen and so do not resemble the architecture of *in vivo* acini

[Lang et al., 2001b]. In support of this work, very similar solid spheroids have also been generated from both normal [Hudson et al., 2000; Kogan et al., 2006; van Leenders et al., 2000] and immortalised primary epithelial cells [Kogan et al., 2006].

In order to overcome this problem, Lang and colleagues experimented with different culture conditions. Previously, there has been little focus on the culture conditions required to generate 3D structures, and there is, as yet, no consensus between different research groups on what media and additive supplements are optimal. It has been possible to generate 3D structures of the mammary gland in both high [Darcy et al., 2000; Debnath et al., 2002; Debnath et al., 2003; Gudjonsson et al., 2002; Petersen et al., 1992; Weaver et al., 2002; Weaver et al., 1997] and low [Blatchford et al., 1999] calcium media, with [Debnath et al., 2002; Debnath et al., 2003; Gudjonsson et al., 2002] or without [Blatchford et al., 1999; Darcy et al., 2000; Petersen et al., 1992; Weaver et al., 2002; Weaver et al., 1997] the addition of serum. In contrast, most of the experiments using prostate cells have tended to use low calcium keratinocyte serum free media (KSFM). While this has proved ideal for the culture of acinus-like spheroids using RWPE-1 cells [Bello-DeOcampo et al., 2001; Webber et al., 1997], it has not been successful for the culture of primary epithelial cells, inducing the formation of solid spheroids [Kogan et al., 2006; Lang et al., 2001b]. In addition, NbMC- 2 rat epithelial cells cultured in high calcium containing DMEM/HamsF12 supplemented with 5% serum [Freeman et al., 1994] and primary prostate cells cultured in DMEM (Lang unpublished data), all form solid spheroids.

Research has shown that estrogen, androgen (in the form of dihydrotestosterone) and 2% serum are required to induce glandular epithelial differentiation [Blanchere et al., 1998; Collins et al., 1991; Fong et al., 1991; Foster et al., 2002; Harkonen and Makela, 2004; Lipschutz et al., 1997; Shekhar et al., 2000]. The addition of these factors to primary prostate epithelial cells in Matrigel® led to the formation of less dense spheroids that were regular in shape. Moreover, the spheroids architecturally resembled *in vivo* acini because they contained a lumen surrounded by one or two layers of columnar and cuboidal epithelial cells (see figure 1B). Microvilli were present at the luminal edge, a basal lamina was observed, but there were few junctional complexes. Phenotypically, the cells were intermediate and luminal, expressing luminal (cytokeratin 18) and basal (cytokeratins 1, 5, 10 and 14) markers, however, a distinct basal layer was lost [Lang et al., 2001b].

As with the collagen raft culture, Matrigel® is also ideally suited to co-culture of both epithelial and stromal cells. This can be achieved by three different methods. The stromal and epithelial cell can be mixed and plated together as a single cell suspension within the Matrigel® [Shekhar et al., 2001]. Alternatively, the two cultures can be plated separately either by seeding stroma within a collagen gel with epithelial cells plated on top in Matrigel® [Yang et al., 2002] or by plating the stromal cells in a tissue culture insert balanced on top of epithelial cells seeded within Matrigel® [Lang and Maitland, 2004; Lang et al., 2001b] (or vice versa). The latter method is often preferable as the two cell types can be easily isolated from each other for forward analysis, but direct cell-cell contact is lost. The effect of stromal co-culture on 3D glandular structure and function has been clearly demonstrated by Lang and colleagues [2001b]. The addition of stroma to the primary epithelial cells grown in KSFM supplemented with 2% serum, estrogen and dihydrotestosterone, produced spheroids that were phenotypically identical to those grown with serum, however they showed a much higher degree of polarisation. The microvilli, golgi and secretory vesicles were organised towards the lumen whereas the nuclei were located predominately towards the basal membrane. β1-integrin expression was also strongly basal whilst PSA expression was luminal. The addition of stroma doubled the spheroid forming efficiency, suggesting that stroma can recruit more epithelia to form spheroids. In summary, primary epithelial and stromal co-culture, in the presence of serum, estrogen and DHT, produces acinus-like spheroids that show both functional and morphological differentiation. Unfortunately, during the process of differentiation, the distinct basal layer of epithelial cells seen in serum free culture is lost. It has been suggested that the basal layer of the prostate is the proliferative compartment, containing the stem cell population, which can differentiate into secretory epithelium [Long et al., 2005; Wang et al., 2001]. It was hypothesised that the majority of spheroids are derived from progenitor cells and only a few from the stem cell population. Whilst the stem cells would have the capacity to generate the full range to prostate epithelial phenotypes, the spheroids derived from progenitors would only have the capacity to differentiate into intermediate and luminal cells but not to replace the basal population, which would therefore be lost in the absence of stem cells [Lang et al., 2001b].

There is considerable evidence for the ability of stromal cells to direct the growth and differentiation of prostate and mammary epithelial cells via the production of soluble growth factors [Condon and Bosland, 1999; Darcy et al., 2000; Hayward et al., 1997]. The flexibility of Matrigel® culturing techniques has led to the identification of possible candidate factors, which can then be added to the culture media to facilitate the faithful reproduction *in vitro* the architecture and phenotype of tissue. One example is hepatocyte growth factor/Scatter factor (HGF), which is expressed by stromal cells and known to stimulate motility, proliferation and morphogenesis of epithelial cells via signalling through the c.Met receptor on epithelial cells [Niranjan et al., 1995; Rosen et al., 1994]. In the mammary gland, the formation of branching tubules could be induced by the addition of HGF to the culture medium of mammary epithelial cells grown in Matrigel® [Niemann et al., 2000; Niemann et al., 1998]. HGF has also induced branching of lung [Nakamura et al., 1997; Ohmichi et al., 1998], kidney [Khoury et al., 2005; Williams and Clark, 2003; Yu et al., 2003], and epidermal cells [Choi and Fuchs, 1990; Ozbun and Meyers, 1996]. However, similar experiments using normal prostate cell lines have not supported this effect (Lang and Pearson, unpublished data). Another candidate factor, also tested by Neimann and co-workers [2000; 1998], is Neuregulin. This has the opposite effect to HGF and instead induced alveoli-like spheroids by binding through the c-erB2 tyrosine kinase receptors.

An additional aspect that is worth consideration is the importance calcium plays in the development of 3D structures. The formation of acini is often associated with an increase in cell-cell contact, primarily through calcium-dependant junctional proteins [Denker and Nigam, 1998; Garrod et al., 1996; Knudsen and Wheelock, 2005; Matter et al., 2005]. In the prostate, in order to form 3D structures resembling *in vivo* acini, epithelial cells are routinely cultured in KSFM media [Bello-DeOcampo et al., 2001; Webber et al., 1997], which has very low levels of calcium (<0.05 mM). BPH-1 cells cultured in KSFM supplemented with 2% serum formed acini with a lumen. However, if cultured in the presence of calcium (1.8 mM) they formed solid spheroids (see figure 2). Moreover, the incubation of fully grown acini with calcium for 24 hours resulted in the loss of the previously formed lumen (Pearson and

Lang, unpublished observations). In contrast, 3D mammary epithelial spheroids are generally cultured in DMEM or HamsF12 [Darcy et al., 2000; Debnath et al., 2002; Debnath et al., 2003; Gudjonsson et al., 2002; Petersen et al., 1992; Weaver et al., 2002; Weaver et al., 1997], both of which contain much higher levels of calcium (1.8 mM and 0.2 mM respectively) compared to KSFM.

This work has highlighted that the addition of growth factors, hormones, serum and calcium can have dramatic effects on the architecture and phenotype of 3D *in vitro* acini and therefore their resemblance to *in vivo* tissue. However, the effect of these factors has not been fully investigated, moreover, there are many unknown factors in serum and bovine pituitary factor (a supplement added to KSFM) which have not been addressed at this stage.

A) Low Calcium **B) High Calcium**

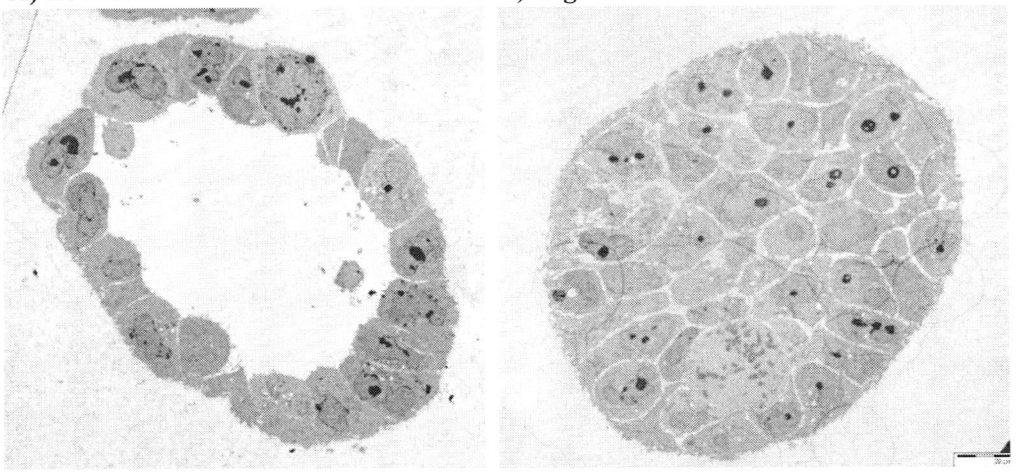

Figure 2. Transmission electron microscopy of BPH-1 epithelial cells in Matrigel® with A) low calcium (<0.05mM) KSFM supplemented with dihydrotestosterone, estrogen and 2% serum and B) high calcium (1.8mM) KSFM supplemented with dihydrotestosterone, estrogen and 2% serum. Bar represents 20 μM.

Synthetic Matrices

Another technique for generating a matrix is the use of synthetic peptide scaffolds. Scaffolds are made from interweaving self assembling peptides such as polylactic acid, polyglycolic acid, hyaluronic acid and alginate, to generate micro-fibres which can encapsulate cells, forming a truly 3D environment resembling the ECM. Cells are able to attach to and migrate along the intertwined scaffold that provides physical support to the developing 3D structures. The biodegradable pre-engineered scaffolds have been used extensively for a variety of different tissues; these include skin [Lutolf and Hubbell, 2005; Lutolf et al., 2003], bone, cartilage, tendon, ligament and nerve [Freed et al., 1994; Freed et al., 1993; Kim, 2005; Wang et al., 2006]. However, the use of these scaffolds to recapitulate glandular structures is still very much in its infancy and one of the few examples has been demonstrated by Semino et al. [2003] using hepatocyte cells. In monolayer, Lig-8 hepatocyte cells express markers of endoderm and hepatocyte lineage but lack the expression of mature hepatocyte markers. When cultured in a 3D peptide scaffold, the Lig-8 cells were able to organise into clusters of spheroids. These spheroids displayed several characteristics of mature hepatocyte behaviour indicating that the peptide scaffold was able to provide a microenvironment that was conducive to enhanced cell differentiation, enabling progenitor cells to form functional hepatocyte-like spheroids.

Whilst artificial matrices have the potential to offer many of the same benefits of Matrigel®, the synthetic peptides form a well-defined ECM, in which the effects of individual matrix molecules can be studied, thus making it possible to clearly define their role in differentiation and morphogenesis. It will be interesting to see this promising area of research develop.

3D Modelling of Tumours

3D Matrigel® models provide an important cell-based assay that offers researchers an invaluable tool to study normal and abnormal differentiation. Compared to the acinus-like structures formed by normal breast epithelial cells in Matrigel®, epithelial cells isolated from breast carcinoma biopsies and tumourigenic cell lines formed solid spheroids with neither correct polarity nor deposition of a basement membrane [Petersen et al., 1992]. This inability of carcinoma cells to form spheroids resembling *in vivo* acini has been supported by observations using the non-malignant S-1 and the tumourigenic T4-2 breast cell lines. S-1 cells formed polarised acini showing expression of β-Catenin, β4 and α6 integrins and laminin 5 at the basal to Matrigel® interface. In contrast, T4-2 cells formed solid acini with no obvious polarisation [Weaver et al., 2002; Weaver et al., 1997]. It was therefore proposed that culturing mammary epithelial cells in Matrigel® could form the basis of a rapid assay for discriminating between normal human breast epithelial cells and their malignant counterpart [Petersen et al., 1992; Weaver et al., 2002; Weaver et al., 1997].

Initially, this observation was also supported by researchers modelling prostate epithelium. The normal RWPE-1 cell line forms polarised acinus-like structures with a distinct lumen. However, the RWPE-1 malignant counterparts (RWPE-2 and WPE1-NB26) and the DU145 carcinoma cell line form amorphous aggregates without organisation or lumen. This led to the conclusion that normal cells can undergo acinar morphogenesis but tumourigenic cells have lost this ability [Bello-DeOcampo et al., 2001; Webber et al., 1997]. However, work by Lang and colleagues disputes this conclusion and suggests that the view that only normal cells are able to form acini is too simplistic. This has arisen from observation of the normal PNT2C2 and metastatic PC-3 cell lines in monolayer and 3D. PNT2C2 cells display a highly differentiated phenotype in monolayer (expression of luminal cytokeratin, PSA and PSMA, both markers of functional differentiation), and form solid spheroids in Matrigel® which have the same phenotype. In contrast, the metastatic PC-3 cells line displays an undifferentiated basal-like phenotype in monolayer (expressing basal cytokeratins and CD44 but no expression of PSA or PSMA), and in Matrigel® forms polarised acini composed of a lumen surrounded by a single layer of cells expressing low levels of basal and luminal cytokeratins but higher levels of PSA and PSMA expression (See figure 3A). It was hypothesised that the PNT2C2, because of its highly differentiated phenotype, could not be induced to undergo further differentiation, unlike the PC-3 cell line. It was therefore proposed that formation of acinus-like structures in Matrigel® is dependant on the differentiation state of a particular cell and not its tumourigenic state [Lang et al., 2001a]. Further experiments have reinforced this theory. In 2006, Lang et al. tested the ability of the non-tumour (ShMac 2 and 6) and tumour (ShMac 5 and P4E6) prostate cell lines to form acinus-like structures. In monolayer, the cell lines showed an intermediate phenotype expressing luminal and basal cytokeratins, CD44 and PSA but not the androgen receptor. Of the four cell lines, ShMac 5 cells displayed the most undifferentiated phenotype. Interestingly the tumourigenic ShMac 5 cell line was also the only cell line to produce acinus-like spheroids in Matrigel®, which, with the addition of stromal cells induced androgen receptor expression and epithelial polarisation (see figure 3B). In contrast, the ShMac 2 and 6 and P4E6 cells all formed multi-layered spheroids without any evidence of a lumen [Lang et al., 2006]. These experiments are the first to directly compare the phenotype and differentiation state of cells in monolayer with that in 3D and, as a result, they have cast doubt on the concept that growth in Matrigel® can distinguish between tumour and non-tumour cells. Such studies, which did not consider the phenotype and state of differentiation of epithelial cells prior to 3D culture, may therefore be making a comparison of the effect Matrigel® induces on differentiated versus undifferentiated cells rather than on the behaviour of tumour and non-tumour cells.

A) PC-3 **B) ShMac 5**

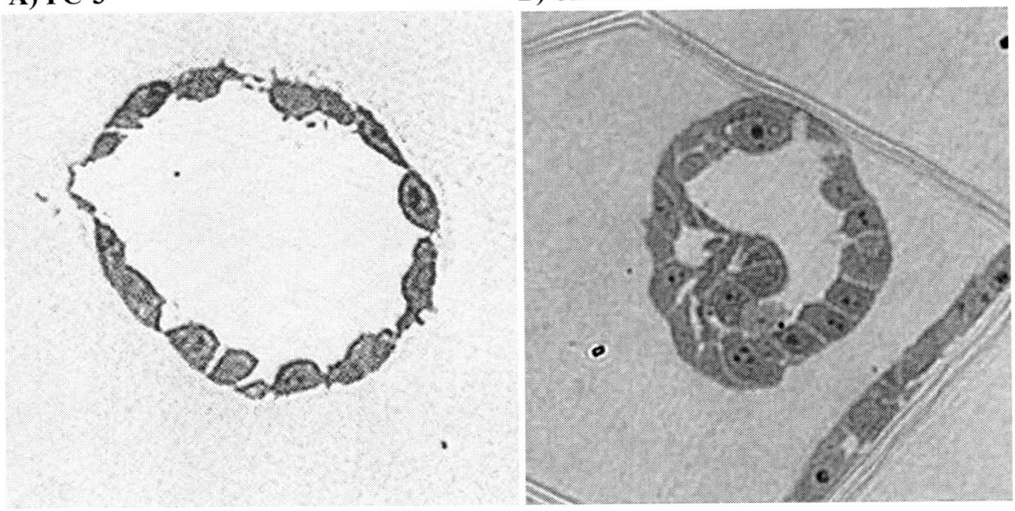

Figure 3. Phase contrast 1 μm sections at X10 objective magnification of A) PC-3 and B) ShMac 5 tumour prostate cell line spheroids grown in Matrigel® and KSFM, supplemented with dihydrotestosterone, estrogen and 2% serum, in the presence of stroma.

USES OF 3D MODELS

Gel matrix 3D modelling of epithelial cells has not only been instrumental in increasing our understanding of the process of differentiation of glandular organs but is also slowly revealing the steps involved in the formation of lumen and ductal structures. As mentioned previously, 3D prostatic acini form as aggregates which develop a lumen after 2-3 days of culture by the fusion of the vacuoles of neighbouring cells [Ma et al., 1997]. This mechanism is similar to that proposed in the assembly of endothelial tubes during the development of a functional circulatory system [Kamei et al., 2006]. In the prostate at least, this theory of lumen formation is unchallenged. However, there is compelling evidence from work using breast and liver cells cultured within Matrigel® that indicates that the lumen is formed by the process of selective apoptosis of internal luminal cells. MDCK kidney cells cultured in Matrigel® rely on a two part model of epithelial morphogenesis. Firstly, polarised lumen-enclosing epithelial cysts resembling prostatic acini develop, followed by the outgrowth of tubules. The cysts develop initially as solid spheroids, consisting of inner luminal cells surrounded by basal cells, which then hollow out by the apoptosis of the luminal cells to generate a lumen. Once this process is complete, morphogens such as HGF induce transient de-differentiation to enable the outer basal cells to protrude into the ECM, forming duct-like structures [Debnath and Brugge, 2005; O'Brien et al., 2002]. A very similar process has also been observed in the formation of mammary acini. MCF-10A spheroids initially develop as solid acini in Matrigel® and only start to become hollow after 6-8 days of culture when the inner subset of poorly polarised cells die off [Debnath and Brugge, 2005; Debnath et al., 2002; Debnath et al., 2003]. Other groups have also observed apoptosis-driven lumen formation in the development of *in vitro* mammary acini [Blatchford et al., 1999; Mailleux et al., 2007; Park et al., 2006; Weaver and Bissell, 1999; Weaver et al., 2002]. Whilst evidence of apoptosis within prostatic acini has not been identified, there is as yet no conclusive evidence to support or refute the apoptotic mechanism of lumen formation in the prostate.

For the first time, collagen and Matrigel® based gel-matrix assays have enabled researchers to generate 3D glandular ductal/acinar structures *in vitro* which closely resemble both the architecture and function of *in vivo* tissue. These model systems have been invaluable in enabling scientists to explore the course of differentiation and tissue formation, in particular the role of the stroma and ECM in directing this process.

3D MODELLING AND THE STEM CELL NICHE

Adult Stem Cells and their Niche

Stem cells are characterised by their ability to self-renew, proliferate indefinitely and to differentiate into multiple cellular progenies or lineages. There are broadly three types of stem cells: embryonic, germinal and adult stem cells. Embryonic stem cells are derived from the first 5 or 6 divisions of the fertilised egg and have the capacity to generate all the cells of the adult organs. Germinal stem cells produce egg and sperm in the adult and are responsible for reproduction. In contrast, adult stem cells are more limited in their potential but are essential in normal tissue renewal to replenish cells as they die, either by natural wear and tear or by injury. Multipotent stem cells are very few in number and generally divide infrequently as most of the cellular renewal is accomplished by tissue-determined transit amplifying cells. Transit-amplifying cells are formed when the stem cell divides asymmetrically; one daughter cell remains a stem cell, in order to continue the process of regeneration, while the other daughter cell will commit to differentiate into a transit amplifying cell. This transit amplifying cell will ultimately differentiate and expand into the full repertoire of cell types found in a particular organ and so start the process of normal tissue renewal. Although transit amplifying cells have low rates of self-renewal and restricted differentiation, they proliferate more rapidly than stem cells and so form a large population of committed progenitors between stem cells and all their terminally differentiated daughter cells [Diaz-Flores et al., 2006].

Given the life-long importance of adult stem cells, it is important that they are protected within the tissues in what has now become known as the stem cell niche. These niches are composed of stem cells and also a diverse gathering of neighbouring differentiated cells, which generate extrinsic factors that control stem cell number, proliferation and fate determination. While these signals have not been fully elucidated, many developmental regulatory signals, including hedgehog, Wnt, bone morphogenic protein (BMP), Notch, fibroblast growth factor (FGF), epidermal growth factor (EGF), stem cell factor (SCF) and transforming growth factor β (TGFβ) have all been shown to play roles in controlling stem cell renewal and differentiation. The niche maintains the stem cell primarily in a quiescent, non-proliferating state by providing signals that inhibit proliferation (such as TGFβ and BMP), and it is only on receipt of stimulatory signals (such as Wnt and Notch) that the stem cells are induced to divide and differentiate [Calabrese et al., 2007; Diaz-Flores et al., 2006; Dontu et al., 2004; Dontu and Wicha, 2005; Fuchs et al., 2004; Kondo et al., 2004; Lee et al., 2006; Li and Neaves, 2006; Patrawala et al., 2005; Reynolds and Weiss, 1996; Robey, 2007; Sun et al., 2006]. Regulating this balance between proliferative and anti-proliferative signals is key to maintaining stem cells and their niche and consequently the ability of the tissue to renew cells and repair damage. It is also important for adult stem cells to be physically anchored within their protective niches; this is achieved by binding to neighbouring differentiated cells and to the ECM. So far adherins junctions, primarily cadherins and catenins, have been identified between stem cells and neighbouring epithelial and stromal cells. while integrin adhesion mediates binding to the ECM and the basal lamina. These adhesion molecules are also important in recruiting stem cells, a process termed 'homing', as once these niches are established they are able to survive, at least transiently, as signalling centres to attract other stem cells [Diaz-Flores et al., 2006; Fuchs et al., 2004].

Recapitulation of this niche in 3D is now at the forefront of scientific research. It will lead to increased understanding of the signals involved in maintaining the stem cell niche as well as the control of pluripotency and self renewal. In addition, it will enable researchers to artificially induce and track the process of differentiation, ideally leading to the reconstruction of 3D glandular structures representing the full repertoire of the different phenotypes present in tissues. There has been a plethora of data showing isolation of stem cells from different tissues including brain [Lee et al., 2006], central nervous system [Bazan et al., 2004; Morshead and van der Kooy, 2004], skin [Bianco and Robey, 2001], corneal epithelium [Papini et al., 2005], embryos [Aberdam, 2004; Chaudhry et al., 2004; Taylor et al., 2006], and bone marrow [Kirouac and Zandstra, 2006; Li and Neaves, 2006]. In addition, different 3D modelling techniques have been used to recapitulate *in vivo* tissue; such as non-adherent agar and Matrigel® for glandular structures [Locke et al., 2005; Tang et al., 2007], non-adherent culture flasks to induce skin formation [Aberdam, 2004], peptide scaffolds to reproduce bone nodules [Bianco and Robey, 2001; Chaudhry et al., 2004; Li et al., 2005], and collagen for reconstruction of the corneal epithelium [Papini et al., 2005]. However, since this field of research is in its infancy, the definition of a stem cell can vary greatly between different research laboratories. Indeed, the method of isolation and characterisation of stem cells differs between different tissues but can also vary for the same tissue between different laboratories. This leads to concerns that cells defined as a stem cell may possess some stem-like properties but are possibly a more differentiated population of early progenitor transit amplifying cells. One of the main challenges facing stem cell research is therefore the ability to isolate and phenotype pure populations of stem cells without inducing cellular differentiation. The definition of a stem cell is that it can self renew and that it is multipotent, therefore both abilities must be proven to ensure that the isolated cells are stem cell-like. The following section will concentrate solely on the identification and isolation of stem cells from the human prostate and mammary glands and their use in 3D modelling.

Breast and Prostate Stem Cell Markers

Using stem cells markers found in the skin and the brain, stem-like cells have been isolated from the mouse mammary gland on the basis of the cell surface markers, $Lin^-/CD29^{hi}/CD24^+$. These cells have the capacity to reconstitute a complete mammary gland *in vivo* from a single cell. When plated into Matrigel®, single cells can also form tissue-like ducts and alveoli. These abilities indicate that this population of cells are multipotent. $Lin^-/CD29^{hi}/CD24^+$cells can also self renew, since the primary outgrowths from these cells can be re-transplanted and form secondary and tertiary outgrowths [Shackleton et al., 2006; Stingl et al., 2006]. Other researchers have used the ability of stem cells to exclude Hoechst dye as a method of isolating breast and prostate stem cells [Bhatt et al., 2003; Kalirai and Clarke, 2006]. The increased dye exclusion can identify a sub-population of cells, termed the side population, after FACS analysis. In the breast, these cells show increased expression of stem cell markers ($p21^{CIP1}$, Msi1) and can regenerate a functional gland *in vivo* and in Matrigel®. However, self renewal has not yet been tested. In the prostate, the cells of the side population remain to be characterised for self renewal and multipotency. However, there has been considerable success in identifying and culturing prostate stem-like cells using the cd133 marker. Research led by A. Collins has shown that a population of prostate stem-like cells can be distinguished from their differentiating progeny by high levels of α2β1 integrin and CD133 expression, previously identified as a haematopoietic stem cell marker. Only 1% of basal epithelial cells are $\alpha2\beta1^{hi}/CD133^+$, and these cells display many of the characteristics of adult stem cells. $\alpha2\beta1^{hi}/CD133^+$ cells are generally quiescent. However, when grown in long term serial culture, they display 4.5-fold greater colony forming efficiency then non-selected basal cells, indicating high *in vitro* self renewal capacity. Moreover, these cells form fully differentiated and functional acini when grafted immediately into the flanks of immuno-compromised mice [Collins et al., 2001; Richardson et al., 2004]. A further marker, Sca-1, has also been used to identify murine stem-like cells in the prostate. These have the ability to regenerate prostate-like tissues *in vivo* but self-renewal was not tested [Xin et al., 2005].

Propagation/Maintenance as Monolayer and Protospheres

A variety of methods exist for maintaining stem cells in culture. In prostate research, stem–like cells ($\alpha2\beta1^{hi}$/$CD133^{+}$) are plated onto collagen coated petri dishes with irradiated mouse fibroblast cells (STO), termed 'feeder cells', enabling the stem cells to make contacts with both the ECM and neighbouring cells. The stem cells are cultured in Keratinocyte Serum Free Media (KSFM), supplemented with EGF, bovine pituitary extract, stem cell factor, cholera toxin, GMCSF and Lif. There is usually a lag time of 5 days to 2 weeks before the initiation of colony growth [Richardson et al., 2004]. In the breast, maintenance of breast stem–like cells has been carried out in suspension. Analogous to primary neural cells [Reynolds and Weiss, 1996; Weiss et al., 1996], they are grown on non-adherent plates in serum free conditions (mammary epithelial growth media, supplemented with B27, EGF and βFGF), leading to the formation of mammospheres after 5-7 days [Dontu et al., 2003]. These mammospheres are composed of solid balls of cells and so are not representative of the acinar/ductal morphology of breast tissue. However, they are enriched in early progenitor/stem cells and are able to differentiate along all three mammary epithelial lineages and to clonally generate complex functional structures in reconstituted 3D Matrigel® culture systems. Cells derived from mammospheres can self renew in culture [Dontu et al., 2003], but as yet the ability to regenerate breast tissues *in vivo* has not been shown. Stem-like/progenitor cells derived from prostatic cell lines can also grow as colonies in agar and free floating 'prostaspheres' in liquid medium [Tokar et al., 2005].

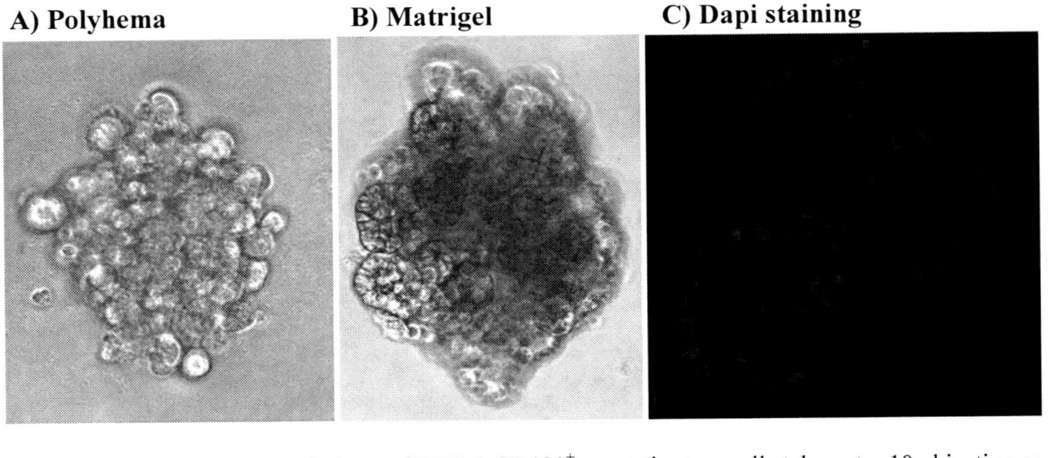

Figure 4. Phase contrast morphology of BPH-1 $CD133^{+}$ prostatic stem cells taken at x 10 objective as A) aggregates on polyhema which are then plated in to B) Matrigel® leading to the formation of large budding structures composed of multiple hollow acini shown by C) nuclear staining by DAPI, x 20 objective.

Growth of Stem Cells in 3D Culture and Reproduction of the Stem Cell Niche

Breast stem-like cells ($Lin^-/CD29^{hi}/CD24^+$) can grow into acinus-like structures in Matrigel® from single cells [Shackleton et al., 2006]. However, this is not the case for prostate stem-like cells ($CD133^+$) which cannot initiate growth of spheroids in Matrigel® from single cells (Lang unpublished data). They are also incapable of initiating colony growth in suspension culture from a single cell. Instead, isolated $CD133^+$ cells must be aggregated on non-adherent tissue culture plates before growth as spheroids in Matrigel® can be achieved (see figure 4). At present, it is difficult to determine why there are differences in the behaviour of stem cells derived from the prostate and the breast. It may indicate differences in the biological activity of the two tissues. Breast tissue has to undergo periods of remodelling during lactation whilst the prostate is slow growing and does not need to deviate from a fixed pattern of tissue growth and renewal. These differences may be reflected in the different abilities of the two stem cell populations. It may also indicate differences in the purity of the stem cell isolations, since in the prostate $CD133^-$ cells can initiate growth of spheroids in Matrigel® from single cells, perhaps the isolated population of $Lin^-/CD29^{hi}/CD24^+$ breast cells still contain progenitor cells. Aggregation of stem cells may also be necessary for the mammosphere cultures to initiate proliferation. In the stem cell niche, direct contact of the stem cell with neighbouring cells and matrix is essential for the correct control of homeostasis. Therefore, aggregation of stem cells with homotypic or heterotypic cells or extracellular matrix would lead to the identification of the correct conditions for reproducing the niche. Aggregated cultures could be used to analyse the ability of cells or matrix to induce proliferation or differentiation in stem cells, such experiments have been elegantly carried out by Dontu et al. [2004] and Liu et al. [2006].

Large aggregated spheroids or mammospheres do, however, lack the correct tissue morphology. Since cell architecture and the correct positioning of cells within a tissue is important for functional differentiation [Ingber, 2005; McBeath et al., 2004], such aggregated cultures may be limited in their ability to reflect true differentiation. Therefore, the culture of stem cells in 3D gels or scaffolds which can reproduce tissue-like architecture remains an important technique to reproduce true tissue differentiation.

Research Discoveries using 3D Modelling of Stem Cells

To determine the role of notch signalling in mammary stem cell fate, Dontu and colleagues [2004] used their 3D mammosphere culture system. Notch signalling was achieved using exogenous ligands or inhibited with antagonists. The activation of Notch promotes self renewal of stem cells as shown by a 10 fold increase in secondary mammosphere formation. Notch also stimulates myoepithelial lineage-specific commitment and proliferation and branching morphogenesis in Matrigel®. In contrast, Notch had no effects on committed cells. Also using the mammosphere culture system, Liu et al. [2006] investigated hedgehog signalling in mammary stem cells. They discovered that the hedgehog pathway was highly active in mammospheres compared to differentiated cells grown as monolayers on collagen. Activation of hedgehog signalling increased mammosphere initiation, suggesting that these pathways promote stem cell renewal.

Both studies show how these models can be applied to increase our understanding of stem cell fate. In our opinion, this will be increased further by combining aggregation techniques with 3D gels and scaffolds to induce tissue architecture. It could also be improved by aggregating cells of known stem cell origin to be sure of the precise nature of the cells present in the aggregate.

Conclusion

Research over the last 5 years has indicated that cell shape and its position within a tissue are important for differentiation and function. Therefore, recapitulation of tissue architecture by 3D cell culture remains the major advantage of this system over 2D cultures. Research in this area is increasing our understanding of how cell:cell contacts and ECM influence tissue structure and epithelial differentiation. This review has discussed a number of basic 3D cell culture models which better mimic the phenotypic and morphological behaviour of cells *in vivo*. However, much more research is required to understand why different cell lines form different structures in 3D i.e. single or multi-layered acini, with or without a lumen, and what the signals are to initiate true acinus-like spheroids. These experiments are essential to properly understand the results that have been generated so far. Further development of *in vitro* 3D culture systems will be invaluable in modelling epithelial differentiation, cancer progression and the testing of therapeutics in a biologically relevant context.

References

Aberdam D. (2004): Derivation of keratinocyte progenitor cells and skin formation from embryonic stem cells. *Int J Dev Biol 48*:203-6.

Atula S., Grenman R., Syrjanen S. (1997): Fibroblasts can modulate the phenotype of malignant epithelial cells in vitro. *Exp Cell Res 235*:180-7.

Barcellos-Hoff M. H., Aggeler J., Ram T. G., Bissell M. J. (1989): Functional differentiation and alveolar morphogenesis of primary mammary cultures on reconstituted basement membrane. *Development 105*:223-35.

Bayne C. W., Ross M., Donnelly F., Chapman K., Buck C., Bollina P., Habib F. K. (1998): Selective interactions between prostate fibroblast and epithelial cells in co-culture maintain the BPH phenotype. *Urol Int 61*:1-7.

Bazan E., Alonso F. J., Redondo C., Lopez-Toledano M. A., Alfaro J. M., Reimers D., Herranz A. S., Paino C. L., Serrano A. B., Cobacho N., Caso E., Lobo M. V. (2004): In vitro and in vivo characterization of neural stem cells. *Histol Histopathol 19*:1261-75.

Bello-DeOcampo D., Kleinman H. K., Deocampo N. D., Webber M. M. (2001): Laminin-1 and alpha6beta1 integrin regulate acinar morphogenesis of normal and malignant human prostate epithelial cells. *Prostate 46*:142-53.

Bello D., Webber M. M., Kleinman H. K., Wartinger D. D., Rhim J. S. (1997): Androgen responsive adult human prostatic epithelial cell lines immortalized by human papillomavirus 18. *Carcinogenesis 18*:1215-23.

Bhatt R. I., Brown M. D., Hart C. A., Gilmore P., Ramani V. A., George N. J., Clarke N. W. (2003): Novel method for the isolation and characterisation of the putative prostatic stem cell. *Cytometry A 54*:89-99.

Bianco P., Robey P. G. (2001): Stem cells in tissue engineering. *Nature 414*:118-21.

Blanchere M., Berthaut I , Portois M. C., Mestayer C., Mowszowicz I. (1998): Hormonal regulation of the androgen receptor expression in human prostatic cells in culture. *J Steroid Biochem Mol Biol 66*:319-26.

Blatchford D. R., Quarrie L. H., Tonner E., McCarthy C., Flint D. J., Wilde C. J. (1999): Influence of microenvironment on mammary epithelial cell survival in primary culture. *J Cell Physiol 181*:304-11.

Calabrese C., Poppleton H., Kocak M., Hogg T. L., Fuller C., Hamner B., Oh E. Y., Gaber M. W., Finklestein D., Allen M., Frank A., Bayazitov I. T., Zakharenko S. S., Gajjar A., Davidoff A., Gilbertson R. J. (2007): A perivascular niche for brain tumor stem cells. *Cancer Cell 11*:69-82.

Chaudhry G. R., Yao D., Smith A., Hussain A. (2004): Osteogenic Cells Derived From Embryonic Stem Cells Produced Bone Nodules in Three-Dimensional Scaffolds. *J Biomed Biotechnol* 2004:203-210.

Choi Y., Fuchs E. (1990): TGF-beta and retinoic acid: regulators of growth and modifiers of differentiation in human epidermal cells. *Cell Regul 1*:791-809.

Clejan S., O'Connor K., Rosensweig N. (2001): Tri-dimensional prostate cell cultures in simulated microgravity and induced changes in lipid second messengers and signal transduction. *J Cell Mol Med 5*:60-73.

Collins A. T., Habib F. K., Maitland N. J., Neal D. E. (2001): Identification and isolation of human prostate epithelial stem cells based on alpha(2)beta(1)-integrin expression. *J Cell Sci 114*:3865-72.

Collins J. E., Legan P. K., Kenny T. P., MacGarvie J., Holton J. L., Garrod D. R. (1991): Cloning and sequence analysis of desmosomal glycoproteins 2 and 3 (desmocollins): cadherin-like desmosomal adhesion molecules with heterogeneous cytoplasmic domains. *J Cell Biol 113*:381-91.

Condon M. S. (2005): The role of the stromal microenvironment in prostate cancer. *Semin Cancer Biol 15*:132-7.

Condon M. S., Bosland M. C. (1999): The role of stromal cells in prostate cancer development and progression. In Vivo 13:61-5.

Cunha G. R., Hayward S. W., Wang Y. Z. (2002): Role of stroma in carcinogenesis of the prostate. *Differentiation 70*:473-85.

Cunha G. R., Hayward S. W., Wang Y. Z., Ricke W. A. (2003): Role of the stromal microenvironment in carcinogenesis of the prostate. *Int J Cancer 107*:1-10.

Cunha G. R., Ricke W., Thomson A., Marker P. C., Risbridger G., Hayward S. W., Wang Y. Z., Donjacour A. A., Kurita T. (2004): Hormonal, cellular, and molecular regulation of normal and neoplastic prostatic development. *J Steroid Biochem Mol Biol 92*:221-36.

Darcy K. M., Zangani D., Shea-Eaton W., Shoemaker S. F., Lee P. P., Mead L. H., Mudipalli A., Megan R., Ip M. M. (2000): Mammary fibroblasts stimulate growth, alveolar morphogenesis, and functional differentiation of normal rat mammary epithelial cells. *In Vitro Cell Dev Biol Anim 36*:578-92.

Debnath J., Brugge J. S. (2005): Modelling glandular epithelial cancers in three-dimensional cultures. *Nat Rev Cancer 5*:675-88.

Debnath J., Mills K. R., Collins N. L., Reginato M. J., Muthuswamy S. K., Brugge J. S. (2002): The role of apoptosis in creating and maintaining luminal space within normal and oncogene-expressing mammary acini. *Cell 111*:29-40.

Debnath J., Muthuswamy S. K., Brugge J. S. (2003): Morphogenesis and oncogenesis of MCF-10A mammary epithelial acini grown in three-dimensional basement membrane cultures. *Methods 30*:256-68.

Denker B. M., Nigam S. K. (1998): Molecular structure and assembly of the tight junction. *Am J Physiol 274*:F1-9.

Diaz-Flores L., Jr., Madrid J. F., Gutierrez R., Varela H., Valladares F., Alvarez-Arguelles H., Diaz-Flores L. (2006): Adult stem and transit-amplifying cell location. *Histol Histopathol 21*:995-1027.

Dontu G., Abdallah W. M., Foley J. M., Jackson K. W., Clarke M. F., Kawamura M. J., Wicha M. S. (2003): In vitro propagation and transcriptional profiling of human mammary stem/progenitor cells. *Genes Dev 17*:1253-70.

Dontu G., Jackson K. W., McNicholas E., Kawamura M. J., Abdallah W. M., Wicha M. S. (2004): Role of Notch signaling in cell-fate determination of human mammary stem/progenitor cells. *Breast Cancer Res 6*:R605-15.

Dontu G., Wicha M. S. (2005): Survival of mammary stem cells in suspension culture: implications for stem cell biology and neoplasia. *J Mammary Gland Biol Neoplasia 10*:75-86.

Edelman D. B., Keefer E. W. (2005): A cultural renaissance: in vitro cell biology embraces three-dimensional context. *Exp Neurol 192*:1-6.

Emerman J. T., Enami J., Pitelka D. R., Nandi S. (1977): Hormonal effects on intracellular and secreted casein in cultures of mouse mammary epithelial cells on floating collagen membranes. *Proc Natl Acad Sci U S A 74*:4466-70.

Emerman J. T., Pitelka D. R. (1977): Maintenance and induction of morphological differentiation in dissociated mammary epithelium on floating collagen membranes. *In Vitro 13*:316-28.

Fong C. J., Sherwood E. R., Sutkowski D. M., Abu-Jawdeh G. M., Yokoo H., Bauer K. D., Kozlowski J. M., Lee C. (1991): Reconstituted basement membrane promotes morphological and functional differentiation of primary human prostatic epithelial cells. *Prostate 19*:221-35.

Foster C. S., Dodson A., Karavana V., Smith P. H., Ke Y. (2002): Prostatic stem cells. *J Pathol 197*:551-65.

Freed L. E., Grande D. A., Lingbin Z., Emmanual J., Marquis J. C., Langer R. (1994): Joint resurfacing using allograft chondrocytes and synthetic biodegradable polymer scaffolds. *J Biomed Mater Res 28*:891-9.

Freed L. E., Marquis J. C., Nohria A., Emmanual J., Mikos A. G., Langer R. (1993): Neocartilage formation in vitro and in vivo using cells cultured on synthetic biodegradable polymers. *J Biomed Mater Res 27*:11-23.

Freeman M. R., Bagli D. J., Lamb C. C., Guthrie P. D., Uchida T., Slavin R. E., Chung L. W. (1994): Culture of a prostatic cell line in basement membrane gels results in an enhancement of malignant properties and constitutive alterations in gene expression. *J Cell Physiol 158*:325-36.

Fuchs E., Tumbar T., Guasch G. (2004): Socializing with the neighbors: stem cells and their niche. *Cell 116*:769-78.

Garlick J. A. (2007): Engineering skin to study human disease--tissue models for cancer biology and wound repair. *Adv Biochem Eng Biotechnol 103*:207-39.

Garrod D., Chidgey M., North A. (1996): Desmosomes: differentiation, development, dynamics and disease. *Curr Opin Cell Biol 8*:670-8.

Goodwin T. J., Prewett T. L., Spaulding G. F., Becker J. L. (1997): Three-dimensional culture of a mixed mullerian tumor of the ovary: expression of in vivo characteristics. *In Vitro Cell Dev Biol Anim 33*:366-74.

Goodwin T. J., Prewett T. L., Wolf D. A., Spaulding G. F. (1993): Reduced shear stress: a major component in the ability of mammalian tissues to form three-dimensional assemblies in simulated microgravity. *J Cell Biochem 51*:301-11.

Gudjonsson T., Ronnov-Jessen L., Villadsen R., Rank F., Bissell M. J., Petersen O. W. (2002): Normal and tumor-derived myoepithelial cells differ in their ability to interact with luminal breast epithelial cells for polarity and basement membrane deposition. *J Cell Sci 115*:39-50.

Hagios C., Lochter A., Bissell M. J. (1998): Tissue architecture: the ultimate regulator of epithelial function? *Philos Trans R Soc Lond B Biol Sci 353*:857-70.

Hall J. A., Maitland N. J., Stower M., Lang S. H. (2002): Primary prostate stromal cells modulate the morphology and migration of primary prostate epithelial cells in type 1 collagen gels. *Cancer Res 62*:58-62.

Harkonen P. L., Makela S. I. (2004): Role of estrogens in development of prostate cancer. *J Steroid Biochem Mol Biol 92*:297-305.

Hayward S. W., Rosen M. A., Cunha G. R. (1997): Stromal-epithelial interactions in the normal and neoplastic prostate. *Br J Urol 79 Suppl 2*:18-26.

Howlett A. R., Hodges G. M., Rowlatt C. (1986): Epithelial-stromal interactions in the adult bladder: urothelial growth, differentiation, and maturation on culture facsimiles of bladder stroma. *Dev Biol 118*:403-15.

Hudson D. L. (2004): Epithelial stem cells in human prostate growth and disease. *Prostate Cancer Prostatic Dis 7*:188-94.

Hudson D. L., O'Hare M., Watt F. M., Masters J. R. (2000): Proliferative heterogeneity in the human prostate: evidence for epithelial stem cells. *Lab Invest 80*:1243-50.

Hull B. E., Sher S. E., Rosen S., Church D., Bell E. (1983): Structural integration of skin equivalents grafted to Lewis and Sprague-Dawley rats. *J Invest Dermatol 81*:429-36.

Ingber D. E. (2005): Mechanical control of tissue growth: function follows form. *Proc Natl Acad Sci U S A 102*:11571-2.

Kalirai H., Clarke R. B. (2006): Human breast epithelial stem cells and their regulation. *J Pathol 208*:7-16.

Kamei M., Saunders W. B., Bayless K. J., Dye L., Davis G. E., Weinstein B. M. (2006): Endothelial tubes assemble from intracellular vacuoles in vivo. *Nature 442*:453-6.

Keely P. J., Fong A. M., Zutter M. M., Santoro S. A. (1995): Alteration of collagen-dependent adhesion, motility, and morphogenesis by the expression of antisense alpha 2 integrin mRNA in mammary cells. *J Cell Sci 108 (Pt 2)*:595-607.

Khoury H., Naujokas M. A., Zuo D., Sangwan V., Frigault M. M., Petkiewicz S., Dankort D. L., Muller W. J., Park M. (2005): HGF converts ErbB2/Neu epithelial morphogenesis to cell invasion. *Mol Biol Cell 16*:550-61.

Kim H. S., Lee B. L., Bae S. I., Kim Y. I., Park J. G., Kleinman H. K., Kim W. H. (1998): Differentiation of a colon cancer cell line on a reconstituted basement membrane in vitro. *Int J Exp Pathol 79*:443-51.

Kim J. B. (2005): Three-dimensional tissue culture models in cancer biology. *Semin Cancer Biol 15*:365-77.

Kirouac D. C., Zandstra P. W. (2006): Understanding cellular networks to improve hematopoietic stem cell expansion cultures. *Curr Opin Biotechnol 17*:538-47.

Knudsen K. A., Wheelock M. J. (2005): Cadherins and the mammary gland. *J Cell Biochem 95*:488-96.

Kogan I., Goldfinger N., Milyavsky M., Cohen M., Shats I., Dobler G., Klocker H., Wasylyk B., Voller M., Aalders T., Schalken J. A., Oren M., Rotter V. (2006): hTERT-immortalized prostate epithelial and stromal-derived cells: an authentic in vitro model for differentiation and carcinogenesis. *Cancer Res 66*:3531-40.

Kondo T., Setoguchi T., Taga T. (2004): Persistence of a small subpopulation of cancer stem-like cells in the C6 glioma cell line. *Proc Natl Acad Sci U S A 101*:781-6.

Lam J. S., Reiter R. E. (2006): Stem cells in prostate and prostate cancer development. *Urol Oncol 24*:131-40.

Lang S. H., Maitland N. J. (2004): Coculture of prostate cancer cells. *Methods Mol Med 88*:295-302.

Lang S. H., Sharrard R. M., Stark M., Villette J. M., Maitland N. J. (2001a): Prostate epithelial cell lines form spheroids with evidence of glandular differentiation in three-dimensional Matrigel cultures. *Br J Cancer 85*:590-9.

Lang S. H., Smith J., Hyde C., Macintosh C., Stower M., Maitland N. J. (2006): Differentiation of prostate epithelial cell cultures by matrigel/ stromal cell glandular reconstruction. *In Vitro Cell Dev Biol Anim 42*:273-80.

Lang S. H., Stark M., Collins A., Paul A. B., Stower M. J., Maitland N. J. (2001b): Experimental Prostate Epithelial Morphogenesis in Response to Stroma and Three-Dimensional Matrigel Culture. *Cell Growth Differ 12*:631-640.

Lang S. H., Stower M., Maitland N. J. (2000): In vitro modelling of epithelial and stromal interactions in non-malignant and malignant prostates. *Br J Cancer 82*:990-7.

Langlois A. J., Holder W. D., Jr., Iglehart J. D., Nelson-Rees W. A., Wells S. A., Jr., Bolognesi D. P. (1979): Morphological and biochemical properties of a new human breast cancer cell line. *Cancer Res 39*:2604-13.

Lee E. Y., Parry G., Bissell M. J. (1984): Modulation of secreted proteins of mouse mammary epithelial cells by the collagenous substrata. *J Cell Biol 98*:146-55.

Lee J., Kotliarova S., Kotliarov Y., Li A., Su Q., Donin N. M., Pastorino S., Purow B. W., Christopher N., Zhang W., Park J. K., Fine H. A. (2006): Tumor stem cells derived from glioblastomas cultured in bFGF and EGF more closely mirror the phenotype and genotype of primary tumors than do serum-cultured cell lines. *Cancer Cell 9*:391-403.

Li L., Neaves W. B. (2006): Normal stem cells and cancer stem cells: the niche matters. *Cancer Res 66*:4553-7.

Li W. J., Tuli R., Huang X., Laquerriere P., Tuan R. S. (2005): Multilineage differentiation of human mesenchymal stem cells in a three-dimensional nanofibrous scaffold. *Biomaterials 26*:5158-66.

Lipschutz J. H., Foster B. A., Cunha G. R. (1997): Differentiation of rat neonatal ventral prostates grown in a serum-free organ culture system. *Prostate 32*:35-42.

Liu A. Y., True L. D., LaTray L., Nelson P. S., Ellis W. J., Vessella R. L., Lange P. H., Hood L., van den Engh G. (1997): Cell-cell interaction in prostate gene regulation and cytodifferentiation. *Proc Natl Acad Sci U S A 94*:10705-10.

Liu S., Dontu G., Mantle I. D., Patel S., Ahn N. S., Jackson K. W., Suri P., Wicha M. S. (2006): Hedgehog signaling and Bmi-1 regulate self-renewal of normal and malignant human mammary stem cells. *Cancer Res 66*:6063-71.

Locke M., Heywood M., Fawell S., Mackenzie I. C. (2005): Retention of intrinsic stem cell hierarchies in carcinoma-derived cell lines. *Cancer Res 65*:8944-50.

Long R. M., Morrissey C., Fitzpatrick J. M., Watson R. W. (2005): Prostate epithelial cell differentiation and its relevance to the understanding of prostate cancer therapies. *Clin Sci (Lond) 108*:1-11.

Lutolf M. P., Hubbell J. A. (2005): Synthetic biomaterials as instructive extracellular microenvironments for morphogenesis in tissue engineering. *Nat Biotechnol 23*:47-55.

Lutolf M. P., Lauer-Fields J. L., Schmoekel H. G., Metters A. T., Weber F. E., Fields G. B., Hubbell J. A. (2003): Synthetic matrix metalloproteinase-sensitive hydrogels for the conduction of tissue regeneration: engineering cell-invasion characteristics. *Proc Natl Acad Sci U S A 100*:5413-8.

Ma Y. L., Fujiyama C., Masaki Z., Sugihara H. (1997): Reconstruction of prostatic acinus-like structure from ventral and dorsolateral prostatic epithelial cells of the rat in three-dimensional collagen gel matrix culture. *J Urol 157*:1025-31.

Mailleux A. A., Overholtzer M., Schmelzle T., Bouillet P., Strasser A., Brugge J. S. (2007): BIM regulates apoptosis during mammary ductal morphogenesis, and its absence reveals alternative cell death mechanisms. *Dev Cell 12*:221-34.

Marker P. C., Donjacour A. A., Dahiya R., Cunha G. R. (2003): Hormonal, cellular, and molecular control of prostatic development. *Dev Biol 253*:165-74.

Martin I., Miot S., Barbero A., Jakob M., Wendt D. (2007): Osteochondral tissue engineering. *J Biomech 40*:750-65.

Matter K., Aijaz S., Tsapara A., Balda M. S. (2005): Mammalian tight junctions in the regulation of epithelial differentiation and proliferation. *Curr Opin Cell Biol 17*:453-8.

McBeath R., Pirone D. M., Nelson C. M., Bhadriraju K., Chen C. S. (2004): Cell shape, cytoskeletal tension, and RhoA regulate stem cell lineage commitment. *Dev Cell 6*:483-95.

Mitrofanova E., Hagan C., Qi J., Seregina T., Link C., Jr. (2003): Sodium iodide symporter/radioactive iodine system has more efficient antitumor effect in three-dimensional spheroids. *Anticancer Res 23*:2397-404.

Morshead C. M., van der Kooy D. (2004): Disguising adult neural stem cells. *Curr Opin Neurobiol 14*:125-31.

Mueller-Klieser W. (1997): Three-dimensional cell cultures: from molecular mechanisms to clinical applications. *Am J Physiol 273*:C1109-23.

Nakamura T., Matsumoto K., Kiritoshi A., Tano Y., Nakamura T. (1997): Induction of hepatocyte growth factor in fibroblasts by tumor-derived factors affects invasive growth of tumor cells: in vitro analysis of tumor-stromal interactions. *Cancer Res 57*:3305-13.

Niemann C., Brinkmann V., Birchmeier W. (2000): Hepatocyte growth factor and neuregulin in mammary gland cell morphogenesis. *Adv Exp Med Biol 480*:9-18.

Niemann C., Brinkmann V., Spitzer E., Hartmann G., Sachs M., Naundorf H., Birchmeier W. (1998): Reconstitution of mammary gland development in vitro: requirement of c-met and c-erbB2 signaling for branching and alveolar morphogenesis. *J Cell Biol 143*:533-45.

Niranjan B., Buluwela L., Yant J., Perusinghe N., Atherton A., Phippard D., Dale T., Gusterson B., Kamalati T. (1995): HGF/SF: a potent cytokine for mammary growth, morphogenesis and development. *Development 121*:2897-908.

O'Brien L. E., Zegers M. M., Mostov K. E. (2002): Opinion: Building epithelial architecture: insights from three-dimensional culture models. *Nat Rev Mol Cell Biol 3*:531-7.

Ohmichi H., Koshimizu U., Matsumoto K., Nakamura T. (1998): Hepatocyte growth factor (HGF) acts as a mesenchyme-derived morphogenic factor during fetal lung development. *Development 125*:1315-24.

Ozbun M. A., Meyers C. (1996): Transforming growth factor beta1 induces differentiation in human papillomavirus-positive keratinocytes. *J Virol 70*:5437-46.

Papini S., Rosellini A., Nardi M., Giannarini C., Revoltella R. P. (2005): Selective growth and expansion of human corneal epithelial basal stem cells in a three-dimensional-organ culture. *Differentiation 73*:61-8.

Park C. C., Zhang H., Pallavicini M., Gray J. W., Baehner F., Park C. J., Bissell M. J. (2006): Beta1 integrin inhibitory antibody induces apoptosis of breast cancer cells, inhibits growth, and distinguishes malignant from normal phenotype in three dimensional cultures and in vivo. *Cancer Res 66*:1526-35.

Parry G., Lee E. Y., Farson D., Koval M., Bissell M. J. (1985): Collagenous substrata regulate the nature and distribution of glycosaminoglycans produced by differentiated cultures of mouse mammary epithelial cells. *Exp Cell Res 156*:487-99.

Patrawala L., Calhoun T., Schneider-Broussard R., Zhou J., Claypool K., Tang D. G. (2005): Side population is enriched in tumorigenic, stem-like cancer cells, whereas ABCG2+ and ABCG2- cancer cells are similarly tumorigenic. *Cancer Res 65*:6207-19.

Peehl D. M. (2004): Are primary cultures realistic models of prostate cancer? *J Cell Biochem 91*:185-95.

Peehl D. M. (2005): Primary cell cultures as models of prostate cancer development. *Endocr Relat Cancer 12*:19-47.

Petersen O. W., Ronnov-Jessen L., Howlett A. R., Bissell M. J. (1992): Interaction with basement membrane serves to rapidly distinguish growth and differentiation pattern of normal and malignant human breast epithelial cells. *Proc Natl Acad Sci U S A 89*:9064-8.

Poumay Y., Coquette A. (2007): Modelling the human epidermis in vitro: tools for basic and applied research. *Arch Dermatol Res 298*:361-9.

Reynolds B. A., Weiss S. (1996): Clonal and population analyses demonstrate that an EGF-responsive mammalian embryonic CNS precursor is a stem cell. *Dev Biol 175*:1-13.

Richardson G. D., Robson C. N., Lang S. H., Neal D. E., Maitland N. J., Collins A. T. (2004): CD133, a novel marker for human prostatic epithelial stem cells. *J Cell Sci 117*:3539-45.

Robey R. (2007): The environment matters. *Nat Rev Cancer 7*:160.

Robinson E. J., Neal D. E., Collins A. T. (1998): Basal cells are progenitors of luminal cells in primary cultures of differentiating human prostatic epithelium. *Prostate 37*:149-60.

Rosen E. M., Nigam S. K., Goldberg I. D. (1994): Scatter factor and the c-met receptor: a paradigm for mesenchymal/epithelial interaction. *J Cell Biol 127*:1783-7.

Schoop V. M., Mirancea N., Fusenig N. E. (1999): Epidermal organization and differentiation of HaCaT keratinocytes in organotypic coculture with human dermal fibroblasts. *J Invest Dermatol 112*:343-53.

Semino C. E., Merok J. R., Crane G. G., Panagiotakos G., Zhang S. (2003): Functional differentiation of hepatocyte-like spheroid structures from putative liver progenitor cells in three-dimensional peptide scaffolds. *Differentiation 71*:262-70.

Shackleton M., Vaillant F., Simpson K. J., Stingl J., Smyth G. K., Asselin-Labat M. L., Wu L., Lindeman G. J., Visvader J. E. (2006): Generation of a functional mammary gland from a single stem cell. *Nature 439*:84-8.

Shekhar M. P., Werdell J., Santner S. J., Pauley R. J., Tait L. (2001): Breast stroma plays a dominant regulatory role in breast epithelial growth and differentiation: implications for tumor development and progression. *Cancer Res 61*:1320-6.

Shekhar M. P., Werdell J., Tait L. (2000): Interaction with endothelial cells is a prerequisite for branching ductal-alveolar morphogenesis and hyperplasia of preneoplastic human breast epithelial cells: regulation by estrogen. *Cancer Res 60*:439-49.

Smalley K. S., Lioni M., Herlyn M. (2006): Life isn't flat: taking cancer biology to the next dimension. *In Vitro Cell Dev Biol Anim 42*:242-7.

Stark H. J., Baur M., Breitkreutz D., Mirancea N., Fusenig N. E. (1999): Organotypic keratinocyte cocultures in defined medium with regular epidermal morphogenesis and differentiation. *J Invest Dermatol 112*:681-91.

Stingl J., Eirew P., Ricketson I., Shackleton M., Vaillant F., Choi D., Li H. I., Eaves C. J. (2006): Purification and unique properties of mammary epithelial stem cells. *Nature 439*:993-7.

Streuli C. H., Bissell M. J. (1990): Expression of extracellular matrix components is regulated by substratum. *J Cell Biol 110*:1405-15.

Sun L., Hui A. M., Su Q., Vortmeyer A., Kotliarov Y., Pastorino S., Passaniti A., Menon J., Walling J., Bailey R., Rosenblum M., Mikkelsen T., Fine H. A. (2006): Neuronal and glioma-derived stem cell factor induces angiogenesis within the brain. *Cancer Cell 9*:287-300.

Sutherland R. M., Inch W. R., McCredie J. A., Kruuv J. (1970): A multi-component radiation survival curve using an in vitro tumour model. *Int J Radiat Biol Relat Stud Phys Chem Med 18*:491-5.

Tang D. G., Patrawala L., Calhoun T., Bhatia B., Choy G., Schneider-Broussard R., Jeter C. (2007): Prostate cancer stem/progenitor cells: identification, characterization, and implications. *Mol Carcinog 46*:1-14.

Taylor R. A., Cowin P. A., Cunha G. R., Pera M., Trounson A. O., Pedersen J., Risbridger G. P. (2006): Formation of human prostate tissue from embryonic stem cells. *Nat Methods 3*:179-81.

Tokar E. J., Ancrile B. B., Cunha G. R., Webber M. M. (2005): Stem/progenitor and intermediate cell types and the origin of human prostate cancer. *Differentiation 73*:463-73.

van Leenders G., Dijkman H., Hulsbergen-van de Kaa C., Ruiter D., Schalken J. (2000): Demonstration of intermediate cells during human prostate epithelial differentiation in situ and in vitro using triple-staining confocal scanning microscopy. *Lab Invest 80*:1251-8.

Wang R., Xu J., Juliette L., Castilleja A., Love J., Sung S. Y., Zhau H. E., Goodwin T. J., Chung L. W. (2005): Three-dimensional co-culture models to study prostate cancer growth, progression, and metastasis to bone. *Semin Cancer Biol 15*:353-64.

Wang Y., Hayward S., Cao M., Thayer K., Cunha G. (2001): Cell differentiation lineage in the prostate. *Differentiation 68*:270-9.

Wang Y., Kim H. J., Vunjak-Novakovic G., Kaplan D. L. (2006): Stem cell-based tissue engineering with silk biomaterials. *Biomaterials 27*:6064-82.

Watt F. M. (1991): Cell culture models of differentiation. *Faseb J 5*:287-94.

Watt F. M. (2002): Role of integrins in regulating epidermal adhesion, growth and differentiation. *Embo J 21*:3919-26.

Weaver V. M., Bissell M. J. (1999): Functional culture models to study mechanisms governing apoptosis in normal and malignant mammary epithelial cells. *J Mammary Gland Biol Neoplasia 4*:193-201.

Weaver V. M., Fischer A. H., Peterson O. W., Bissell M. J. (1996): The importance of the microenvironment in breast cancer progression: recapitulation of mammary tumorigenesis using a unique human mammary epithelial cell model and a three-dimensional culture assay. *Biochem Cell Biol 74*:833-51.

Weaver V. M., Lelievre S., Lakins J. N., Chrenek M. A., Jones J. C., Giancotti F., Werb Z., Bissell M. J. (2002): beta4 integrin-dependent formation of polarized three-dimensional architecture confers resistance to apoptosis in normal and malignant mammary epithelium. *Cancer Cell 2*:205-16.

Weaver V. M., Petersen O. W., Wang F., Larabell C. A., Briand P., Damsky C., Bissell M. J. (1997): Reversion of the malignant phenotype of human breast cells in three-dimensional culture and in vivo by integrin blocking antibodies. *J Cell Biol 137*:231-45.

Webber M. M., Bello D., Kleinman H. K., Hoffman M. P. (1997): Acinar differentiation by non-malignant immortalized human prostatic epithelial cells and its loss by malignant cells. *Carcinogenesis 18*:1225-31.

Webber M. M., Quader S. T., Kleinman H. K., Bello-DeOcampo D., Storto P. D., Bice G., DeMendonca-Calaca W., Williams D. E. (2001): Human cell lines as an in vitro/in vivo model for prostate carcinogenesis and progression. *Prostate 47*:1-13.

Weiss S., Reynolds B. A., Vescovi A. L., Morshead C., Craig C. G., van der Kooy D. (1996): Is there a neural stem cell in the mammalian forebrain? *Trends Neurosci 19*:387-93.

Whitehead R. H., Jones J. K., Gabriel A., Lukies R. E. (1987): A new colon carcinoma cell line (LIM1863) that grows as organoids with spontaneous differentiation into crypt-like structures in vitro. *Cancer Res 47*:2683-9.

Williams M. J., Clark P. (2003): Microscopic analysis of the cellular events during scatter factor/hepatocyte growth factor-induced epithelial tubulogenesis. *J Anat 203*:483-503.

Wozniak M. A., Keely P. J. (2005): Use of three-dimensional collagen gels to study mechanotransduction in T47D breast epithelial cells. *Biol Proced Online 7*:144-61.

Xin L., Lawson D. A., Witte O. N. (2005): The Sca-1 cell surface marker enriches for a prostate-regenerating cell subpopulation that can initiate prostate tumorigenesis. *Proc Natl Acad Sci U S A 102*:6942-7.

Yang H., Han S., Kim H., Choi Y. M., Hwang K. J., Kwo H. C., Kim S. K., Cho D. J. (2002): Expression of integrins, cyclooxygenases and matrix metalloproteinases in three-dimensional human endometrial cell culture system. *Exp Mol Med 34*:75-82.

Yu W., O'Brien L. E., Wang F., Bourne H., Mostov K. E., Zegers M. M. (2003): Hepatocyte growth factor switches orientation of polarity and mode of movement during morphogenesis of multicellular epithelial structures. *Mol Biol Cell 14*:748-63.

Zahir N., Weaver V. M. (2004): Death in the third dimension: apoptosis regulation and tissue architecture. *Curr Opin Genet Dev 14*:71-80.

Zhau H. E., Goodwin T. J., Chang S. M., Baker T. L., Chung L. W. (1997): Establishment of a three-dimensional human prostate organoid coculture under microgravity-simulated conditions: evaluation of androgen-induced growth and PSA expression. *In Vitro Cell Dev Biol Anim 33*:375-80.

Zhu Y. K., Umino T., Liu X. D., Wang H. J., Romberger D. J., Spurzem J. R., Rennard S. I. (2001): Contraction of fibroblast-containing collagen gels: initial collagen concentration regulates the degree of contraction and cell survival. *In Vitro Cell Dev Biol Anim 37*:10-6.

In: New Cell Differentiation Research Topics
Editor: Hitoshi Saitama, pp. 213-227

ISBN: 978-1-60021-937-5

Chapter VI

REGULATION OF MAST CELL DEVELOPMENT AND APOPTOSIS BY TH1 AND TH2 CYTOKINES

Zhi-Qing Hu*, Wei-Hua Zhao and Tadakatsu Shimamura
Department of Microbiology and Immunology, Showa University School of Medicine, Tokyo, Japan.

ABSTRACT

SCF and c-*kit* signaling is essential for both human and murine mast cell development, while IL-3 is required for murine mast cell hyperplasia that occurs in response to various stimuli. Relatively stable numbers of mast cells within connective and mucosal tissues are maintained by the constitutive expression of SCF in its membrane-bound and soluble forms by stromal cells (such as fibroblasts, epithelial cells, endothelial cells, bronchial smooth muscle cells) and by the expression of c-*kit* on mast cells at all stages of differentiation. Besides SCF and IL-3, the Th2 cytokines IL-4, IL-9, IL-10 and IL-13, also play critical roles in mast cell development by synergistically promoting the actions of SCF and IL-3 on mast cell proliferation and differentiation. However, the Th2 cytokines alone are unable to support neither the proliferation nor survival of mast cells. In contrast, the Th1 cytokine IFN-γ suppresses mast cell development from various murine and human sources. Interestingly, the above generally-accepted roles of the Th1 and Th2 cytokines on mast cell development are currently being challenged by controversial findings showing that mast cell development is up-regulated by IFN-γ and down-regulated by IL-4, IL-10 and IL-13. This review summarizes the recent progress in our understanding of how the Th1 and Th2 cytokines regulate mast cell development either in a positive or negative manner, with a particular focus on IL-4-induced mast cell apoptosis.

INTRODUCTION

Mast cell-committed precursor cells are derived from $CD34^{+}$/c-kit^{+} multipotent hematopoietic progenitor cells and very small numbers of them are present in bone marrow, blood, spleen and other lymphoid organs. Such precursor cells are unable to be identified as mast cells by their morphology or histochemistry because they lack high-affinity IgE receptor and have few secretory granules. The precursor cells migrate into virtually all tissues throughout the body, where they undergo final maturation in response to local microenvironmental factors [1-4].

By using the conditioned medium derived from mitogen-stimulated lymphocytes, apparently pure populations of mast cells can be developed *in vitro* from murine hematopoietic cells [5-9]. The cytokine responsible for this proliferation is interleukin (IL-) 3 [10-13]. Research into the mechanism of mast cell deficiency in Sl/Sl^{d} and W/W^{v} mice [14-18] has led to the finding of another crucial mast cell growth factor: stem cell factor (SCF) and its receptor, c-kit [19-25]. Withdrawal of SCF from both human and murine mast cell cultures, and withdrawal of IL-3 from the latter, can result in mast cell apoptosis [26,27]. However, in IL-3-knockout mice, the number of tissue mast cells appears unaltered [28], and under *in vitro* conditions, mast cells can develop from hematopoietic cells of Sl/Sl^{d} and W/W^{v} mice in the presence of IL-3 [29]. It is now clear that SCF and c-*kit* signaling is indispensable for both human and murine mast cell development, while IL-3 is required for murine mast cell hyperplasia that occurs in response to various stimuli. Relatively stable numbers of mast cells within connective and mucosal tissues are maintained by the constitutive expression of SCF in its membrane-bound and soluble forms by stromal cells (such as fibroblasts, epithelial cells, endothelial cells, bronchial smooth muscle cells) and by the expression of c-*kit* on mast cells at all stages of differentiation.

Since the discovery of type 1 T helper and type 2 T helper cells in mice and humans [30-34], there has been much research interest in the influence of Th1 and Th2 cytokines on mast cell development and survival. Th1 and Th2 cells are characterized by the patterns of specific cytokines they produce following activation. Th1 cells produce IFN-γ and tumor necrosis factor (TNF) β, whereas Th2 cells produce IL-4, IL-5, IL-6, IL-9, IL-10 and IL-13. Th1 cells are involved in cell-mediated inflammatory reactions. Th2 cytokines can stimulate B-cell proliferation and antibody production, and are therefore considered necessary for the full maturation of the humoral immune system. Particularly, IL-4 is well known for its role in class switching of IgE responses. Evidence suggests that Th2 cytokines play a triggering role in the activation and/or recruitment of IgE antibody-producing B cells, mast cells and eosinophils involved in allergic inflammation. The characteristic cytokines of Th1 and Th2 are mutually inhibitory for the differentiation and effector functions of the reciprocal Th1 and Th2 phenotypes. IFN-γ inhibits the differentiation and functions of Th2 cells, and can lead to a dominant Th1 response. IL-12 produced by professional antigen presenting cells, such as activated macrophages and dendritic cells, stimulates the differentiation of Th1 cells. Conversely, IL-4 stimulates the development of Th2 cells and promotes the production of

* Correspondence concerning this article should be addressed to: Dr. Zhi-Qing Hu, Department of Microbiology and Immunology, Showa University School of Medicine, 1-5-8 Hatanodai, Shinagawa-ku, Tokyo 142-8555,

Th2 cytokines including IL-4 itself. Futhermore, IL-4, IL-10 and IL-13 inhibit Th1 cell proliferation and oppose the effects of IFN-γ on macrophages.

Th1 and Th2 cytokines exert divergent influences on mast cell development. The Th2 cytokines IL-4, IL-9, IL-10 and IL-13, play a critical role in mast cell development by synergistically promoting the actions of SCF and IL-3 on mast cell proliferation and differentiation, although the Th2 cytokines on their own are unable to support neither the proliferation nor survival of mast cells. Accordingly, IL-4 [35-38], IL-9 [39], IL-10 [38,40] and IL-13 [41] are also called mast cell growth factors. The appearance of Th2 cytokines at sites of Th2-mediated immunity amplifies the proliferation and differentiation of mast cell progenitors. Under normal conditions, there are few mast cells in gastrointestinal, genitourinary and respiratory mucosal epithelial surfaces, but their numbers increase markedly in mucosal inflammation. For example, intestinal helminthic infections elicit a Th2-polarized adaptive immune response and give rise to a profound intraepithelial mastocytosis [42]. In contrast, the Th1 cytokine IFN-γ suppresses murine [43-45] and human [46-48] mast cell development and activation [49-52]. Interestingly, the above generally-accepted roles of the Th1 and Th2 cytokines on mast cell development are currently being challenged by controversial findings showing that mast cell development is up-regulated by IFN-γ and down-regulated by IL-4, IL-10 and IL-13. This review summarizes the recent progress in our understanding of how the Th1 and Th2 cytokines regulate mast cell development either in a positive or negative manner, with a particular focus on IL-4-induced mast cell apoptosis.

Before addressing these topics, we will begin by describing properties of *in vitro* systems that are typically used for mast cell development.

CHARACTERISTICS OF *IN VITRO* MAST DEVELOPMENT SYSTEMS

The regulation of mast cell development *in vivo* is particularly complex. Culture conditions required for maturation and the correct function of mast cells include a cytokine milieu and a tissue-specific microenvironment consisting of various cellular types, such as T cells, fibroblasts, endothelial cells, macrophages and even mast cells themselves. As summarized in Table 1, mast cell development involves not only mast cell growth factors but also many mediators released in the inflammatory response, such as PGE, IFN-γ, IL-6, IL-12, TNF-α and TGF-β, as confirmed in a series of experiments in our laboratory [53-60] and others [61-81].

Various sources, such as bone marrow cells, splenic cells, peritoneal cells, peripheral/cord blood cells and purified cells with specific markers such as $CD34^+$/c-kit^+, can be used for mast cell development *in vitro*. The systems to develop mast cells can be roughly divided into either a liquid culture version or a semi-solid culture version using methylcellulose. Different culture systems, cell sources, mouse strains and even initial cell numbers are critical and may result in different and even contradictory conclusions. The effect of a cytokine on undifferentiated mast cells may be considerably different from its

Japan. Tel: 81-3-3784-8131; Fax: 81-3-3784-3069; Email: zqhu@med.showa-u.ac.jp.

Table 1. Cytokines regulating mast cell development

Cytokine	Effect*	Brief description and references
GM-SCF	D	Promotes macrophage proliferation and indirectly inhibits mast cell development [61]
IFN-γ	U	Promotes mast cell development from mouse splenic cells [57,59] and promotes human mast cell survival [62]
	D	Inhibits mast cell development [43-48] and activation [49-51] and induces mast cell apoptosis [52]
IL-1α	U	Triggers PGE_2 production and then indirectly promotes mast cell development in cooperation with membrane-bound SCF expressed by fibroblasts [63]
IL-3	U	Crucial growth factor for murine mast cells [10-13]
IL-4	U	Confirmed mast cell growth factor [35-38]
	D	Inhibits mast cell development [48,57,64-67] and induces mast cell apoptosis [60,64,68-71]
IL-5	U	Cofactor for proliferation of human mast cells [4,46]
IL-6	U	Triggers mast cell development from mouse splenic cells [58] and promotes human mast cell development as cofactor of SCF [4,46,61]
	D	Reduces SCF-dependent human mast cell development from $CD34^+$ cord blood cells [72]
IL-9	U	Confirmed mast cell growth factor [39]
IL-10	U	Confirmed mast cell growth factor [38,40]
	D	Induces mast cell apoptosis [68,71,73]
IL-12	U	Promotes murine mast cell development in an IFN-γ-dependent [57,59] and -independent way [74]
IL-13	U	Confirmed mast cell growth factor [41]
	D	Induces mast cell apoptosis [60]
IL-15	U	Prevents mouse mast cell apoptosis [75]
NGF	U	Promotes IL-3-dependent murine mast cell development [76] and promotes human mast cell survival synergistically with SCF [77]
SCF	U	Crucial growth factor for human and murine mast cells [19-25]
TGF-β	D	Suppresses mast cells proliferation [60] and induces mast cell apoptosis [78]
TNF-α	U	Triggers mast cell development from mouse splenic cells [58] and bone marrow cells [79]
TPO	U	Promotes SCF-dependent human mast cell development [80]
TRAIL	D	Induces human mast cell apoptosis [81]

* D: down-regulation; U: up-regulation.

effect on mature mast cells. Furthermore, human mast cells and murine mast cells may respond differently to a given cytokine. The effects of a factor on mast cell development can also be divided into either direct or indirect action, and the final result may depend on the balance between them. Generally, the direct effect of a stimulant can be ascertained from data derived from semi-solid culture systems seeded with limiting numbers of pure cell populations. Conversely, data need to be analyzed carefully when derived from liquid cultures seeded with relatively large numbers of cells of mixed populations, as results are typically confounded by diverse, indirect effects caused by the stimulant on mast cell development.

Most research has focused on examining the direct effect of a cytokine on mast cells or their precursors, while the indirect effects of microenvironmental factors and cytokines have been largely neglected. In addition, indirect effects are commonly misinterpreted as direct effects. A thorough understanding of the characteristics of each *in vitro* system for mast cell development is essential to clarify such confusing and even contradictory results.

Up- and Down-Regulation of Mast Cell Development by the T_H1 Cytokine IFN-γ

It is well known that the Th1 cytokine IFN-γ suppresses murine [43-45] and human [46-48] mast cell development and activation [49-51]. Recently, it has been reported that IFN-γ induced apoptosis in developing mast cells in a STAT-1-dependent manner, when IFN-γ was added into the culture of murine bone marrow cells in the presence of IL-3 and SCF [52]. In contrast to the above reports, we found that mast cell development from mouse splenic cell cultures was directly proportional to IFN-γ levels in the supernatants recovered on days 2 and 4. In the cultures without exogenous IL-3 and SCF, rIFN-γ induced mast cell development in a dose-dependent manner. Time course studies showed the importance of adding IFN-γ into the cultures during the early phase of culture (i.e. on days 0 and 2 of a 12-day culture). When endogenous IFN-γ (present at the early phase of culture) was neutralized by an anti-IFN-γ antibody, all stimulants, including LPS [53,54], rIFN-γ [57], rIL-10 [58], PCE_2 [56] and dibutyryl cAMP [56], failed to induce mast cell development from splenic cells. The above results reveal that IFN-γ is crucial for the survival and/or differentiation of splenic mast cell precursors, although IFN-γ itself is not a mast cell growth factor [57]. This conclusion was further supported by our investigation into the mechanisms of strain-dependent development of mast cells from mouse splenocytes [59].

In a culture of pure human mast cells generated from cultures of umbilical cord blood mononuclear cells, in the presence of SCF and IL-6, Yanagida *et al.* showed that mast cells underwent apoptosis after withdrawal of SCF and IL-6. The addition of IFN-γ into the cultures suppressed apoptosis and prolonged cell survival in a dose-dependent manner. This survival-promoting effect of IFN-γ was blocked by neutralizing antibodies against both IFN-γ and IFN-γ receptor. The mast cells were further confirmed to bear IFN-γ receptors on their surface [43,62]. These findings suggest that IFN-γ may directly act on mast cells via specific receptors in certain aspects of inflammatory reactions.

The exact mechanisms for the discrepancy in reports related to the effects of IFN-γ on mast cell development are not clear. Different culture systems, cell sources and the balance between direct and indirect effects of IFN-γ on mast cell precursors may be responsible for the contradictory conclusions.

T_H2 Cytokine IL-4 Functions as both a Mast Cell Growth Factor and an Apoptosis-Inducing Factor

IL-4 is a multifunctional Th2 cell-associated cytokine that, through STAT6-dependent signaling, plays a critical role in the regulation of immune responses [82-84]. The role of IL-4 in allergic inflammation is not limited to its capacity to induce the production of allergen-specific IgE antibodies via the class-switch effect. IL-4 is also a well-known mast cell growth factor and is responsible, at least in part, for mast cell hyperplasia. IL-4 and IL-13 exhibit a functional overlap that can be explained by the sharing of a common receptor component and STAT6-dependent signaling [85]. Since initial reports [35,36], the ability of IL-4 to promote both proliferation and differentiation of either murine or human mast cells has been well documented and reviewed [86-89].

We unexpectedly found that the number of mast cells that developed from mouse spleen cells was inversely proportional to IL-4 levels in cultures. Addition of rIL-4 dose-dependently suppressed mast cell development induced by rIFN-γ, rIL-10, LPS, PCE and dibutyryl cAMP. The suppressive action of IL-4 was completely diminished by the addition of neutralizing anti-IL-4 antibody [57]. To eliminate the possibility that IL-4 specifically inhibits mast cell development from spleen cells, we tested the effect of IL-4 on mast cell development from peritoneal and bone marrow cells. In the presence of IL-3 and SCF, large numbers of mast cells developed from such cells during 8 to 12 days of culture, while the addition of IL-4 into the cultures dose- and time-dependently suppressed mast cell development. These results indicate that suppression of mast cell development is a general property of IL-4 and is not specific to mast cell development from spleen cells [60]. Reports from other laboratories also support our conclusion; IL-4 inhibits SCF-dependent mast cell development from various sources of progenitors especially in the early phase of culture [48,64-68].

One possible mechanism is that IL-4 directly acts on mast cells and their precursors at various stages of their differentiation and results in suppression of their development. It has been reported that IL-4 decreased c-*kit* expression of CD34+ progenitors but had no effect on mature mast cell c-*kit* expression [48,65,67].

Human rIL-4 has been reported to induce apoptosis of cord blood-derived mast cells [64,69]. In murine systems, IL-4 alone, or in combination with IL-10, has been reported to directly elicit apoptosis of developing mast cells from bone marrow cells [68,70,71]. Importantly, however, the cells used in these studies were from mixed populations, and the influence of indirect effects of IL-4 was not considered.

To evaluate if IL-4 directly induces apoptosis of mast cells or their precursors, we separated c-*kit*$^+$ cells (including mast cell precursors) from spleen cells and seeded them in a semi-solid culture to observe mast cell development. In the presence of IL-3 and SCF, mast cell development from the c-*kit*$^+$ cells was significantly promoted by IL-4 [57]. IL-4 also enhanced the growth of purified mast cells and a cell line representing immature mast cells (IC-2 cells) in liquid culture [60]. Thus, in agreement with most previous reports, IL-4, acting as a synergistic growth factor with IL-3 and SCF, is able to promote the growth of mast cells and their precursors.

To evaluate whether the suppressive effect of IL-4 was an indirect process, different densities of peritoneal cells were cultured in semi-solid medium with IL-3 and SCF, in the presence of IL-4. IL-4 promoted the development of mast cells from peritoneal cells seeded at a density of 10^4 cells per ml, per culture, but completely suppressed mast cell development when the peritoneal cells were incubated at 5 x 10^4 cells per ml, per culture. These findings suggest that this IL-4-induced suppression of mast cell development may be dependent on the presence of other peritoneal cells of non-mast cell lineage. To evaluate this possibility further, we removed macrophages from the cultured populations. Mast cell development in the presence of rIL-3 and rSCF was not affected, but the suppressive activity of rIL-4 was greatly reduced after the removal of macrophages. Replacing the macrophages completely restored the ability of the peritoneal cells to exhibit IL-4-induced suppression of mast cell development. When peritoneal cells from macrophage-deficient *op/op* mice were stimulated with rIL-3, rSCF and rIL-4, mast cell development was not suppressed by IL-4, but was greatly promoted. These results demonstrate that IL-4-induced suppression of mast cell development is an indirect process and requires the presence of macrophages [60].

Next we tested whether IL-4 can stimulate macrophages to produce a soluble factor that can in turn suppress mast cell development. Purified mast cells or mast cell lines were cultured in supernatants collected from the cultures of peritoneal cells. Compared to supernatants from cultures maintained with two cytokines (IL-3 and SCF), supernatants from cultures with three cytokines (IL-3, IL-4 and SCF) showed suppressive activity on IC-2 cells, MC/9 cells and purified mast cells. A direct action of IL-4 was not responsible for this suppressive activity, since neutralizing antibody against IL-4 did not abolish their suppressive activity. The supernatants from IL-4-stimulated peritoneal cells of macrophage-deficient *op/op* mice did not show any suppressive activity on IC-2 cells, MC/9 cells and purified mast cells [60].

The IL-4-induced suppressive factor in the supernatants was further confirmed to be an apoptosis-inducing factor since mast cells underwent apoptosis when exposed to such supernatants despite the presence of high concentrations of mast cell growth factors including IL-3, IL-4 and SCF. This effect of IL-4 is dependent on STAT6 signaling because IL-4 and IL-13 did not suppress mast cell development from the spleen and peritoneal cells of STAT6$^{-/-}$ mice. Moreover, supernatants recovered from cultures of IL-4- and IL-13-stimulated peritoneal cells of STAT6$^{-/-}$ mice did not exhibit apoptosis-inducing activity [60].

The immune system relies on cell death to help maintain homeostasis and avoid disease. While apoptosis induced by various apoptotic factors plays a central role in shaping the repertoire of mature cells of the immune system, little is known about apoptosis-inducing factors that down-regulate mast cells. Our studies have characterized a novel IL-4-dependent mechanism that can contribute to mast cell homeostasis and/or to changes in mast cell populations during Th2-associated responses. The fact that IL-4 induces the production of this factor from macrophages offers new insight into why IL-4 can promote mast cell growth under certain circumstances but can also promote mast cell apoptosis and suppress mast cell development in other settings. The precise molecular identification of this mast cell apoptosis-inducing factor may be of considerable interest.

Contradictory Actions of the Th2 Cytokine IL-10 on Mast Cell Development

As one of the Th2 cytokines, IL-10 was first recognized as a cytokine-synthesis inhibitory factor capable of inhibiting T cells, monocytes and macrophages and blocking various inflammatory responses [90]. On mast cells, however, IL-10 is confirmed to be a mast cell growth factor [38,40,86-89]. In contrast, IL-10 was recently reported to be able to inhibit expression and function of c-*kit* and the high affinity receptor for IgE [67,91]. Combined stimulation with IL-4 and IL-10 induced apoptosis of differentiated mast cells [68,71]. The same group also reported that IL-10 alone could induce mast cell apoptosis [73]. In the system, mixed populations of mast cells and macrophages were developed when bone marrow cells were cultured with IL-3 plus SCF. The addition of IL-10 greatly decreased the population expansion due to the induction of apoptosis [73]. In the system we use to develop mast cells from splenic cells, however, IL-10 dose-dependently promoted mast cell development in the presence of endogenous IL-3, but IL-4 suppressed mast cell development [57]. Therefore, the mechanism of mast cell apoptosis induced by IL-10 [73] is clearly different from the apoptosis induced by IL-4 [60]. Further research is needed to clarify the above inconsistencies and the mechanism of the down-regulation of mast cell development by IL-10.

Conclusion

Mast cell-deficient W/W^v mice show a significantly increased mortality when subjected to acute bacterial peritonitis or gram-negative pneumonia because they lack the capacity to efficiently recruit neutrophils to the infected tissue. The neutrophil recruitment and survival rates are restored after a reconstitution of W/W^v mice by mast cell transplantation [92,93]. Mast cells are potential immunoregulatory cells which contribute to the initiation, development, expression, and regulation of acquired immune responses by producing multiple cytokines [94-98]. Mast cell hyperplasia is a characteristic feature of various pathologies such as allergic diseases, chronic inflammatory processes, fibrotic disorders, wound healing and even neoplastic tissue transformation [99]. IFN-γ, IL-4, IL-10 and IL-13 directly or indirectly orchestrate the recruitment, activation and apoptosis of mast cells via both up- and down-regulation. Therefore, Thl/Th2 cytokines and the Thl/Th2 balance are extremely important and may determine whether mast cell-mediated immune responses are appropriate or lead to severe immunopathologies.

REFERENCES

[1] Kitamura, Y. Matsuda, H. & Hatanaka, K. (1979). Clonal nature of mast-cell clusters formed in *W/W^v* mice after bone marrow transplantation. *Nature, 281*, 154-155.

[2] Kirshenbaum, A. S. Kessler, S. W. Goff, J. P. & Metcalfe, D. D. (1991). Demonstration of the origin of human mast cells from CD34$^+$ bone marrow progenitor cells. *J. Immunol., 146*, 1410-1415.

[3] Rodewald, H. R. Dessing, M. Dvorak, A. M. & Galli, S. J. (1996). Identification of a committed precursor for the mast cell lineage. *Science, 271*, 818-822.

[4] Kirshenbaum, A. S. Goff, J. P. Semere, T. Foster, B. Scott, L. M. & Metcalfe, D. D. (1999). Demonstration that mast cells arise from a progenitor cell population that is CD34$^+$, c-kit$^+$, and expresses aminopeptidase N (CD13). *Blood, 94*, 2333-2342.

[5] Nabel, G. Galli, S. J. Dvorak, A. M. Dvorak, H. F. & Cantor, H. (1981). Inducer T lymphocytes synthesize a factor that stimulates proliferation of cloned mast cells. *Nature, 291*, 332-334.

[6] Nagao, K. Yokoro, K. & Aaronson, S. A. (1981). Continuous lines of basophil/mast cells derived from normal mouse bone marrow. *Science, 212*, 333-335.

[7] Razin, E. Cordon-Cardo, C. & Good, R. A. (1981). Growth of a pure population of mouse mast cells *in vitro* with conditioned medium derived from concanavalin A-stimulated splenocytes. *Proc. Natl. Acad. Sci. USA, 78*, 2559-2561.

[8] Schrader, J. W. (1981). In *in vitro* production and cloning of the P cell, a bone marrow-derived null cell that expresses H-2 and Ia-antigens, has mast cell-like granules, and is regulated by a factor released by activated T cells. *J. Immunol., 126*, 452-458.

[9] Tertian, G. Yung, Y. P. Guy-Grand, D. & Moore, M. A. (1981). Long term *in vitro* culture of murine mast cells. I. Description of a growth factor dependent culture technique. *J. Immunol., 127*, 788-794.

[10] Ihle, J. N. Keller, J. Oroszlan, S. Henderson, L. E. Copeland, T. D. Fitch, F. Prystowsky, M. B. Goldwasser, E. Schrader, J. W. Palaszynski, E. Dy, M. & Label, B. (1983). Biologic properties of homogeneous interleukin 3. I. Demonstration of WEHI-3 growth factor activity, mast cell growth factor activity, p cell-stimulating factor activity, colony-stimulating factor activity, and histamine-producing cell-stimulating factor activity. *J. Immunol., 131*, 282-287.

[11] Razin, E. Ihle, J. N. Seldin, D. Mencia-Huerta, J. M. Katz, H. R. LeBlanc, P. A. Hein, A. Caulfield, J. P. Austen, K. F. & Stevens, R. L. (1984). Interleukin 3: a differentiation and growth factor for the mouse mast cell that contains chondroitin sulfate E proteoglycan. *J. Immunol., 132*, 1479-1486.

[12] Galli, S. J. (1990). New insights into "the riddle of the mast cells": microenvironmental regulation of mast cell development and phenotypic heterogeneity. *Lab. Invest., 62*, 5-33.

[13] Denburg, J. A. (1992). Basophil and mast cell lineages *in vitro* and *in vivo*. *Blood, 79*, 846-860.

[14] Kitamura, Y. Go, S. & Hatanaka, K. (1978). Decrease of mast cells in *W/W^v* mice and their increase by bone marrow transplantation. *Blood, 52*, 447-452.

[15] Kitamura, Y. & Go, S. (1979). Decreased production of mast cells in *Sl/Sl^d* anemic mice. *Blood*, *53*, 492-497.

[16] Fujita, J. Nakayama, H. Onoue, H. Kanakura, Y. Nakano, T. Asai, H. Takeda, S. Honjo, T. & Kitamura, Y. (1988). Fibroblast dependent growth of mouse mast cells *in vitro*: duplication of mast cell depletion in mutant mice of *W/W^v* genotype. *J. Cell. Physiol.*, *134*, 78-84.

[17] Chabot, B. Stephenson, D. A. Chapman, V. M. Besmer, P. & Bernstein, A. (1988). The proto-oncogene c-*kit* encoding a transmembrane tyrosine kinase receptor maps to the mouse *W* locus. *Nature*, *335*, 88-89.

[18] Geissler, E. N. Ryan, M. A. & Housman, D. E. (1988). The dominant-white spotting (*W*) locus of the mouse encodes the c-kit proto-oncogene. *Cell*, *55*, 185-192.

[19] Williams, D. E. Eisenman, J. Baird, A. Rauch, C. Van Ness, K. March, C. J. Park, L. S. Martin, U. Mochizuki, D. Y. Boswell, H. S. Burgess, G. S. Cosman, D. & Lyman, S. D. (1990). Identification of a ligand for the c-*kit* proto-oncogene. *Cell*, *63*, 167-174.

[20] Copeland, N. G. Gilbert, D. J. Cho, B. C. Donovan, P. J. Jenkins, N. A. Cosman, D. Anderson, D. Lyman, S. D. & Williams, D. E. (1990). Mast cell growth factor maps near the steel locus on mouse chromosome 10 and is deleted in a number of steel alleles. *Cell*, *63*, 175-183.

[21] Flanagan, J. G. & Leder, P. (1990). The kit ligand: a cell surface molecule altered in steel mutant fibroblasts. *Cell*, *63*, 185-194.

[22] Zsebo, K. M. Wypych, J. McNiece, I. K. Lu, H. S. Smith, K. A. Karkare, S. B. Sachdev, R. K. Yuschenkoff, V. N. Birkett, N. C. Williams, L. R. Satyagal, V. N. Tung, W. Bosselman, R. A. Mendiaz, E. A. & Langley, K. E. (1990). Identification, purification, and biological characterization of hematopoietic stem cell factor from buffalo rat liver-conditioned medium. *Cell*, *63*, 195-201.

[23] Huang, E. Nocka, K. Beier, D. R. Chu, T. Y. Buck, J. Lahm, H. W. Wellner, D. Leder, P. & Besmer, P. (1990). The hematopoietic growth factor KL is encoded by the Sl locus and is the ligand of the c-kit receptor, the gene product of the W locus. *Cell*, *63*, 225-233.

[24] Anderson, D. M. Lyman, S. D. Baird, A. Wignall, J. M. Eisenman. J. Rauch, C. March, C. J. Boswell, H. S. Gimpel, S. D. Cosman, D. & Williams, D. E. (1990). Molecular cloning of mast cell growth factor, a hematopoietin that is active in both membrane bound and soluble forms. *Cell*, *63*, 235-243.

[25] Valent, P. Spanblochl, E. Sperr, W. R. Sillaber, C. Zsebo, K. M. Agis, H. Strobl, H. Geissler, K. Bettelheim, P. & Lechner, K. (1992). Induction of differentiation of human mast cells from bone marrow and peripheral blood mononuclear cells by recombinant human stem cell factor *kit*-ligand in long-term culture. *Blood*, *80*, 2237-2245.

[26] Iemura, A. Tsai, M. Ando, A. Wershil, B. K. & Galli, S. J. (1994). The c-*kit* ligand, stem cell factor, promotes mast cell survival by suppressing apoptosis. *Am. J. Pathol.*, *144*, 321-328.

[27] Mekori, Y. A. Oh, C. K. & Metcalfe, D. D. (1993). IL-3-dependent murine mast cells undergo apoptosis on removal of IL-3. Prevention of apoptosis by c-*kit* ligand. *J. Immunol.*, *151*, 3775-3784.

[28] Lantz, C. S. Boesiger, J. Song, C. H. Mach, N. Kobayashi, T. Mulligan, R. C. Nawa, Y. Dranoff, G. & Galli, S. J. (1998). Role for interleukin-3 in mast-cell and basophil development and in immunity to parasites. *Nature*, *392*, 90-93.

[29] Kitamura, Y. Oboki, K. & Ito, A. (2006). Molecular mechanisms of mast cell development. *Immunol. Allergy Clin. North Am.*, *26*, 387-405.

[30] Mosmann, T. R. Cherwinski, H. Bond, M. W. Giedlin, M. A. & Coffman R. L. (1986). Two types of murine helper T cell clone. I. Definition according to profiles of lymphokine activities and secreted proteins. *J. Immunol.*, *136*, 2348-2357.

[31] Cherwinski, H. M. Schumacher, J. H. Brown, K. D. & Mosmann, T. R. (1987). Two types of mouse helper T cell clone. Ⅲ. Further differences in lymphokine synthesis between Th1 and Th2 clones revealed by RNA hybridization, functionally monospecific bioassays, and monoclonal antibodies. *J. Exp. Med.*, *166*, 1229-1244.

[32] Del Prete, G. F. De Carli, M. Mastromauro, C. Biagiotti, R. Macchia, D. Falagiani, P. Ricci, M. & Romagnani, S. (1991). Purified protein derivative of *Mycobacterium tuberculosis* and excretory-secretory antigen(s) of *Toxocara canis* expand in vitro human T cells with stable and opposite (type 1 T helper or type 2 T helper) profile of cytokine production. *J. Clin. Invest.*, *88*, 346-350.

[33] Mosmann, T. R. & Coffman, R. L. (1989). TH1 and TH2 cells: different patterns of lymphokine secretion lead to different functional properties. *Annu. Rev. Immunol.*, *7*, 145-173.

[34] Romagnani, S. (1994). Lymphokine production by human T cells in disease states. *Annu. Rev. Immunol.*, *12*, 227-257.

[35] Mosmann, T. R. Bond, M. W. Coffman, R. L. Ohara, J. & Paul, W. E. (1986). T-cell and mast cell lines respond to B-cell stimulatory factor 1. *Proc. Natl. Acad. Sci. USA*, *83*, 5654-5658.

[36] Smith, C. A. & Rennick, D. M. (1986). Characterization of a murine lymphokine distinct from interleukin 2 and interleukin 3 (IL-3) possessing a T-cell growth factor activity and a mast-cell growth factor activity that synergizes with IL-3. *Proc. Natl. Acad. Sci. USA*, *83*, 1857-1861.

[37] Hamaguchi, Y. Kanakura, Y. Fujita, J. Takeda, S. Nakano, T. Tarui, S. Honjo, T. & Kitamura, Y. (1987). Interleukin 4 as an essential factor for in vitro clonal growth of murine connective tissue-type mast cells. *J. Exp. Med.*, *165*, 268-273.

[38] Rennick, D. Hunte, B. Holland, G. & Thompson-Snipes, L. (1995). Cofactors are essential for stem cell factor-dependent growth and maturation of mast cell progenitors: comparative effects of interleukin-3 (IL-3), IL-4, IL-10, and fibroblasts. *Blood*, *85*, 57-65.

[39] Hültner, L. Druez, C. Moeller, J. Uyttenhove, C. Schmitt, E. Rüde, E. Dörmer, P. & Van Snick, J. (1990). Mast cell growth-enhancing activity (MEA) is structurally related and functionally identical to the novel mouse T cell growth factor P40/TCGFIII (interleukin 9). *Eur. J. Immunol.*, *20*, 1413-1416.

[40] Thompson-Snipes, L. Dhar, V. Bond, M. W. Mosmann, T. R. Moore, K. W. & Rennick, D. M. (1991). Interleukin 10: a novel stimulatory factor for mast cells and their progenitors. *J. Exp. Med.*, *173*, 507-510.

[41] He, Y. W. & Malek, T. R. (1995). The IL-2 receptor γ c chain does not function as a subunit shared by the IL-4 and IL-13 receptors. Implication for the structure of the IL-4 receptor. *J. Immunol.*, *155*, 9-12.

[42] Finkelman, F. D. Shea-Donohue, T. Goldhill, J. Sullivan, C. A. Morris, S. C. Madden, K. B. Gause, W. C. & Urban, J. F. Jr. (1997).Cytokine regulation of host defense against parasitic gastrointestinal nematodes: lessons from studies with rodent models. *Annu. Rev. Immunol.*, *15*, 505-533.

[43] Nafziger, J. Arock, M. Guillosson, J. J. & Wietzerbin, J. (1990). Specific high-affinity receptors for interferon-gamma on mouse bone marrow-derived mast cells: inhibitory effect of interferon-gamma on mast cell precursors *Eur. J. Immunol.*, *20*, 113-117.

[44] Takagi, M. Koike, K. & Nakahata, T. (1990). Antiproliferative effect of IFN-γ on proliferation of mouse connective tissue-type mast cells. *J. Immunol.*, *145*, 1880-1884.

[45] Rottem, M. Hull, G. & Metcalfe, D. D. (1994). Demonstration of differential effects of cytokines on mast cells derived from murine bone marrow and peripheral blood mononuclear cells. *Exp. Hematol.*, *22*, 1147-1155.

[46] Ochi, H. Hirani, W. M. Yuan, Q. Friend, D. S. Austen, K. F. & Boyce, J. A. (1999). T helper cell type 2 cytokine-mediated comitogenic responses and CCR3 expression during differentiation of human mast cells *in vitro*. *J. Exp. Med.*, *190*, 267–280.

[47] Kirshenbaum, A. S. Worobec, A. S. Davis, T. A. Goff, J. P. Semere, T. & Metcalfe, D. D. (1998). Inhibition of human mast cell growth and differentiation by interferon γ-1b. *Exp. Hematol.*, *26*, 245–251.

[48] Kulka, M. & Metcalfe, D. D. (2005). High-resolution tracking of cell division demonstrates differential effects of TH1 and TH2 cytokines on SCF-dependent human mast cell production *in vitro*: correlation with apoptosis and *kit* expression. *Blood*, *105*, 592–599.

[49] Coleman, J. W. Buckley, M. G. Holliday, M. R. & Morris, A. G. (1991). Interferon-γ inhibits serotonin release from mouse peritoneal mast cells. *Eur. J. Immunol.*, *21*, 2559–2564.

[50] Bissonnette, E. Y. Chin, B. & Befus, A. D. (1995). Interferons differentially regulate histamine and TNF-α in rat intestinal mucosal mast cells. *Immunology*, *86*, 12–17.

[51] Coleman, J. W. & Morris, A. G. (1991). Identification of a spleen cell-derived factor that inhibits sensitization of murine peritoneal mast cells as interferon-γ. *Int. Arch. Allergy Appl. Immunol.*, *94*, 179–180.

[52] Mann-Chandler, M. N. Kashyap, M. Wright, H. V. Norozian, F. Barnstein, B. O. Gingras, S. Parganas, E. & Ryan, J. J. (2005). IFN-γ induces apoptosis in developing mast cells. *J. Immunol.*, *175*, 3000-3005.

[53] Hu, Z-Q. Yoshida, T. & Shimamura, T. (1992). Induction and identification of mast cells from long-term culture of mouse spleen cells without conditioned medium. *J. Immunol. Methods*, *149*, 173-181.

[54] Hu, Z-Q. Asano, K. Yamazaki, T. & Shimamura, T. (1994). Effect of lipopolysaccharide on mouse mast cell induction by a splenic cell culture system. *Infect. Immun.*, *62*, 3844-3849.

[55] Hu, Z-Q. Yamazaki, T. Cai, Z. Yoshida, T. & Shimamura, T. (1994). Mast cells display natural suppressor activity partially by releasing transforming growth factor-β. *Immunology*, *82*, 482-486.

[56] Hu, Z-Q. Asano, K. Seki, H. & Shimamura, T. (1995). An essential role of prostaglandin E on mouse mast cell induction. *J. Immunol.*, *155*, 2134-2142.

[57] Hu, Z-Q. Zenda, N. & Shimamura, T. (1996). Down-regulation by IL-4 and up-regulation by IFN-γ of mast cell induction from mouse spleen cells. *J. Immunol.*, *156*, 3925-3931.

[58] Hu, Z-Q. Kobayashi, K. Zenda, N. & Shimamura, T. (1997). Tumor necrosis factor-α- and interleukin-6-triggered mast cell development from mouse spleen cells. *Blood*, *89*, 526-533.

[59] Hu, Z-Q. Zhao, W-H. & Shimamura, T. (2006). Mechanisms of strain-dependent development mast cells from mouse splenocytes. *Immunol. Cell Biol.*, *84*, 184–191.

[60] Hu, Z-Q. Zhao, W-H. Shimamura, T. & Galli, S. J. (2006). Interleukin 4-triggered, STAT6-dependent production of a factor that induces mouse mast cell apoptosis. *Eur. J. Immunol.*, *36*, 1275-1284.

[61] Saito, H. Ebisawa, M. Tachimoto, H. Shichijo, M. Fukagawa, K. Matsumoto, K. Iikura, Y. Awaji, T. Tsujimoto, G. Yanagida, M. Uzumaki, H. Takahashi, G. Tsuji, K. & Nakahata, T. (1996). Selective growth of human mast cells induced by *Steel* factor, IL-6, and prostaglandin E2, from cord blood mononuclear cells. *J. Immunol.*, *157*, 343-350.

[62] Yanagida, M. Fukamachi, H. Takei, M. Hagiwara, T. Uzumaki, H. Tokiwa, T. Saito, H. Iikura, Y. & Nakahata, T. (1996). Interferon-γ promotes the survival and FcεRI-mediated histamine release in cultured human mast cells. *Immunology*, *89*, 547-552.

[63] Kameyoshi, Y. Morita, E. Tanaka, T. Hiragun, T. & Yamamoto, S. (2000). Interleukin-1 alpha enhances mast cell growth by a fibroblast-dependent mechanism. *Arch. Dermatol. Res.*, *292*, 240-247.

[64] Oskeritzian, C. A. Wang, Z. Kochan, J. P. Grimes, M. Du, Z. Chang, H-W. Grant, S. & Schwartz, L. B. (1999). Recombinant human (rh) IL-4-mediated apoptosis and recombinant human IL-6-mediated protection of recombinant human stem cell factor-dependent human mast cells derived from cord blood mononuclear cell progenitors. *J. Immunol.*, *163*, 5105-5115.

[65] Nilsson, G. Miettinen, U. Ishizaka, T. Ashman, L. K. Irani, A. M. & Schwartz, L. B. (1994). Interleukin-4 inhibits the expression of Kit and tryptase during stem cell factor-dependent development of human mast cells from fetal liver cells. *Blood*, *84*, 1519-1527.

[66] Sillaber, C. Sperr, W. R. Agis, H. Spanblochl, E. Lechner, K. & Valent, P. (1994). Inhibition of stem cell factor dependent formation of human mast cells by interleukin-3 and interleukin-4. *Int Arch. Allergy Immunol.*, *105*, 264-268.

[67] Mirmonsef, P. Shelburne, C. P. Yeatman, F. C. 2nd. Chong, H. J. & Ryan, J. J. (1999). Inhibition of Kit expression by IL-4 and IL-10 in murine mast cells: role of STAT6 and phosphatidylinositol 3'-kinase. *J. Immunol.*, *163*, 2530-2539.

[68] Yeatman, C. F. 2nd. Jacobs-Helber, S. M. Mirmonsef, P. Gillespie, S. R. Bouton, L. A. Collins, H. A. Sawyer, S. T. Shelburne, C. P. & Ryan, J. J. (2000). Combined

stimulation with the T helper cell type 2 cytokines interleukin (IL)-4 and IL-10 induces mouse mast cell apoptosis. *J. Exp. Med.*, *192*, 1093-1103.

[69] Oskeritzian, C. A. Zhao, W. Pozez, A. L. Cohen, N. M. Grimes, M. & Schwartz, L. B. (2004). Neutralizing endogenous IL-6 renders mast cells of the MCT type from lung, but not the MCTC type from skin and lung, susceptible to human recombinant IL-4-induced apoptosis. *J. Immunol.*, *172*, 593-600.

[70] Bailey, D. P. Kashyap, M. Mirmonsef, P. Bouton, L. A. Domen, J. Zhu, J. Dessypris, E. N. & Ryan, J. J. (2004). Interleukin-4 elicits apoptosis of developing mast cells via a Stat6-dependent mitochondrial pathway. *Exp. Hematol.*, *32*, 52-59.

[71] Bouton, L. A. Ramirez, C. D. Bailey, D. P. Yeatman, C. F. Yue, J. Wright, H. V. Domen, J. Rosato, R. R. Grant, S. Fischer-Stenger, K. & Ryan, J. J. (2004). Costimulation with interleukin-4 and interleukin-10 induces mast cell apoptosis and cell-cycle arrest: the role of p53 and the mitochondrion. *Exp. Hematol.*, *32*, 1137-1145.

[72] Kinoshita, T. Sawai, N. Hidaka, E. Yamashita, T. & Koike, K. (1999). Interleukin-6 directly modulates stem cell factor-dependent development of human mast cells derived from $CD34^+$ cord blood cells. *Blood*, *94*, 496-508.

[73] Bailey, D. P. Kashyap, M. Bouton, L. A. Murray, P. J. & Ryan, J. J. (2006). Interleukin-10 induces apoptosis in developing mast cells and macrophages. *J. Leukoc. Biol.*, *80*, 581-589.

[74] Helmby, H. & Grencis, R. K. (2003). IFN-gamma-independent effects of IL-12 during intestinal nematode infection. *J. Immunol.*, *171*, 3691-3696.

[75] Masuda, A. Matsuguchi, T. Yamaki, K. Hayakawa, T. & Yoshikai, Y. (2001). Interleukin-15 prevents mouse mast cell apoptosis through STAT6-mediated Bcl-xL expression. *J. Biol. Chem.*, *276*, 26107-26113.

[76] Matsuda, H. Kannan, Y. Ushio, H. Kiso, Y. Kanemoto, T. Suzuki, H. & Kitamura, Y. (1991). Nerve growth factor induces development of connective tissue-type mast cells *in vitro* from murine bone marrow cells. *J. Exp. Med.*, *174*, 7-14.

[77] Kanbe, N. Kurosawa, M. Miyachi, Y. Kanbe, M. Saitoh, H. & Matsuda, H. (2000). Nerve growth factor prevents apoptosis of cord blood-derived human cultured mast cells synergistically with stem cell factor. *Clin. Exp. Allergy*, *30*, 1113-1120.

[78] Norozian, F. Kashyap, M. Ramirez, C. D. Patel, N. Kepley, C. L. Barnstein, B. O. & Ryan, J. J. (2006). TGFβ1 induces mast cell apoptosis. *Exp. Hematol.*, *34*, 579–587.

[79] Wright, H. V. Bailey, D. Kashyap, M. Kepley, C. L. Drutskaya, M. S. Nedospasov, S. A. & Ryan, J. J. (2006). IL-3-mediated TNF production is necessary for mast cell development. *J. Immunol.*, *176*, 2114–2121.

[80] Sawai, N. Koike, K. Mwamtemi, H. H. Kinoshita, T. Kurokawa, Y. Sakashita, K. Higuchi, T. Takeuchi, K. Shiohara, M. Kamijo, T. Ito, S. Kato, T. Miyazaki, H. Yamashita, T. & Komiyama, A. (1999). Thrombopoietin augments stem cell factor-dependent growth of human mast cells from bone marrow multipotential hematopoietic progenitors. *Blood*, *93*, 3703-3712.

[81] Berent-Maoz, B. Piliponsky, A. M. Daigle, I. Simon, H. U. & Levi-Schaffer, F. (2006). Human mast cells undergo TRAIL-induced apoptosis. *J. mmunol.*, *176*, 2272-2278.

[82] Kaplan, M. H. Schindler, U. Smiley, S. T. & Grusby, M. J. (1996). Stat6 is required for mediating responses to IL-4 and for development of Th2 cells. *Immunity*, *4*, 313-319.

[83] Shimoda, K. van Deursen, J. Sangster, M. Y. Sarawar, S. R. Carson, R. T. Tripp, R. A. Chu, C. Quelle, F. W. Nosaka, T. Vignali, D. A. Doherty, P. C. Grosveld, G. Paul, W. E. & Ihle, J. N. (1996). Lack of IL-4-induced Th2 response and IgE class switching in mice with disrupted Stat6 gene. *Nature*, *380*, 630-633.

[84] Takeda, K. Tanaka, T. Shi, W. Matsumoto, M. Minami, M. Kashiwamura, S. Nakanishi, K. Yoshida, N. Kishimoto, T. & Akira, S. (1996). Essential role of Stat6 in IL-4 signaling. *Nature*, *380*, 627-630.

[85] Zurawski, S. M. Vega, F. Jr. Huyghe, B. & Zurawski, G. (1993). Receptors for interleukin-13 and interleukin-4 are complex and share a novel component that functions in signal transduction. *EMBO J.*, *12*, 2663-2670.

[86] Boyce, J. A. (2003). Mast cells: beyond IgE. *J. Allergy Clin. Immunol.*, *111*, 24-32.

[87] Shelburne, C. P. & Ryan, J. J. (2001). The role of Th2 cytokines in mast cell homeostasis. *Immunol. Rev.*, *179*, 82–93.

[88] Shiohara, M. & Koike, K. (2005). Regulation of mast cell development. *Chem. Immunol. Allergy*, *87*, 1-21.

[89] Okayama, Y. & Kawakami, T. (2006). Development, migration, and survival of mast cells. *Immunol. Res.*, *34*, 97-115.

[90] Moore, K. W. de Waal Malefyt, R. Coffman, R. L. & O'Garra, A. (2001). Interleukin-10 and the interleukin-10 receptor. *Annu. Rev. Immunol.*, *19*, 683-765.

[91] Gillespie, S. R. DeMartino, R.R. Zhu, J. Chong, H. J. Ramirez, C. Shelburne, C. P. Bouton, L. A. Bailey, D. P. Gharse, A. Mirmonsef, P. Odom, S. Gomez, G. Rivera, J. Fischer-Stenger, K. & Ryan, J. J. (2004). IL-10 inhibits FcεRI expression in mouse mast cells. *J. Immunol.*, *172*, 3181-3188.

[92] Echtenacher, B. Mannel, D. N. & Hultner, L. (1996). Critical protective role of mast cells in a model of acute septic peritonitis. *Nature*, *381*, 75-77.

[93] Malaviya, R. Ikeda, T. Ross, E. & Abraham, S. N. (1996). Mast cell modulation of neutrophil influx and bacterial clearance at sites of infection through TNF-alpha. *Nature*, *381*, 77-80.

[94] Gordon, J. R. & galli, S. J. (1990). Mast cells as a source of both preformed and immunologically inducible TNF-alpha/cachectin. *Nature*, *346*, 274-276.

[95] Gordon, J. R. Burd, P. R. & Galli, S. J. (1990). Mast cells as a source of multifunctional cytokines. *Immunol. Today*, *11*, 458-464.

[96] Wedemeyer, J. Tsai, M. & Galli, S. J. (2000). Roles of mast cells and basophils in innate and acquired immunity. *Curr. Opin. Immunol.*, *12*, 624-631.

[97] Galli, S. J. Kalesnikoff, J. Grimbaldeston, M. A. Piliponsky, A. M. Williams, C. M. & Tsai, M. (2005). Mast cells as "tunable" effector and immunoregulatory cells: recent advances. *Annu. Rev. Immunol.*, *23*, 749-786.

[98] Grimbaldeston, M. A. Metz, M. Yu, M. Tsai, M. & Galli, S. J. (2006). Effector and potential immunoregulatory roles of mast cells in IgE-associated acquired immune responses. *Curr. Opin. Immunol.*, *18*, 751–760.

[99] Bischoff, S. C. & Sellge, G. (2002). Mast cell hyperplasia: role of cytokine. *Int. Arch. Allergy Immunol.*, *127*, 118-122.

In: New Cell Differentiation Research Topics
Editor: Hitoshi Saitama, pp. 229-246
ISBN: 978-1-60021-937-5

Chapter VII

THE BIG BROTHER OF BLOOD: THE NEED FOR CONSTANT VIDEO SURVEILLANCE OF INDIVIDUAL CELLS

Erin Drew and Timm Schroeder*
Hematopoiesis Group, Institute of Stem Cell Research, GSF - National Research Center for Environment and Health, Neuherberg (Munich), Germany.

ABSTRACT

Cellular differentiation is the commitment of cells in response to internal and external signals, and is often accompanied by morphological changes. Hematopoiesis is a well-understood differentiation system that leads to the formation of the various blood cell types, which are all derived from the hematopoietic stem cell (HSC). Research about this process has exploded in the last few decades with the discovery of cell surface antigen-based isolation methods and cytokine response characterizations. However, the kinetics of these responses and the specific fates of individual cells remain unclear. This is because the data are based on the average of cell populations and/or conditions that observe cell behaviour outside of their normal environment. Therefore, new technologies must be developed and utilized, following single cells, to dissect the exact nature of this differentiation process.

It is becoming increasingly clear that the bone marrow has supporting cells that influence and control blood differentiation. Research has only begun to understand what cells types and factors are involved. The data so far have recognized that some "niche cells" allow HSC to self-renew and differentiate in a controlled manner that, in turn, leads to the mature adult hematopoietic system.

Although a few components of this complex niche are now recognized, the exact nature of the HSC-niche relationship is still unclear, as technical obstacles have yet to be

* Correspondence concerning this article should be addressed to: Dr. Timm Schroeder, PhD, Institute of Stem Cell Research, GSF - National Research Center for Environment and Health, Ingolstaedter Landstr. 1, D-85764 Neuherberg, Germany. TEL: +49 (0)89 3187 3758; FAX: +49 (0)89 3187 3761; email: timm.schroeder@gsf.de.

overcome. Much of the data are based on fixed tissues that show the localization and behaviour of a HSC at a single point in time. This neglects the temporal component to this dynamic relationship between the HSC and its niche. Furthermore, researchers have had to measure the presence of HSC by indirect functional readouts for its progeny, thus ignoring the individual characteristics and behaviours of these cells. So far, HSC have not been directly observed as live single cells, with the individual characterization of their behaviour and differentiation within their normal niche.

To understand the temporal and spatial relationships the HSC has with the various supporting cell types, HSC must be visualized within their niche over time, under constant video surveillance. Only then will we be able to understand the HSC-microenvironment relationship. This review will discuss the need for new technologies that will allow for a greater understanding of the HSC-niche relationship that influences normal blood differentiation.

INTRODUCTION

Hematopoiesis refers to the process of blood development that leads to the formation of at least ten mature lineages. This includes the differentiation of red and white blood cells, the latter of which can be sub-divided into lymphoid and myeloid cells. Lymphoid cells include B, T, lymphoid dendritic and natural killer (NK) cells, which are essential for the adaptive immune system. Myeloid cells include granulocytes, macrophages, myeloid dendritic cells, megakaryocytes or erythrocytes. Granulocytes (neutrophils, basophils, eosinophils and mast cells) and macrophages have roles in the innate immune system, megakaryocytes give rise to platelets that are essential for blood clotting and red blood cells are essential in carbon dioxide/oxygen gas exchange process (Janeway, 2001).

Depending on the stage of development in mammals, hematopoiesis occurs at different anatomical sites. For example, in the developing embryo, primitive hematopoiesis takes place in the yolk sac and in the AGM region (aorta-gonad-mesonephros). Later on, it moves to the fetal liver and then right before birth hematopoiesis transits to the bone marrow where it remains the primary site of blood formation throughout the rest of the life of the animal (Dzierzak, 2003).

Within these different sites of hematopoiesis, blood cells can arise from the hematopoietic stem cell (HSC). This cell is defined by its ability to self-renew and to differentiate into all of the various blood cell lineages. Over the past decades, the combination of several technologies has allowed researchers to isolate and study the function of these rare cells so that today, HSC are the most well-characterized stem cell population in mammals (Takano et al., 2004).

HSC RESEARCH TECHNIQUES: IN THE BEGINNING

The concept of the HSC was first recognized in 1909 by the Russian aristocrat Alexander Maximow: "In the mammalian organism exists one cell type, [that]... produce by homoplastic proliferation again and again lymphocytes...[and] circulates with the blood- and lymph stream throughout the organism and is able to regain after a sufficient time-lag of inactivity its full developmental potential" (Fliedner, 1998). It is generally accepted that the first demonstration of the existence of this cell was 50 years later by Till and McCulloch (Becker et al., 1963; Siminovitch et al., 1963; Till and McCulloch, 1961). However, it is worth noting that in 1945, when Ray Owen was studying immune tolerance, he observed that twin calves shared blood cell types, demonstrating that *in utero*, each calf had obtained HSC from their twin (Weissman, 2002).

Till and McCulloch showed that transplanted bone marrow cells were capable of producing spleen forming units in irradiated mice (Till and McCulloch, 1961). Furthermore, they induced random chromosomal breaks in the transplanted cells to demonstrate that the spleen colonies were clonally derived, since these showed the same pattern of breaks (Becker et al., 1963). Importantly, some of these cells were able to produce secondary spleen colonies and reconstitute hematopoiesis in irradiated mice (Siminovitch et al., 1963). These results demonstrated the existence of an HSC, however, at this time, the identification and purification of these cells was not possible.

Identification and isolation of the HSC

In the 1960s and 70s technologies were developed that eventually led to the to identification and isolation of various hematopoietic cells, including the HSC. Namely, these were the ability to sort cells using flow cytometry and the production of monoclonal antibodies.

Flow cytometric Sorting

Flow cytometric cell sorting of viable cells was pioneered by Herzenberg *et al* (Herzenberg and Sweet, 1976). This technique allows researchers to sort cells based on physical characteristics (size and granularity) and/or by specific expression of fluorescence, either from specific dyes or fluorescent proteins. This has led to the current capabilities of researchers to simultaneously sort cells using more than ten colours. It has become an indispensable tool for any lab that wants to study pure population of cells, not only in hematopoiesis but also in other areas of cell biology.

Monoclonal Antibodies

With the advent of monoclonal antibody generation using hybridomas (Kohler and Milstein, 1975), researchers could generate antibodies against specific cell surface proteins and subsequently combine these with fluorescent markers and use them to sort cells. Weissman and colleagues were the first to isolate different populations of hematopoietic cells when they were studying immune responses of B and T cells (Weissman, 2002). At that time, they found that the progenitor cells lay within the lineage negative fraction of bone marrow and began to add other markers to purify this population. They finally showed that Lineage$^-$ Thy-1^{lo}c-kit$^+$Sca-1$^+$ bone marrow cells were capable of reconstituting the entire hematopoietic system in irradiated mice and followed their self-renewal in secondary recipients to establish their self-renewal potential (Ikuta et al., 1992; Uchida and Weissman,

1992). Thus, the very rare HSC could be enriched and was functionally shown to maintain its self-renewal and differentiation capacities.

Other groups use slightly different combinations of cell surface proteins for the purification of HSC, including Lineage$^-$c-kit$^+$Sca-1$^+$CD34$^-$ (Osawa et al., 1996). More recently the CD150$^+$CD48$^-$CD41$^-$ "SLAM" signature has been shown to highly enrich for HSC (Kiel et al., 2005). Importantly, the latter profile allows for visualization of these cells *in vivo*, whereas Lineage$^-$CD34$^-$c-kit$^+$Sca-1$^+$ and other marker combinations are too complex to examine histochemically.

Dye exclusion has also been successfully used to enrich for HSC. Used dyes include Hoechst 33342 and Rhodamine-123 (Rho), both of which are effluxed by HSC (Goodell et al., 2004; Kim et al., 1998). Hoechst is detected with two different wavelengths using flow cytometry and HSC fall into a specific so-called Hoechst^{low} "side population" (SP). These dyes have been combined with antigen staining combinations to isolate highly purified HSC populations (Goodell et al., 2004; Matsuzaki et al., 2004)

HSC are arguable the rarest population of hematopoietic cells. Therefore, their identification and isolation has been one of the most difficult. One problem in HSC biology today is the variety of antigen combinations used by different researchers as a handle to purify these cells. There is currently controversy over which method is optimal and it is unclear if the observations by one group can be applied to HSC sorted using different criteria, since they may be looking at slightly different populations.

Single Cell Functional Analysis of Isolated HSC

As mentioned before, HSC are defined by their self-renewal capacity and their ability to produce all blood cell lineages over the lifetime of an organism. Therefore, currently the only assay to positively identify HSC is their transplantation into stem cell deficient mice or irradiated wild type recipients, with subsequent transfer of the produced blood cells (after several weeks) into a secondary recipient to measure their renewal (Scadden, 2006). Once blood is taken from these secondary recipients, some weeks further on, and genetically marked blood cells from the original cell is detected within all blood lineages, the original cell can then be retrospectively identified as a multi-potent, self-renewing HSC (Scadden, 2006).

Several groups have performed this assay at the single cell level for purified HSC (Ema et al., 2000; Osawa et al., 1996; Wagers et al., 2002). In these experiments, Lineage$^-$CD34$^{-/lo}$c-kit$^+$Sca-1$^+$ BM cells were sorted and single cells were transplanted into lethally irradiated mice. Of the mice that received a single cell, 21-40% showed long-term reconstitution. Similarly, it has been shown that single cell transplantation of "SLAM-code" sorted cells (CD150$^+$CD48$^-$Sca-1$^+$Lineage$^-$c-kit$^+$) resulted in long-term reconstitution in 47% of recipients, putatively reflecting a higher purity of HSC (Kiel et al., 2005).

These percentages of engrafted transplants suggest that antibody staining and sorting can enrich for HSC but that these may not be pure populations. Serial transplantation of single cells is still currently the only robust assay to test HSC function. However, its drawback is that it is retrospective, and for this reason it is difficult to prospectively isolate and study HSC directly.

Studying Hematopoiesis: The Variables to Consider

HSC Heterogeneity

Hematopoiesis is most commonly thought of as a series of changes that a cell predictably undergoes, passing through defined stages until it becomes a mature blood cell (Wagers et al., 2002). However, this hierarchical model has come into question with the recent observations made using individually sorted HSC. It was shown that after one division, an HSC can produce one multi-lineage blood cell and a sister cell that is restricted to a specific blood lineage (Takano et al., 2004). However, whether this occurs in the absence of micromanipulation under conditions where these cells interact with niche cells, remains an open question.

Likewise, more recently, Sieburg *et al.* (Sieburg et al., 2006) have shown that the reconstituting patterns of blood populations from a single HSC differ, suggesting HSC have an intrinsic predisposition in their repopulating kinetics. They have shown that this is inherent to individual HSC, since they give rise to daughter HSC with a similar pattern of behaviour. They also found that isolation of Lineage^{-}Rho^{-}SP cells selects for certain HSC reconstitution patterns, suggesting that some HSC are excluded when these parameters are used.

Therefore, even within highly enriched populations of HSC, or by assaying the function of HSC, independent of isolation, it is obvious that HSC have heterogeneous behaviours. These different behaviours may be a result of various external and/or internal signals. However, this remains difficult to ascertain due to the rarity of these cells. Therefore, technology is required that can analyze the cells at the single cell level so that sub-populations may be able to be identified prospectively.

Intrinsic Influences

As cells mature, chromatin becomes more condensed to form heterochromatin, accompanied by DNA methylation and histone deacetylation, so that genes become inaccessible to the transcription machinery and are no longer expressed. This process is generally thought to be irreversible so that differentiation is a uni-linear process that does not normally allow more differentiated cells to "de-differentiate" and cells committed to a certain lineage cannot "trans-differentiate" to form a different blood lineage (Wagers et al., 2002). How cell fate is directed is not well understood, however, it is thought to involve an orchestration of transcription factor levels, response to cytokines and other cell-to-cell interactions.

It has been demonstrated that pluripotent cell lines "promiscuously" express low levels of lineage-restricted transcription factors and markers at the single cell level (Hu et al., 1997). Likewise, it has been shown that myeloid restricted progenitors co-express myeloid and erythroid genes and lymphoid restricted progenitors co-express B and T cell specific genes, before they have committed to differentiate along either of these lineages (Miyamoto et al., 2002). Therefore, it has been proposed that this is a means of priming cells to respond to other influences, until this leads to mono-lineage commitment, thus combining internal and external influences (Cross and Enver, 1997).

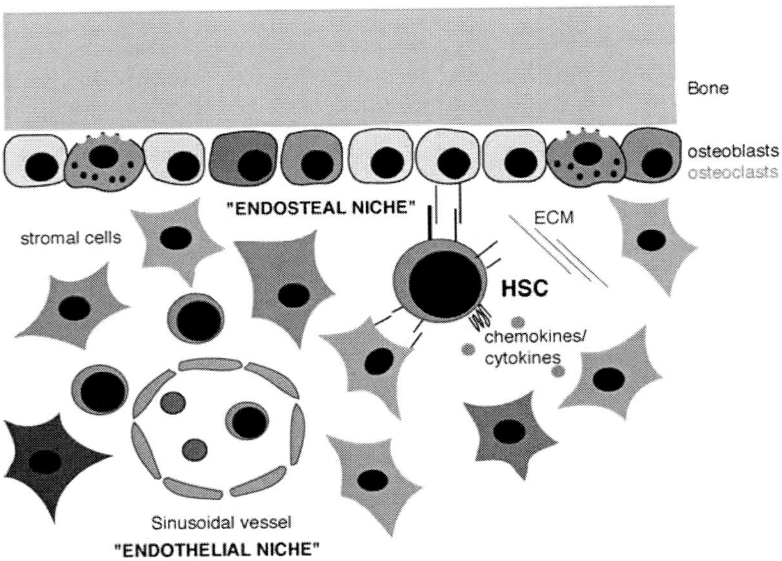

Figure 1. A proposed model of the bone marrow niche. Various HSC niche components are depicted that include osteoblasts, osteoclasts, stromal cells, sinusoidal endothelium, extracellular matrix (ECM) and chemokines and/or cytokines. These interact with the HSC through a variety of proteins. Various colours/shades of osteoblasts and stromal cells are shown to depict the heterogeneity of these populations. Currently there are two niches thought to interact with HSC and are commonly referred to the "endosteal niche" near osteoblasts and the "endothelial niche" near sinusoidal endothelium.

External Influences: The HSC Niche

A niche is defined *ecologically* as a place where an organism can reside and reproduce. Similarly, a *cellular* niche protects a population from death and proliferation control that both protects the population from inappropriate cues that could lead to aberrant growth. In the case of hematopoiesis, the latter could lead to malignancies such as leukemia (Scadden, 2006).

The concept of the HSC niche was coined in 1978 by Schoefield, "the stem cell is seen in association with other cells which determine its behaviour...Its maturation is prevented and, as a result, its continued proliferation as a stem cell is assured. Its progeny, unless they can occupy a similar stem cell 'niche'...proliferate and mature to acquire a high probability of differentiation" (Schofield, 1978). The stem cell niche is thus a functioning entity likely made up of a complex array of cell types and extracellular components that induce self-renewal of stem cells and are able to release these cells for their subsequent differentiation into more mature blood cells.

Just as the ecological niche of an organism changes as it develops, a hematopoietic niche is thought to change as a cell becomes more mature. There have been a few studies recently that have given specific clues about some components of the HSC niche, however it is still not fully understood how this differs from the niches of other hematopoietic cells and whether there is movement of blood cells from one niche to the other, as they mature.

In adult mammals, HSC are found mostly in the bone marrow. Within the marrow there are many factors that allow for their survival, quiescence and/or and proliferation. Here, an HSC is surrounded by hematopoietic and non-hematopoietic cells as well as the ECM components produced by those cells (Arai et al., 2004; Rattis et al., 2004; Whetton and Graham, 1999) (Figure 1). The cumulative influence of these factors is often collectively referred to as the bone marrow HSC niche.

It was not until recently that data have accumulated to verify the HSC niche hypothesis. There are now several lines of evidence to support that osteoblasts, endothelial cells and stromal cells, among others, contribute to the HSC niche (Scadden, 2006).

Osteoblasts

Osteoblasts are immature bone cells, some of which line the bone surface and are responsible for generating new bone by producing and secreting bone matrix (Cho et al., 2006). Due to the close proximity of immature hematopoietic cells to the endosteum (Lord, 1990; Nilsson et al., 1997) and the secretion of hematopoietic-supporting factors by osteoblasts (Taichman, 2005) it was long thought that there was a functional relationship between blood and bone.

Taichman *et al.* were the first to show that human osteoblasts functionally support hematopoiesis (Taichman and Emerson, 1994). Then, in mice, Nilsson *et al.* observed that HSC transplanted into non-irradiated mice home to the endosteal region of the bone marrow (Nilsson et al., 1997). It was also shown that ablation of osteoblastic cells using toxic thymidine kinase under control of the Col2.3 promoter caused the loss of hematopoietic cells, including HSC (Visnjic et al., 2001). Although this study saw an increase of HSC over time, it was the first line of evidence in mice that suggested that osteoblasts were functionally important in supporting hematopoiesis.

In 2003, two back-to-back papers demonstrated the ability of osteoblasts to support HSC (Calvi et al., 2003; Zhang et al., 2003). Using two different genetic approaches, they both showed an increased number of osteoblasts and a specific increase in the number of HSC in the bone marrow, without significant increases to other hematopoietic populations, suggesting that osteoblasts regulate the number of HSC. More recently, one of these groups has shown therapeutic targeting of the HSC osteoblastic niche that in turn, improved hematopoietic recovery after irradiation (Adams et al., 2007), suggesting that the endosteal niche can be manipulated to affect the behaviour of HSC.

In vivo imaging of bone marrow has recently become possible within the mouse femur and scapula. Using this fluorescence microscopy and *Gata-2*-GFP knock-in mice, it was observed that at least some HSC reside within an endosteal niche, there is only one HSC per niche and that they are relatively stationary over time (Suzuki et al., 2006). These results further suggest that HSC localize to the endosteum and their observed behaviour reflects what one would expect to see from a quiescent HSC.

Although the osteoblastic niche appears to regulate HSC, it is not clear whether all osteoblasts have this function. Furthermore, this cell type cannot be the only cell type with this function since osteoblasts are not found within the fetal liver or the spleen, which are also active sites of hematopoiesis throughout development (Kiel et al., 2005; Wilson and Trumpp, 2006).

Endothelial Cells

Since endothelial cells also reside within the fetal liver and the spleen, it has been hypothesized that these cells may also be able to act as a stem cell niche. This is supported by other studies that have shown that endothelial cells can maintain HSC in culture (Li et al., 2004; Ohneda et al., 1998) and blocking of the endothelial-specific VE-cadherin causes loss of hematopoiesis (Avecilla et al., 2004). With the recent discovery of the "SLAM" ($CD150^{+}CD48^{-}CD41^{-}$) signature on HSC, it has become possible to study their whereabouts histochemically and it was reported that 67% of purified populations containing HSC are localized adjacent to sinusoids in the bone marrow and spleen (Kiel et al., 2005). Although these results do not functionally prove that they regulate a niche *in vivo*, their localization suggests that they also may have an HSC niche function.

Stromal Cells

Ten years ago, Moore et al. (Moore et al., 1997) published a study of immortalizing fetal liver stromal cells in an effort to find the molecular identity of an HSC supporting line. They isolated fetal livers from day 14.5 embryos since it is known that there is a high rate of HSC self-renewal in this organ at this stage of development. They found that one out of 200 lines they generated that was able to support robust maintenance of long-term HSC *in vitro* for several weeks (Hackney et al., 2002; Moore et al., 1997; Wineman et al., 1996). This suggested that these supporting lines, as with HSC, are rare cells and indicated high heterogeneity of the largely undefined stromal population within the fetal liver. This also demonstrated that certain stromal cells were functionally sufficient for HSC support. However, it remains controversial whether this reflects a property of the transformation of the line, and whether a stromal niche for HSC maintenance represents a developmentally specific phenomenon *in vivo*.

More recently, it has been demonstrated that there are reticular stromal cells that express SDF-1 located near the endosteum and sinusoidal endothelium in the bone marrow, and that these interact with HSC. This may explain the observation of HSC near both these cell types and provides evidence that stromal cells are indeed important in the HSC niche in the bone marrow (Sugiyama et al., 2006).

Other Hematopoietic Niches

With the recent hype about the HSC niche, the obvious is sometimes forgotten: 1) there are undefined niches that harbour other, more mature types of blood cells 2) the majority of cells within the bone marrow are blood cells and these can affect other blood cells through cell-to-cell contact and the secretion of different factors and 3) blood cells within a niche, may affect the niche components, which in turn can affect their own behaviour.

One recent study has demonstrated that developing B cells change their location within the bone marrow between different types of stromal cells, depending on their stage of maturation. They showed that pre-pro-B cells interact with SDF-1$^+$stroma and then more mature pro-B cells interacted with IL-7R$^+$ stroma (Tokoyoda et al., 2004). Furthermore, they showed that mature plasma cells return to SDF-1$^+$ stroma. These results demonstrate the migration to different niches as B cell maturation progresses and may serve as an example for the study of other hematopoietic niches within the bone marrow.

It is currently not well understood how blood cells affect other blood cells. However, one example that is known is the formation of erythrocytic islands by macrophages in the bone marrow where erythrocytes are produced (Nakamura et al., 1999). More studies need to be done to evaluate these important interactions and the influence they have on the hematopoietic process. It is also not understood how a cell within a niche change its own microenvironment. Molecules involved in blood cell-niche interactions are discussed elsewhere (Taichman, 2005).

In summary, some components of the HSC niche and other hematopoietic niches have been identified within the bone marrow and other tissues. However, the nature of these interactions in real-time is not well understood. Until hematopoietic cells and defined niche components are studied with high temporal resolution, their exact interactions and influence on hematopoietic cell behaviour will remain unclear.

Cytokine Responses and Hematopoietic Differentiation

The niche cells surrounding hematopoietic cells, and hematopoietic cells themselves produce factors that influence the behaviour of other cells. These include the family of proteins called cytokines that have been extensively studied for their effects of hematopoietic differentiation (Janeway, 2001). Each has a corresponding receptor on blood cells and causes cell signaling activation that can lead to the generation of a specific blood cell lineage such as granulocytic-colony stimulating factor (G-CSF), macrophage-stimulating factor (M-CSF), erythropoietin (EPO), interleukin-3 (IL-3), interleukin-7 (IL-7) and thrombopoietin (TPO), among others. These factors can play a role in the differentiation of granulocytes, macrophages, erythrocytes, mast cells, B cells and megakaryocytes, respectively (Metcalf, 1999; Tokoyoda et al., 2004).

However, despite extensive research, it is unclear whether their effects have an instructive or permissive role in cellular differentiation. In other words, do these factors select for the outgrowth of responsive cells (permissive) or do they cause the cells to be pushed along a specific direction of differentiation (instructive)? Some data support a permissive role of some cytokines (Akashi et al., 1997; Goldsmith et al., 1998; Maraskovsky et al., 1997; Stoffel et al., 1999), whereas other studies support an instructive role (Dong et al., 1993; Fukunaga et al., 1993; King et al., 2002; Kondo et al., 2000; Metcalf and Burgess, 1982). Currently, this is a much-debated topic and has not been adequately addressed at the single cell level.

EMERGING TECHNOLOGIES: A BIG BROTHER OF BLOOD

The two main hurdles in hematopoietic research currently are 1) much of the research is based on the study of heterogeneous populations of cells neglecting the individual characteristics and fates of cells and 2) cell populations are analyzed at one or a few time points which doesn't explain how the population of cells dynamically reached its final state.

It is dangerous to draw conclusions from end-point analyses if the whole process cannot be captured to give an accurate picture of what occurred within a population to reach this point. A simple example should demonstrate this point: Given that there are two fates for a cell: division and apoptosis, and there is a maximum of three divisions possible, there are nine possible combinations that the cells could use to reach the same resulting population (Figure 2). Thus, once there are a greater number of divisions possible, and an exponentially greater number of variables to keep track of, it becomes vital to keep track of individual cell behaviour to understand the process.

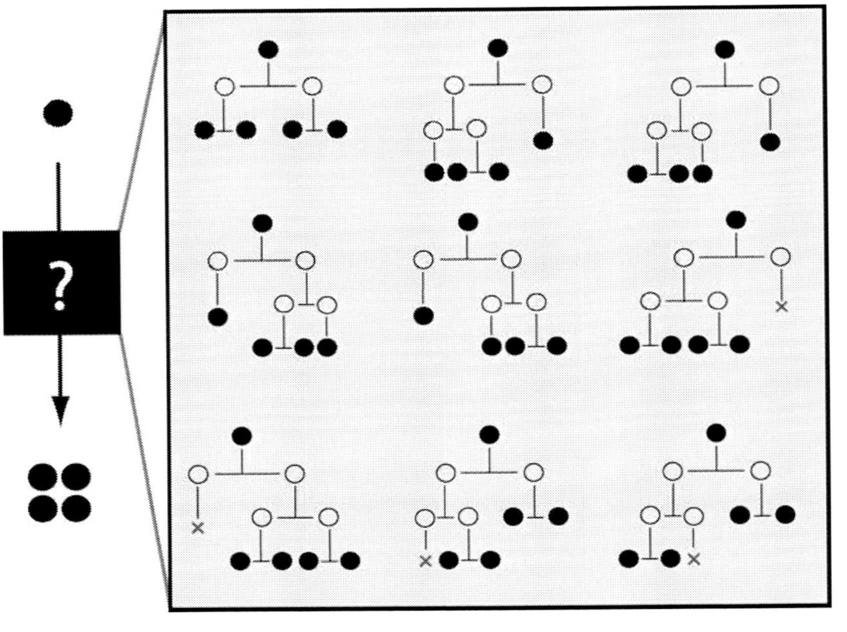

Figure 2. A single cell can follow several possible cellular paths to achieve the same outcome. Given the possibilities of only apoptosis and cell division, and three divisions, nine possible examples are shown that end up with the same end point population of four cells. Thus, single cell fate must be analyzed for a proper understanding of the cellular events that lead to the resulting population.

To this end, a new technology has been developed able to record sequential images of multiple positions for up to a few weeks *in vitro*, that can then be loaded as a video and whereby each individual cell and its progeny can be subsequently analyzed and tracked (Schroeder, 2005). Cells under observation are under controlled atmospheric and temperature conditions, with the possibility of media changes throughout the experiment. Single cell tracking software has been developed to measure the different behaviours of cells so that cellular family trees can be created with the characteristics of each cell being quantified.

These include proliferation times, speed (migration) and size. Thus, individual cell fates can be tracked and the route that the cells took to produce the end population can be understood. Hence, the heterogeneity problem of hematopoietic research is reduced as differential behaviours can be simultaneously observed, recorded and measured.

Fluorescence: a Tool for Tracking the Individual Cell Fates

In addition, various fluorescent proteins can be used with the above technology as a means to measure lineage commitment and/or protein localization. Thus, this provides a powerful technique for non-invasively observing specific hematopoietic lineage maturation and protein segregation.

Since the isolation of GFP (green fluorescent protein) by Shimomura in 1962 (Shimomura et al., 1962) and its cloning in 1992 (Prasher et al., 1992), many other fluorescent proteins have been created and introduced into cells and/or mice. These include Cherry, Tomato, Banana, DsRed, Strawberry, Venus and CFP (cyan fluorescent protein), to name a few (Shaner et al., 2004). Unfortunately, many of these proteins have overlapping emission spectra, which restricts fluorescent combinations. To allow greater distinction of fluorescent signals, sub-cellular restricted expression of fluorescent proteins have recently been generated using corresponding signal peptides, so that the same fluorescent protein can be expressed in the same cell with different sub-cellular localizations to allow distinguishable, simultaneous expression (eg. mitochondria, nucleus, plasma membrane) (Okita et al., 2004).

There are several transgenic and knock-in mice that have been created with fluorescent expression under the control of hematopoietic lineage specific gene promoters. Currently the following mice with hematopoietic lineage specific expression of fluorescence (mostly GFP) are available: Beta-globin (erythrocytes) (Heck et al., 2003), CD2 (T cells) (Singbartl et al., 2001), CD41 (megakaryocytes) (Zhang et al., 2007), Epo-receptor (erythroid progenitors) (Heinrich et al., 2004), GATA-1 (erythrocytes, megakaryocytes, eosinophils and mast cells) (Iwasaki et al., 2005), GATA-2 (early hematopoietic progenitors) (Kobayashi-Osaki et al., 2005), Lysozyme M (granulocytes and monocytes) (Faust et al., 2000), PAX5 (B cells) (Fuxa and Busslinger, 2007), PU.1 (granulocytes and monocytes and others) (Back et al., 2004; Kirstetter et al., 2006), RAG-1 and RAG-2 (lymphoid progenitors, lymphoid cells) (Kuwata et al., 1999; Yu et al., 1999), Stem Cell Antigen-1 (Sca-1) (HSC and multi-potent progenitors and others) (Hanson et al., 2003; Ma et al., 2002). Therefore, cells from these mice can be used to follow the differentiation of these corresponding lineages over real-time *in vitro*. Furthermore, by using destabilized versions of fluorescent proteins, which are degraded shortly after their translation, researchers can precisely visualize when a specific gene is down-regulated (Li et al., 1998).

These tools allow the visualization of different lineage differentiation as blood cells interact with niche cells. Therefore, by using cells from mice that express a fluorescent protein in a certain lineage, one can visualize whether a microenvironment promotes their quiescence or promotes their differentiation. It is also possible to determine how different microenvironments change the behaviour of blood cells in this way. Thus, motility, proliferation and differentiation can be measured simultaneously (as well as other parameters) using constant video surveillance and fluorescent reporters.

In addition to providing an understanding of how different hematopoietic cells interact with various niche cells, this technology allows visualization of the maturation and migration of blood cells from one niche to another in real-time. Thus, it is a powerful tool for understanding what signals and cell-to-cell contacts are essential to allow for hematopoiesis to be maintained and what causes certain lineages to mature (intrinsic versus extrinsic cues).

The understanding of the components promoting HSC survival and propagation without losing their ability to self-renew could be applied to various medical applications. This knowledge would allow researchers to find out how to expand specific blood populations *ex vivo*. For example, autologous stem cell expansion would be useful for the treatment of various hematological disorders and to aid the recovery from chemotherapy.

Future Perspectives: *in vivo* Blood Surveillance

Intravital microscopy is a powerful tool to observe the behaviour of blood cells *in vivo*. Several groups have studied the rolling and adhesion of lymphocytes to endothelium *in vivo* over real time (Kubes, 2002; Sumen et al., 2004). This has also been combined with fluorescent dyes that have allowed visualization of these cells within different organs such as the lymph nodes and the bone marrow (Mazo et al., 2005).

In addition to visualization of blood cells in the context of endothelium or bone marrow, tracking of cell *in vivo* has been attempted in the lymph nodes and stromal cells have been shown to direct the movement of B and T cells within discrete sub-regions of the lymph nodes through the action of chemokines and their receptors (Bajenoff et al., 2006). Not only is this the first step in observing the behaviour of blood cells within their normal environment, or niche, but also demonstrates the power of niche cells and their products to influence the strict localization of blood cells. Since stromal cells can direct the movement of blood cells in this context, it is probable that they also do so in the bone marrow environment.

Intravital microscopy provides a powerful tool for observing the migration and behaviour of cells *in vivo* however, it is limited in the fact that it does not permit one to follow individual cells for extended periods of time. Fluorescent dyes can be used to label cell populations that can be subsequently injected, but there are many cell types (both blood and non-blood) that cannot be readily identified. Therefore, it is difficult to ascertain exact cell interactions and how specific molecules affect cell behaviour.

It will be of great interest to combine the single cell tracking technology developed *in vitro* and intravital microscopy *in vivo* imaging to allow for long-term single cell tracking of cells *in vivo*. Within this setting cells could be non-invasively observed within their normal environment and we would be able to draw firm conclusions about their behaviour and understand precisely how they are influenced to constantly ensure the correct balance of blood cell production.

Conclusion

There is a need for single cell technology to answer long-standing questions that remain in hematopoietic biology. These may be answered through time-lapse imaging of individual cells in combination with fluorescent marking of proteins and/or maturation stages with subsequent single cell tracking. As well, levels of protein expression can be observed, as well as the behaviour a cell undergoes to reach a certain lineage under the influence of cytokines (permissive versus instructive). By following the behaviour of cells and their dynamic interactions with niche cells over time, we can better understand how HSC and other hematopoietic cells are maintained *in vivo*.

The combination of these technologies with *in vivo* imaging will be essential for a full understanding of this complex process. The isolation of HSC was a result of monoclonal antibody generation and flow cytometric cell sorting. There are now other questions in hematopoietic biology that can only be answered as new technology is developed. The knowledge gained as these technologies become available should allow scientists to be able to better manipulate hematopoietic cells for the use in the clinic for various hematological diseases and treatments in the near future.

Acknowledgements

ED was supported by a stipendium from the Alexander von Humboldt Foundation.

References

Adams, G. B., Martin, R. P., Alley, I. R., Chabner, K. T., Cohen, K. S., Calvi, L. M., Kronenberg, H. M., and Scadden, D. T. (2007). Therapeutic targeting of a stem cell niche. *Nat Biotechnol 25*, 238-243.

Akashi, K., Kondo, M., von Freeden-Jeffry, U., Murray, R., and Weissman, I. L. (1997). Bcl-2 rescues T lymphopoiesis in interleukin-7 receptor-deficient mice. Cell *89*, 1033-1041.

Arai, F., Hirao, A., Ohmura, M., Sato, H., Matsuoka, S., Takubo, K., Ito, K., Koh, G. Y., and Suda, T. (2004). Tie2/angiopoietin-1 signaling regulates hematopoietic stem cell quiescence in the bone marrow niche. *Cell 118*, 149-161.

Avecilla, S. T., Hattori, K., Heissig, B., Tejada, R., Liao, F., Shido, K., Jin, D. K., Dias, S., Zhang, F., Hartman, T. E., *et al.* (2004). Chemokine-mediated interaction of hematopoietic progenitors with the bone marrow vascular niche is required for thrombopoiesis. *Nat Med 10*, 64-71.

Back, J., Dierich, A., Bronn, C., Kastner, P., and Chan, S. (2004). PU.1 determines the self-renewal capacity of erythroid progenitor cells. *Blood 103*, 3615-3623.

Bajenoff, M., Egen, J. G., Koo, L. Y., Laugier, J. P., Brau, F., Glaichenhaus, N., and Germain, R. N. (2006). Stromal cell networks regulate lymphocyte entry, migration, and territoriality in lymph nodes. *Immunity 25*, 989-1001.

Becker, A. J., Mc, C. E., and Till, J. E. (1963). Cytological demonstration of the clonal nature of spleen colonies derived from transplanted mouse marrow cells. *Nature 197*, 452-454.

Calvi, L. M., Adams, G. B., Weibrecht, K. W., Weber, J. M., Olson, D. P., Knight, M. C., Martin, R. P., Schipani, E., Divieti, P., Bringhurst, F. R., *et al.* (2003). Osteoblastic cells regulate the haematopoietic stem cell niche. *Nature 425*, 841-846.

Cho, J. Y., Lee, W. B., Kim, H. J., Mi Woo, K., Baek, J. H., Choi, J. Y., Hur, C. G., and Ryoo, H. M. (2006). Bone-related gene profiles in developing calvaria. *Gene 372*, 71-81.

Cross, M. A., and Enver, T. (1997). The lineage commitment of haemopoietic progenitor cells. *Curr Opin Genet Dev 7*, 609-613.

Dong, F., van Buitenen, C., Pouwels, K., Hoefsloot, L. H., Lowenberg, B., and Touw, I. P. (1993). Distinct cytoplasmic regions of the human granulocyte colony-stimulating factor receptor involved in induction of proliferation and maturation. *Mol Cell Biol 13*, 7774-7781.

Dzierzak, E. (2003). Ontogenic emergence of definitive hematopoietic stem cells. *Curr Opin Hematol 10*, 229-234.

Ema, H., Takano, H., Sudo, K., and Nakauchi, H. (2000). In vitro self-renewal division of hematopoietic stem cells. *J Exp Med 192*, 1281-1288.

Faust, N., Varas, F., Kelly, L. M., Heck, S., and Graf, T. (2000). Insertion of enhanced green fluorescent protein into the lysozyme gene creates mice with green fluorescent granulocytes and macrophages. *Blood 96*, 719-726.

Fliedner, T. M. (1998). Prologue to characteristics and potentials of blood stem cells, Stem Cells volume 16, supplement 1. *Stem Cells 16*, 357-360.

Fukunaga, R., Ishizaka-Ikeda, E., and Nagata, S. (1993). Growth and differentiation signals mediated by different regions in the cytoplasmic domain of granulocyte colony-stimulating factor receptor. *Cell 74*, 1079-1087.

Fuxa, M., and Busslinger, M. (2007). Reporter gene insertions reveal a strictly B lymphoid-specific expression pattern of pax5 in support of its B cell identity function. *J Immunol 178*, 3031-3037.

Goldsmith, M. A., Mikami, A., You, Y., Liu, K. D., Thomas, L., Pharr, P., and Longmore, G. D. (1998). Absence of cytokine receptor-dependent specificity in red blood cell differentiation in vivo. *Proc Natl Acad Sci U S A 95*, 7006-7011.

Goodell, M. A., McKinney-Freeman, S., and Camargo, F. D. (2004). Isolation and characterization of side population cells. *Methods Mol Biol 290*, 343-352.

Hackney, J. A., Charbord, P., Brunk, B. P., Stoeckert, C. J., Lemischka, I. R., and Moore, K. A. (2002). A molecular profile of a hematopoietic stem cell niche. *Proc Natl Acad Sci USA 99*, 13061-13066.

Hanson, P., Mathews, V., Marrus, S. H., and Graubert, T. A. (2003). Enhanced green fluorescent protein targeted to the Sca-1 (Ly-6A) locus in transgenic mice results in efficient marking of hematopoietic stem cells in vivo. *Exp Hematol 31*, 159-167.

Heck, S., Ermakova, O., Iwasaki, H., Akashi, K., Sun, C. W., Ryan, T. M., Townes, T., and Graf, T. (2003). Distinguishable live erythroid and myeloid cells in beta-globin ECFP x lysozyme EGFP mice. *Blood 101*, 903-906.

Heinrich, A. C., Pelanda, R., and Klingmuller, U. (2004). A mouse model for visualization and conditional mutations in the erythroid lineage. *Blood 104*, 659-666.

Herzenberg, L. A., and Sweet, R. G. (1976). Fluorescence-activated cell sorting. *Sci Am 234*, 108-117.

Hu, M., Krause, D., Greaves, M., Sharkis, S., Dexter, M., Heyworth, C., and Enver, T. (1997). Multilineage gene expression precedes commitment in the hemopoietic system. *Genes Dev 11*, 774-785.

Ikuta, K., Uchida, N., Friedman, J., and Weissman, I. L. (1992). Lymphocyte development from stem cells. *Annu Rev Immunol 10*, 759-783.

Iwasaki, H., Mizuno, S., Mayfield, R., Shigematsu, H., Arinobu, Y., Seed, B., Gurish, M. F., Takatsu, K., and Akashi, K. (2005). Identification of eosinophil lineage-committed progenitors in the murine bone marrow. *J Exp Med 201*, 1891-1897.

Janeway, C. (2001). *Immunobiology : the immune system in health and disease*, 5th edn (New York, Garland Pub.).

Kiel, M. J., Yilmaz, O. H., Iwashita, T., Terhorst, C., and Morrison, S. J. (2005). SLAM family receptors distinguish hematopoietic stem and progenitor cells and reveal endothelial niches for stem cells. *Cell 121*, 1109-1121.

Kim, M., Cooper, D. D., Hayes, S. F., and Spangrude, G. J. (1998). Rhodamine-123 staining in hematopoietic stem cells of young mice indicates mitochondrial activation rather than dye efflux. *Blood 91*, 4106-4117.

King, A. G., Kondo, M., Scherer, D. C., and Weissman, I. L. (2002). Lineage infidelity in myeloid cells with TCR gene rearrangement: a latent developmental potential of proT cells revealed by ectopic cytokine receptor signaling. *Proc Natl Acad Sci U S A 99*, 4508-4513.

Kirstetter, P., Anderson, K., Porse, B. T., Jacobsen, S. E., and Nerlov, C. (2006). Activation of the canonical Wnt pathway leads to loss of hematopoietic stem cell repopulation and multilineage differentiation block. *Nat Immunol 7*, 1048-1056.

Kobayashi-Osaki, M., Ohneda, O., Suzuki, N., Minegishi, N., Yokomizo, T., Takahashi, S., Lim, K. C., Engel, J. D., and Yamamoto, M. (2005). GATA motifs regulate early hematopoietic lineage-specific expression of the Gata2 gene. *Mol Cell Biol 25*, 7005-7020.

Kohler, G., and Milstein, C. (1975). Continuous cultures of fused cells secreting antibody of predefined specificity. *Nature 256*, 495-497.

Kondo, M., Scherer, D. C., Miyamoto, T., King, A. G., Akashi, K., Sugamura, K., and Weissman, I. L. (2000). Cell-fate conversion of lymphoid-committed progenitors by instructive actions of cytokines. *Nature 407*, 383-386.

Kubes, P. (2002). The complexities of leukocyte recruitment. *Semin Immunol 14*, 65-72.

Kuwata, N., Igarashi, H., Ohmura, T., Aizawa, S., and Sakaguchi, N. (1999). Cutting edge: absence of expression of RAG1 in peritoneal B-1 cells detected by knocking into RAG1 locus with green fluorescent protein gene. *J Immunol 163*, 6355-6359.

Li, W., Johnson, S. A., Shelley, W. C., and Yoder, M. C. (2004). Hematopoietic stem cell repopulating ability can be maintained in vitro by some primary endothelial cells. *Exp Hematol 32*, 1226-1237.

Li, X., Zhao, X., Fang, Y., Jiang, X., Duong, T., Fan, C., Huang, C. C., and Kain, S. R. (1998). Generation of destabilized green fluorescent protein as a transcription reporter. *J Biol Chem 273*, 34970-34975.

Lord, B. I. (1990). The architecture of bone marrow cell populations. *Int J Cell Cloning 8*, 317-331.

Ma, X., Robin, C., Ottersbach, K., and Dzierzak, E. (2002). The Ly-6A (Sca-1) GFP transgene is expressed in all adult mouse hematopoietic stem cells. *Stem Cells 20*, 514-521.

Maraskovsky, E., O'Reilly, L. A., Teepe, M., Corcoran, L. M., Peschon, J. J., and Strasser, A. (1997). Bcl-2 can rescue T lymphocyte development in interleukin-7 receptor-deficient mice but not in mutant rag-1-/- mice. *Cell 89*, 1011-1019.

Matsuzaki, Y., Kinjo, K., Mulligan, R. C., and Okano, H. (2004). Unexpectedly efficient homing capacity of purified murine hematopoietic stem cells. *Immunity 20*, 87-93.

Mazo, I. B., Honczarenko, M., Leung, H., Cavanagh, L. L., Bonasio, R., Weninger, W., Engelke, K., Xia, L., McEver, R. P., Koni, P. A., *et al.* (2005). Bone marrow is a major reservoir and site of recruitment for central memory CD8+ T cells. *Immunity 22*, 259-270.

Metcalf, D. (1999). Stem cells, pre-progenitor cells and lineage-committed cells: are our dogmas correct? *Ann N Y Acad Sci 872*, 289-303; discussion 303-284.

Metcalf, D., and Burgess, A. W. (1982). Clonal analysis of progenitor cell commitment of granulocyte or macrophage production. *J Cell Physiol 111*, 275-283.

Miyamoto, T., Iwasaki, H., Reizis, B., Ye, M., Graf, T., Weissman, I. L., and Akashi, K. (2002). Myeloid or lymphoid promiscuity as a critical step in hematopoietic lineage commitment. *Dev Cell 3*, 137-147.

Moore, K. A., Ema, H., and Lemischka, I. R. (1997). In vitro maintenance of highly purified, transplantable hematopoietic stem cells. *Blood 89*, 4337-4347.

Nakamura, M., Yagi, H., Endo, Y., Kosugi, H., Ishi, T., and Itoh, T. (1999). A time kinetic study of the effect of aminobisphosphonate on murine haemopoiesis. *Br J Haematol 107*, 779-790.

Nilsson, S. K., Dooner, M. S., Tiarks, C. Y., Weier, H. U., and Quesenberry, P. J. (1997). Potential and distribution of transplanted hematopoietic stem cells in a nonablated mouse model. *Blood 89*, 4013-4020.

Ohneda, O., Fennie, C., Zheng, Z., Donahue, C., La, H., Villacorta, R., Cairns, B., and Lasky, L. A. (1998). Hematopoietic stem cell maintenance and differentiation are supported by embryonic aorta-gonad-mesonephros region-derived endothelium. *Blood 92*, 908-919.

Okita, C., Sato, M., and Schroeder, T. (2004). Generation of optimized yellow and red fluorescent proteins with distinct subcellular localization. *Biotechniques 36*, 418-422, 424.

Osawa, M., Hanada, K., Hamada, H., and Nakauchi, H. (1996). Long-term lymphohematopoietic reconstitution by a single CD34- low/negative hematopoietic stem cell. *Science 273*, 242-245.

Prasher, D. C., Eckenrode, V. K., Ward, W. W., Prendergast, F. G., and Cormier, M. J. (1992). Primary structure of the Aequorea victoria green-fluorescent protein. *Gene 111*, 229-233.

Rattis, F. M., Voermans, C., and Reya, T. (2004). Wnt signaling in the stem cell niche. *Curr Opin Hematol 11*, 88-94.

Scadden, D. T. (2006). The stem-cell niche as an entity of action. *Nature 441*, 1075-1079.

Schofield, R. (1978). The relationship between the spleen colony-forming cell and the haemopoietic stem cell. *Blood Cells 4*, 7-25.

Schroeder, T. (2005). Tracking hematopoiesis at the single cell level. *Ann N Y Acad Sci 1044*, 201-209.

Shaner, N. C., Campbell, R. E., Steinbach, P. A., Giepmans, B. N., Palmer, A. E., and Tsien, R. Y. (2004). Improved monomeric red, orange and yellow fluorescent proteins derived from Discosoma sp. red fluorescent protein. *Nat Biotechnol 22*, 1567-1572.

Shimomura, O., Johnson, F. H., and Saiga, Y. (1962). Extraction, purification and properties of aequorin, a bioluminescent protein from the luminous hydromedusan, Aequorea. *J Cell Comp Physiol 59*, 223-239.

Sieburg, H. B., Cho, R. H., Dykstra, B., Uchida, N., Eaves, C. J., and Muller-Sieburg, C. E. (2006). The hematopoietic stem compartment consists of a limited number of discrete stem cell subsets. *Blood 107*, 2311-2316.

Siminovitch, L., McCulloch, E. A., and Till, J. E. (1963). The Distribution of Colony-Forming Cells among Spleen Colonies. *J Cell Physiol 62*, 327-336.

Singbartl, K., Thatte, J., Smith, M. L., Wethmar, K., Day, K., and Ley, K. (2001). A CD2-green fluorescence protein-transgenic mouse reveals very late antigen-4-dependent CD8+ lymphocyte rolling in inflamed venules. *J Immunol 166*, 7520-7526.

Stoffel, R., Ziegler, S., Ghilardi, N., Ledermann, B., de Sauvage, F. J., and Skoda, R. C. (1999). Permissive role of thrombopoietin and granulocyte colony-stimulating factor receptors in hematopoietic cell fate decisions in vivo. *Proc Natl Acad Sci U S A 96*, 698-702.

Sugiyama, T., Kohara, H., Noda, M., and Nagasawa, T. (2006). Maintenance of the hematopoietic stem cell pool by CXCL12-CXCR4 chemokine signaling in bone marrow stromal cell niches. *Immunity 25*, 977-988.

Sumen, C., Mempel, T. R., Mazo, I. B., and von Andrian, U. H. (2004). Intravital microscopy: visualizing immunity in context. *Immunity 21*, 315-329.

Suzuki, N., Ohneda, O., Minegishi, N., Nishikawa, M., Ohta, T., Takahashi, S., Engel, J. D., and Yamamoto, M. (2006). Combinatorial Gata2 and Sca1 expression defines hematopoietic stem cells in the bone marrow niche. *Proc Natl Acad Sci U S A 103*, 2202-2207.

Taichman, R. S. (2005). Blood and bone: two tissues whose fates are intertwined to create the hematopoietic stem-cell niche. *Blood 105*, 2631-2639.

Taichman, R. S., and Emerson, S. G. (1994). Human osteoblasts support hematopoiesis through the production of granulocyte colony-stimulating factor. *J Exp Med 179*, 1677-1682.

Takano, H., Ema, H., Sudo, K., and Nakauchi, H. (2004). Asymmetric division and lineage commitment at the level of hematopoietic stem cells: inference from differentiation in daughter cell and granddaughter cell pairs. *J Exp Med 199*, 295-302.

Till, J. E., and McCulloch, E. A. (1961). A direct measurement of the radiation sensitivity of normal mouse bone marrow cells. *Radiat Res 14*, 213-222.

Tokoyoda, K., Egawa, T., Sugiyama, T., Choi, B. I., and Nagasawa, T. (2004). Cellular niches controlling B lymphocyte behavior within bone marrow during development. *Immunity 20*, 707-718.

Uchida, N., and Weissman, I. L. (1992). Searching for hematopoietic stem cells: evidence that Thy-1.1lo Lin- Sca-1+ cells are the only stem cells in C57BL/Ka-Thy-1.1 bone marrow. *J Exp Med 175*, 175-184.

Visnjic, D., Kalajzic, I., Gronowicz, G., Aguila, H. L., Clark, S. H., Lichtler, A. C., and Rowe, D. W. (2001). Conditional ablation of the osteoblast lineage in Col2.3deltatk transgenic mice. *J Bone Miner Res 16*, 2222-2231.

Wagers, A. J., Christensen, J. L., and Weissman, I. L. (2002). Cell fate determination from stem cells. *Gene Ther 9*, 606-612.

Weissman, I. L. (2002). The road ended up at stem cells. I*mmunol Rev 185*, 159-174.

Whetton, A. D., and Graham, G. J. (1999). Homing and mobilization in the stem cell niche. *Trends Cell Biol 9*, 233-238.

Wilson, A., and Trumpp, A. (2006). Bone-marrow haematopoietic-stem-cell niches. *Nat Rev Immunol 6*, 93-106.

Wineman, J., Moore, K., Lemischka, I., and Muller-Sieburg, C. (1996). Functional heterogeneity of the hematopoietic microenvironment: rare stromal elements maintain long-term repopulating stem cells. *Blood 87*, 4082-4090.

Yu, W., Nagaoka, H., Misulovin, Z., Meffre, E., Suh, H., Jankovic, M., Yannoutsos, N., Casellas, R., Besmer, E., Papavasiliou, F., *et al.* (1999). RAG expression in B cells in secondary lymphoid tissues. *Cold Spring Harb Symp Quant Bio*l *64*, 207-210.

Zhang, J., Niu, C., Ye, L., Huang, H., He, X., Tong, W. G., Ross, J., Haug, J., Johnson, T., Feng, J. Q., *et al.* (2003). Identification of the haematopoietic stem cell niche and control of the niche size. *Nature 425*, 836-841.

Zhang, J., Varas, F., Stadtfeld, M., Heck, S., Faust, N., and Graf, T. (2007). CD41-YFP mice allow in vivo labeling of megakaryocytic cells and reveal a subset of platelets hyperreactive to thrombin stimulation. *Exp Hematol 35*, 490-499.

In: New Cell Differentiation Research Topics
Editor: Hitoshi Saitama, pp. 247-258
ISBN: 978-1-60021-937-5

Chapter VIII

CELL DIFFERENTIATION OF ECTOMESENCHYMAL DENTAL CELLS

C. Morsczeck
Institute of human genetics- University of Regensburg, 93053 Regensburg, Germany.

ABSTRACT

Complex human tissues harbour stem cells and/or precursor cells, which are responsible for tissues' development or tissues' repair. These cells provide new insights into dental development and are important for characterizations of cellular differentiation processes under in vitro conditions. In recent times undifferentiated cells were isolated from dental tissues like the periodontal ligament, the dental papilla or the dental follicle. These cells are realistic sources for the construction for example of an engineered periodontium, which submits the connection between implanted tooth and jaw. Therefore, dental precursor cells are interesting for novel treatments of periodontitis or dental implantation. However, we need more information about their differentiation mechanisms for the evaluation of dental treatments. Dental follicle precursor cells (DFPCs) are derived from an embryonic dental tissue of extracted wisdom teeth. These cells are easy accessible and ideal for testing cell differentiation mechanisms during dental development. However, there are many publications about dental precursor cells or stem cells derived from dental ectomesenchymal tissues. This review gives also a short overview about known ectomesenchymal cells of dental origin and classifies these cells.

INTRODUCTION

Tooth maladies are widespread in industrial countries. For example, approximately two thirds of the German citizens have periodontitis or gum inflammations, which are very frequently the cause for tooth loss. Moreover, there is also an association between presence and severity of periodontitis and coronary artery disease (Geerts et al. 2004). Despite medical progress many periodontal illnesses cannot be entirely repaired with current medical treatments. Since many years new opportunities for periodontal treatments are discussed for example by tissue engineering. The actual use of stem cells for dental implants would be a revolutionary step for dentistry. Recently, an artificial tooth root was grown by dental stem cells in a mini-pig model. Contrary, molecular mechanisms of stem cell differentiation into functional periodontal tissues are not very well determined scientifically. This review summarizes current dental differentiation approaches with isolated, undifferentiated dental stem cells or progenitor cells.

Cell Differentiation of Embryonic Cells into Dental Tissues

In the last years two major categories of stem cells are discussed for cell therapies in medicine: embryonic stem cells and adult stem cells. Embryonic stem cells theoretically have the chance to differentiate into all body cells (pluripotency) and have the characteristic to proliferate infinitely. Human embryonic stem cells (hES) can be isolated from the internal cell mass of blastocytes of an 8-9 days old embryo (Terskikh et al., 2006). These cells express embryonic stem cell markers like OCT-4, NANOG and SOX-2 (Trounson, 2004; Cowan et al., 2004). hES are cultivated on an embryonic fibroblasts cell layer (feeder cells). Embryonic stem cells form compact colonies on this fibroblast cell layer, whereby cloning of a single cell is possible. Subsequently, we can yield high numbers of hES after the initial isolation of single cells. A special characteristic of embryonic stem cells is the formation of teratomas after exogenic implantations into immunosuppressed mice in animal testings. Teratomas belong to germ cell tumors and consist of cells of three different germ layers. This tissue can also form organ like structures (Trounson, 2004). For in vitro differentiation tests hES are growing in a culture medium supplemented with foetal calf serum without feeder cells. Later, pre-differentiated hESs form embryonic bodies, which consist of cells of all three germ layers. For cell therapies special progenitor cells can be derived from embryonic bodies. However, embryonic stem cells form tumors after cell transplantations, therefore they are still not suitable for cell therapies. Recently, attempts were done combining murine embryonic, dental epithelium cells with embryonic stem cells. This combination of cells was afterwards transplanted into a kidney cap. Here a tooth germ like structure was formed, which was created without host cells (Ohazama et al. 2004). However, scientific works with human embryonic stem cells is ethically precarious, since killing and use of unborn human life for research purposes contradicts our moral values (current discussion: Yoshimura et al., 2006). Another category of stem cells are the adult stem cells, whose use is ethically harmless. Adult stem cells have several advantages. For example, adult stem cells can easily be isolated from different human tissues. Furthermore, self-donate adult stem cells circumvent rejections after implantation by the immune system. An important drawback for the use of adult stem cells for cell therapies is the asymmetrical cell division of adult stem cells (Shirley 2002). This means that it is impracticable to increase cell numbers of adult stem cells after isolation of single cells from adult tissues. Therefore, larger quantities of donor tissues are required in order to isolate a sufficient number of stem cells for a therapy. Adult stem cells can be classified on the basis of both origin and differentiation plasticity. There are multi-potent adult stem cells and less potent progenitor cells, which differentiate into a limited number of postmitotic cell types (overview article Terskih et al., 2006). Based on these facts we can conclude that adult stem cells represent a source for cell therapies only for more or less particular treatments.

MESENCHYMAL STEM CELLS AND DENTAL TREAMENT

Today, there are different types of adult stem cells for cell therapies. The hematopoietic stem cells originating from the bone marrow are successfully used for the therapy of leukaemia since many years. Non-hematopoietic mesenchymal stem cells are localized in the bone marrow. These cells - also known as „Bone Marrow Stromal Cells “(BMSCs) - were described decades ago for the first time (Friedenstein and Kuralesova, 1971). BMSCs can be isolated from single cell suspension from bone marrow aspirates via their high plastic adherence and clonogenicity (Pittenger et al., 1999). These cells can migrate into organs such as liver and brain after transplantation into an immuncompromised mouse (Mezey et al. 2003, Huttmann et al. 2003). Here, mesenchymal stem cells are capable of differentiation toward osteoblasts, chondrocytes or retinal cells, which means transdifferentiation into cells of at least two different germ layers (Pittenger et al. 1999, Woodbury et al. 2000, Mezey et al., 2000, Tomita et al. 2002). However, it is controversial whether transplanted stem cells acquire tissue specific markers by transdifferentiation or cell fusion (Prockop et al. 2003). Nonetheless, current studies support the hypothesis that transdifferentiation is possible (Crain et al. 2005). Today, mesenchymal stem cells in combination with dental stem cells are discussed for suitable dental cell therapies (Maria et al. 2007, Ohazama et al. 2004).

DENTAL EPITHELIAL STEM CELLS

The tooth development is based on interactions between the ectodermal and the ectomesenchymal (neural crest) germ layers. The development of the tooth germ begins in the mandibular arch of the embryo. The dental ectoderm builds the surface of the mandibular arch of an embryo and is in close vicinity with the underlying ectomesenchymal tissue. Presently, it is accepted that the dental development starts in the dental ectoderm. In previous studies it was demonstrated that embryonic, dental epithelial tissues were capable to induce the development of a tooth germ in combination with non-dental ectomesenchymal tissues (Ten Cate 1996). By contrast, the dental ectomesoderm in combination with non-dental epithelial tissues did not induce tooth development. Later steps of the dental development are driven by a complex interaction of both the ectodermal tissue and the ectomesenchymal tissue. Two important factors, BMP-4 and FGF-8, produced of the dental ectoderm are significant for the start of tooth development (Modino and Sharpe 2005).

Nevertheless, it is evident that stem cells or progenitor cells are involved in dental development; the identification of dental stem cells was easier said than done. Contrary to humans, which have no regeneration of incisors, rodents regenerate incisors and form new enamel after tooth eruption. Thereby the murine incisor is an ideal model for the development of the tooth crown and especially for the development of ameloblasts. Irma Thesleff's group identified these dental ectodermal stem cells in tissue explants of adult mouse incisors for the first time (Harada et al. 1999). BrdU-labelled stem cells enabled a detailed examination of cell migration and ameloblast development in dental explants. The fibroblast growth factor (FGF), in particular FGF-10 and the activated Notch-Pathway are essential for the undifferentiated status of dental stem cells and for a directed differentiation of stem cells into ameloblasts or in cells of the Stratum intermedium (Harada et al. 2002, 2006). These dental epithelial stem cells are lost, however, after tooth eruption in humans and therefore these dental stem cells are not available for studies of human dental development.

Dental Ectomesenchymal Stem Cells

In contrast to dental epithelial stem cells undifferentiated dental ectomesenchymal cells are not entirely lost after tooth eruption in man. For example, dental ectomesenchymal stem cells were isolated from the dental pulp (DPSCs) of extracted wisdom teeth (Gronthos et al. 2000). Bone marrow derived mesenchymal stem cells and DPSCs have similar features. For example, both cell types are plastic adherent and colony-forming cells. By contrast DPSCs were found to be capable to differentiate into odontoblasts. In a DNA microarray study, DPSCs could be distinguished from bone marrow derived mesenchymal stem cells (Shi et al. 2001). Here, DPSCs differentially expressed cell cycle associated genes. This result corresponds to DFPCs' high proliferation rate compared to the proliferation rate of mesenchymal stem cells (Gronthos et al. 2000). In an additional study DPSCs were also found to be capable of differentiating into adipocytes and neural-like cells (Gronthos et al. 2002). Interestingly, dental stem cells are located perivascular and express the stem cell marker Stro-1 (Shi and Gronthos 2003). Recently, a special stem cell population was isolated from the dental pulp (Kerkis et al. 2006). These stem cells can be induced to undergo uniform differentiation into smooth and skeletal muscles, neurons, cartilage, and bone under chemically defined culture conditions. Furthermore, ectcomesenchymal stem cells SHED (stem cells of human exfoliated deciduous teeth) were isolated from the dental pulp of exfoliated deciduous teeth (Miura et al. 2003). These cells were cultivated either as fibroblast like, plastic-adherent cells or like neural stem cells as neurospheres. SHED could also be differentiated into odontoblast, adipocytes and neural cells (Miura et al. 2003). SHED induced bone formation and produced dentin under in vivo conditions; however, these cells migrated and survived also in the murine brain after introduction into immuncompromised animals (Miura et al. 2003).

A further class of dental ectomesenchymal stem cells are PDL stem cells, which were isolated from the root surface of extracted teeth (Seo et al. 2004). These cells could be isolated as plastic-adherent, colony forming cells with only a low potential for osteogenic differentiation under in vitro conditions (Seo et al. 2004). PDL stem cells differentiate under in vivo and in vitro conditions toward cells or tissues, which are very similar to the periodontium (Seo et al. 2004). Recently, PDL stem cells were also isolated from sheep and pigs (Gronthos et al. 2006, Sonoyama et al. 2006).

A new class of dental stem cells was isolated from the dental papilla of wisdom teeth or from incisors of 4 month old mini-pigs (SCAP, stem cells from apical papilla) (Sonoyama et al. 2006, Jo et al. 2007). This tissue is an embryonic like tissue and becomes dental pulp after tooth maturation. Therefore, SCAPs can only be isolated at a certain stage of tooth development. The SCAPs have a greater capacity for dentin regeneration than DPSCs, because the dental papilla contains a higher number of adult stem cells compared with the maturated dental pulp. In addition, SCAPs' are probably even more undifferentiated than DPSCs, because they are originated from an embryonic-like tissue. Interestingly, it was exclusively a combination of SCAPs and PDL stem cells, which confers the formation of a dental connective tissue for the attachment of an artificial tooth crown in the jaw. It is however unknown what stem cells are due for the creation of an artificial dental connective tissue for tooth attachment in this exciting experiment described by Sonoyama et al. 2006. Moreover, there is no information about the mechanisms of cell differentiation in this particular in vivo differentiation experiment.

The formation of dentin and the formation of periodontal like tissues are the main differentiation potentials of dental pulp associated stem cells and PDL stem cells, respectively. Although PDL-stem cells differentiate toward periodontal cells the dental follicle represents the precursor of the periodontium and plays a crucial role in tooth development. The dental follicle is separated from developing dentin by epithelial cell layers (Hertwig's epithelial root sheath) during early steps of the periodontal development. The current observation is that disintegration of the Hertwig's epithelial root sheath into epithelial fragments allows ectomesenchymal cells from the dental follicle proper to come in contact with the dentin surface where they differentiate into mature cells of the periodontium (Spouge 1980; MacNeil and Thomas 1993). Therefore, this tissue should also contain progenitor cells for the alveolar osteoblasts, PDL-fibroblasts and cementoblasts. In a first experiment, progenitor cells were isolated from dental follicles of bovine. Interestingly, these cells were cultivated in vitro as spheroids only. Surprisingly, differentiation of dental follicle cells toward dental tissues was exclusively possible under in vivo conditions (Handa et al. 2002).

It is obvious that dental ectomesenchymal stem cells can be classified in 2 different groups of stem cells in respect to their major differentiation potentials. The first group is the dental pulp associated group with DPSCs, SHEDs and SCAPs and the second group contains PDL stem cells and dental follicle progenitor cells and is related to the periodontium.

Human Dental Follicle Precursor Cells (DFPC)

The human dental follicle (dental sac) is a tissue of the tooth germ, which can be easily isolated after wisdom tooth extraction. Likewise bovine dental follicles, cells of the human dental sac develop into the mature periodontium consisting of the alveolar bone, the PDL and the tooth cement. The dental follicle contains ectomesenchymal cells which are derived from the neural crest. DFPCs like BMSCs are plastic adherent and colony forming cells (figure 1). These precursor cells can proliferate under in vitro conditions and express the stem cell markers Notch-1 and Nestin. DFPCs differ in their gene expression profile from mesenchymal stem cells (Morsczeck et al. 2005a). However, like mesenchymal stem cells DFPCs differentiated into osteoblast-like cells under in vitro conditions, but differentiated DFPCs also formed sturdy connective tissues beside clusters of mineralized tissue. These connective tissues were stable and built biological diaphragms-like structures. These structures were taken off mechanically for histological investigations from the surface of cell culture flasks (figure 2). In histological investigations the diaphragms turned out as complex connective tissue structures, which were also interrupted, interestingly enough, by vessel-like structures (Morsczeck et al. 2005a). In repeated, independent experiments it was possible to differentiate these precursor cells either more strongly into dental connective tissue-like structures or in bone-like tissues. Here, typical markers were expressed for the PDL and the tooth cement, like e.g. collagen type XII (Col_12) and the cementum attachment protein (CAP). On the other side, the dentin sialophosphoprotein (DSPP) and Ameloblastin, characteristic markers for odontoblasts and ameloblasts, were not detected in differentiated DFPCs (Morsczeck et al. 2005b). These results demonstrate that in vitro formed cells and tissues can be designated as a periodontium-like tissue, which is responsible for the connection between tooth and jaw. DFPCs differentially expressed osteoblast-like markers bone sialoprotein (BS) and osteocalcin (OCN) in a differentiation experiment under in vivo conditions in immuncompromised mice. Nevertheless, human connective tissues were exclusively detectable in histological samples (Morsczeck et al. 2005a). In order to characterize DFPCs more exactly, gene expression profiles were examined and compared with mesenchymal stem cells, osteoblasts and PDL-fibroblasts (Morsczeck et al. 2005a). Differences between different cell types were evident for the expression of insulin like growth factors (IGF)-2, collagen type I (Col_1), bone sialoprotein and osteocalcin, whereby dental cells expressed low levels of osteoblast markers such as OCN and BS and increased levels of IGF-2 and Col_1. In addition, DFPCs could be distinguished from mesenchymal stem cells, osteoblasts and PDL fibroblasts by quantitative gene expression analyses (Morsczeck 2005a).

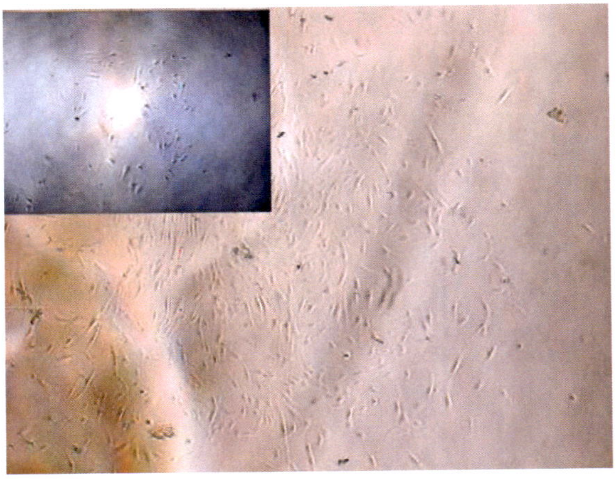

Figure 1. DFPCs are in general plastic adherent cells. The embedded small picture displays a colony forming unit after approximately 1 week. Here, dental cells had heterogenic cell morphologies. After 3 weeks DFPCs assembled to bigger cell clusters (large picture). Interestingly, DFPCs had a more fibroblast-like morphology.

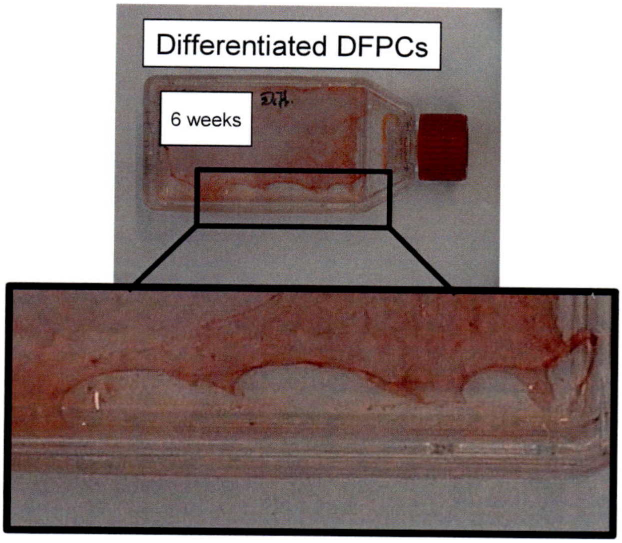

Figure 2. DFPCs differentiated into calcified clusters (alizarin red colouring) and into connective tissue-like multilevel cell layers. Differentiated DFPCs (6 weeks after induction) formed a visible and stable membrane-like structure, which can be taken off mechanically from the surface of the culture flask.

Recently, Salles and colleagues published a study about DFPCs. This group demonstrated that DFPCs express the mesenchymal stem cell marker Stro-1 and confirmed that human DFPCs have multipotential mesenchymal precursor cell properties (Kemoun et al. 2007). Here, DFPCs differentiated toward multiple mesenchymal derived cell types, such as cementoblasts, chondrocytes and adipocytes.

Mechanisms of Osteogenic Differentiation in DFPCs

The exact molecular mechanisms of osteogenic differentiation of dental progenitor cells or dental stem cells are not very well examined. The growth factor BMP-2 is a frequently discussed inducer for the mineralization process in mesenchymal progenitor cells or stem cells (Popowics et al. 2005). On the other side a mix of proteins from dental tissues and special transcription factors like MSX-2 are proposed for the inhibition of osteogenic differentiation in dental cells (Hakki et al. 2001, Yoshizawa et al. 2004). For example the differential expression of MSX-2 in PDL fibroblasts blocks the affinity of runx2 to the OCN promoter and inhibits mineralization of PDL cells (Yoshizawa et al. 2004). PDL fibroblasts synthesize a gum like extracellular matrix in the PDL for a flexible connection between tooth and jaw. This protective effect by MSX-2 seems to be waived by the differential expression of the transcription factor DLX-5 (Ryo et al. 2006). DLX-3 - likewise DLX-5 - is also an important factor for the initiation of osteogenic differentiation in undifferentiated osteoblasts (Hassan et al. 2006). In a current model for osteogenic differentiation the increase of BMP2 expression initiates the osteogenic differentiation process in undifferentiated osteoblasts (Hassan et al. 2006). An increased concentration of BMP2 facilitates a decrease for MSX-2 and an increase of DLX-3. DLX-3 builds complexes with the *runx2* promoter and inhibits the attachment of MSX-2 to this promoter. At this stage of differentiation, the expression of DLX-5 increases and this transcription factor binds to both *runx2* and *OCN* promoters, resulting in an increased gene expression of *runx2* and *OCN* (figure 3). Finally, both promoters bind runx2, which emphasizes *runx2* and *OCN* transcription levels and the differentiation process of osteoblasts. Additionally, Smad 1 and Smad 5 are involved to a lesser extent in the activation of OCN and BSP promoters (Hassan et al. 2006).

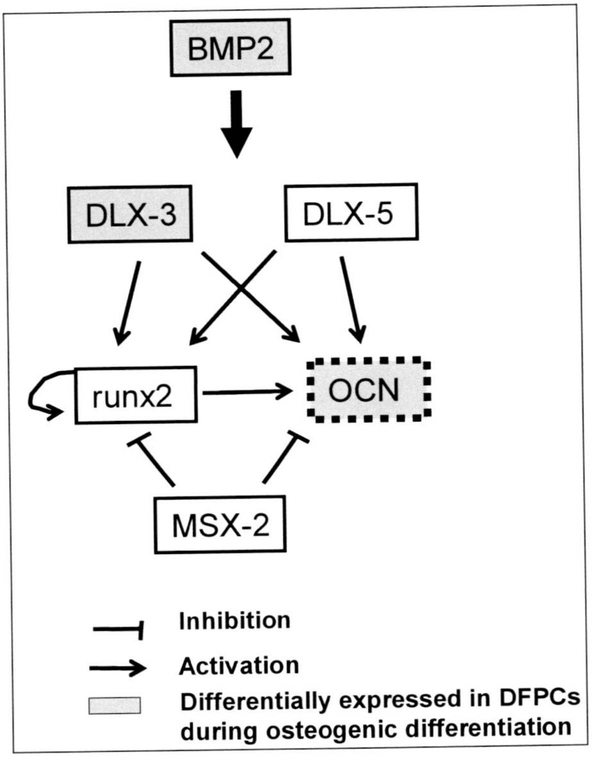

Figure 3. Activation and inhibition of runx2 and OCN promoter activities by transcription factors MSX-2, DLX-3, DLX-5 and runx2 during osteogenic differentiation in osteoblasts. Differentially expressed genes in dental follicle cells during osteogenic differentiation are coloured in grey.

In the light of osteogenic differentiation mechanisms in osteoblasts and PDL-fibroblasts, gene expressions in DFPCs were examined in vitro for osteogenic differentiation associated factors during osteogenic differentiation (Morsczeck et al. 2005b, Morsczeck 2006). Here, it was remarkable that BMP-2, an inducer of osteogenic differentiation in undifferentiatied osteoblasts, was differentially expressed in DFPCs. However, only individual mineralized clusters surrounded by sturdy connective tissue were formed during differentiation (see above and Morsczeck 2006). Recently, Salles and colleagues demonstrated that DFPCs express different BMP-receptors (BMPR)-Ia,-Ib,-II. Moreover, they showed that the alkaline phosphates (AP) activity and the mineralization in DFPCs increased after long term incubation with BMP2 and BMP7 (Kemoun et al. 2007).

Gene expressions of transcription factors give important information about the cellular mechanisms of osteogenic differentiation (Morsczeck 2006). As mentioned above, the initiation of osteogenic differentiation in undifferentiated PDL fibroblasts is regulated by homeodomain proteins like MSX-2 and DLX-5. Contrary to PDL fibroblasts gene expressions of MSX-2 and DLX-5 are nearly constitutive in DFPCs during osteogenic differentiation (Morsczeck 2006). Dental follicle cells seem to use therefore different mechanisms for the regulation of osteogenic differentiation than PDL fibroblasts. Consequently, it is conceivable that DFPCs differentiate into both mineralized and connective tissues like structures after differentiation in vitro. The exact reasons for this are still unknown and have to be examined. In DFPCs – likewise in osteoblasts - DLX-3 was differentially expressed during osteogenic differentiation. This transcription factor, however, was also differentially expressed without an osteogenic inductor like dexamethasone in long-term cultures. Possibly, the osteogenic differentiation in DFPCs is independent of a major inductor for osteogenic differentiation (Morsczeck 2006). Contrary to mesenchymal stem cells or undifferentiated osteoblasts, *DLX-5*, *runx2* and *Osterix* are not differentially expressed in osteogenic differentiated DFPCs (Morsczeck 2006). DFPCs consequently seem to use different differentiation mechanisms than mesenchymal stem cells or osteoblasts. However, there are also similarities with osteogenic differentiation mechanisms in osteoblasts or PDL fibroblasts (figure 3).

Recently, Salles and colleagues reported that signal transduction pathways containing Smad1 and MAPK are activated during mineralization after stimulations of DFPCs with enamel matrix protein (EMD), BMP-2 and BMP-7 (Kemoun et al. 2007).

Conclusion

We do not know much about the mechanisms of cell differentiation. Furthermore, the cell differentiation potential of DFPCs is not yet investigated in details. The neural differentiation potential of the DFPCs for example was not examined yet. It is possible that DFPCs have a promising neural differentiation potential, because these cells express the neural stem cell markers Nestin and Notch-1. Moreover, the differentiation of DFPCs can be also induced by insulin which as well is a supplement for neural differentiation of embryonic stem cells (Morsczeck et al. 2005b). In contrast, neural stem cells proliferate as neurospheres in serum replacement media which is unlikely for DFPCs. Likewise SHEDs (Miura et al. 2003) or neural stem cells DFPCs express the undifferentiated neural stem cell markers Nestin and Notch-1, which mRNAs are also constitutively expressed during osteogenic differentiation in DFPCs (Morsczeck et al. 2005b). In spite of these, the functions for cell proliferation and cell differentiation of Notch-1 in DFPCs are unknown. However, this problem will be tackled in future projects.

REFERENCES

Cowan,C.A., Klimanskaya,I., McMahon,J., Atienza,J., Witmyer,J., Zucker,J.P., Wang,S., Morton,C.C., McMahon,A.P., Powers,D., and Melton,D.A. (2004). Derivation of embryonic stem-cell lines from human blastocysts. *N. Engl. J. Med. 350*, 1353-1356.

Crain B.J., Tran S.D., Mezey E. (2005). Transplanted human bone marrow cells generate new brain cells. *J Neurol Sci. 233*(1-2):121-123

Friedenstein, A. and Kuralesova,A.I. (1971). Osteogenic precursor cells of bone marrow in radiation chimeras. *Transplantation 12*, 99-108

Geerts, S.O., Legrand, V., Charpentier, J., Albert, A., Rompen, E.H. (2004). Further evidence of the association between periodontal conditions and coronary artery disease. *J Periodontol. 75*:1 274-80

Gronthos, S., Mankani, M., Brahim, J., Robey, P.G., Shi, S. (2000). Postnatal human dental pulp stem cells (DPSCs) in vitro and in vivo. *Proc Natl Acad Sci U S A. 97*(25):13625-30

Gronthos, S., Brahim, J., Li, W., Fisher, L.W., Cherman, N., Boyde, A., DenBesten, P., Robey, P.G., Shi, S. (2002). Stem cell properties of human dental pulp stem cells. *J Dent Res. 81*:531-5.

Gronthos, S., Mrozik, K., Shi, S., Bartold, P.M. (2006). Ovine periodontal ligament stem cells: isolation, characterization, and differentiation potential. *Calcif Tissue Int. 79*(5):310-7

Hakki, S.S., Berry, J.E., Somerman, M.J. (2001). The effect of enamel matrix protein derivative on follicle cells in vitro. *J Periodontol. 72*(5):679-87

Handa, K., Saito, M., Yamauchi, M., Kiyono, T., Sato, S., Teranaka, T., Sampath Narayanan, A. (2002). Cementum matrix formation in vivo by cultured dental follicle cells. *Bone. 31*(5):606-11

Harada, H., Ichimori,Y., Yokohama-Tamaki, T., Ohshima, H., Kawano, S., Katsube, K., Wakisaka, S. (2006) Stratum intermedium lineage diverges from ameloblast lineage via Notch signaling. *Biochem Biophys Res Commun. 340*(2):611-6.

Harada, H., Toyono, T., Toyoshima, K., Yamasaki, M., Itoh, N., Kato, S., Sekine, K., Ohuchi, H. (2002). FGF10 maintains stem cell compartment in developing mouse incisors. *Development. 129*(6):1533-41

Harada, H., Kettunen, P., Jung, H.S., Mustonenm T., Wang, Y.A., Thesleff, I. (1999). Localization of putative stem cells in dental epithelium and their association with Notch and FGF signaling. *J Cell Biol. 147*(1):105-20.

Hassan, M.Q., Tare, R.S., Lee, S.H., Mandeville, M., Morasso, M.I., Javed, A., van Wijnen, A.J., Stein, J.L., Stein, G.S., Lian, J.B. (2006). BMP2 commitment to the osteogenic lineage involves activation of Runx2 by DLX3 and a homeodomain transcriptional network. *J Biol Chem. 281*(52):40515-26.

Huttmann,A., Li,C.L., and Duhrsen,U. (2003). Bone marrow-derived stem cells and "plasticity". *Ann. Hematol. 82*, 599-604.

Jo ,Y.Y., Lee, H.J., Kook, S.Y., Choung, H.W., Park, J.Y., Chung, J.H., Choung, Y.H., Kim, E.S., Yang, H.C., Choung, P.H. (2007). Isolation and characterization of postnatal stem cells from human dental tissues. *Tissue Eng. 13*:767-73.

Kemoun, P., Laurencin-Dalicieux, S., Rue, J., Farges, J.C., Gennero, I., Conte-Auriol, F., Briand-Mesange, F., Gadelorge, M., Arzate, H., Narayanan, A.S., Brunel, G., Salles, J.P. (2007). Human dental follicle cells acquire cementoblast features under stimulation by BMP-2/-7 and enamel matrix derivatives (EMD) in vitro. *Cell Tissue Res. Apr 19*; [Epub ahead of print]

Kerkis, I., Kerkis, A., Dozortsev, D., Stukart-Parsons, G.C., Massironi, S.M., Pereira, L.V., Caplan, A.I., Cerruti, H.F. (2006). Isolation and characterization of a population of immature dental pulp stem cells expressing OCT-4 and other embryonic stem cell markers. *Cells Tissues Organs.184*:105-16

Maria, O.M., Khosravi, R., Mezey, E., Tran, S.D. (2007). Cells from bone marrow that evolve into oral tissues and their clinical applications. *Oral Dis. 13*(1):11-6.

MacNeil R.L., Thomas H.F. (1993) Development of the murine periodontium. II. Role of the periodontal attachment. *J Periodontol 64*:285–291

Mezey,E., Key,S., Vogelsang,G., Szalayova,I., Lange,G.D., and Crain,B. (2003). Transplanted bone marrow generates new neurons in human brains. *Proc. Natl. Acad. Sci. U. S. A 100*, 1364-1369.

Miura, M., Gronthos, S., Zhao, M., Lu, B., Fisher, L.W., Robey, P.G., Shi, S. (2003).SHED: stem cells from human exfoliated deciduous teeth. *Proc Natl Acad Sci U S A. 100*(10):5807-12.

Modino, S.A., Sharpem P.T. (2005).Tissue engineering of teeth using adult stem cells. *Arch Oral Biol. 50*(2):255-8.

Morsczeck C, Moehl C, Gotz W, Heredia A, Schaffer TE, Eckstein N, Sippel C, Hoffmann KH (2005a). In vitrodifferentiation of human dental follicle cells with dexamethasone and insulin. *Cell Biol Int. 29*(7),567-75

Morsczeck C, Gotz W, Schierholz J, Zeilhofer F, Kuhn U, Mohl C, Sippel C, Hoffmann KH (2005b). Isolation of precursor cells (PCs) from human dental follicle of wisdom teeth. *Matrix Biol. 24*(2),155-65

Morsczeck C. Gene expression of runx2, osterix, c-fos, DLX-3, DLX-5 & MSX-2 in dental follicle cells during osteogenic differentiation in vitro. *Calcif Tissue Int.* 2006 78(2):98-102

Ohazama A, Modino SA, Miletich I, Sharpe PT (2004). Stem-cell-based tissue engineering of murine teeth. *J Dent Res. 83*(7):518-22

Pittenger,M.F., Mackay,A.M., Beck,S.C., Jaiswal,R.K., Douglas,R., Mosca,J.D., Moorman,M.A., Simonetti,D.W., Craig,S., and Marshak,D.R. (1999). Multilineage potential of adult human mesenchymal stem cells. *Science 284*, 143-147.

Popowics, T., Foster, B.L., Swanson, E.C., Fong, H., Somerman, M.J. (2005). Defining the roots of cementum formation. *Cells Tissues Organs.;181*(3-4):248-57.

Seo, B.M., Miura, M., Gronthos, S., Bartold, P.M., Batouli, S., Brahim, J., Young, M., Robey, P.G., Wang, C.Y., Shi, S. (2004). Investigation of multipotent postnatal stem cells from human periodontal ligament. *Lancet 364*(9429):149-55.

Shi, S., Gronthos, S. (2003). Perivascular niche of postnatal mesenchymal stem cells in human bone marrow and dental pulp. *J Bone Miner Res. 18*(4):696-704.

Shi, S., Robey, P.G., Gronthos, S. (2001).Comparison of human dental pulp and bone marrow stromal stem cells by cDNA microarray analysis. *Bone 29*(6), 532-539

Sherley, J.L. (2002). Asymmetric cell kinetics genes: the key to expansion of adult stem cells in culture. *ScientificWorldJournal. 2*:1906-1921

Sonoyama, W., Liu, Y., Fang, D., Yamaza, T., Seo, B.M., Zhang, C., Liu, H., Gronthos, S., Wang, C.Y., Shi S, Wang S (2006).Mesenchymal stem cell-mediated functional tooth regeneration in Swine. *PLoS ONE. Dec 20*;1:e79.

Spouge J.D. (1980) A new look at the rests of Malassez: a review of their embryological origin, anatomy, and possiblae role in periodontal health and disease. *J Periodontol 51*:437–444

Prockop D.J., Gregory C.A., Spees J.L. (2003). One strategy for cell and gene therapy: harnessing the power of adult stem cells to repair tissues. *Proc Natl Acad Sci U S A. 100* Suppl 1, 11917-23

Ryoo, H.M., Lee, M.H., Kim, Y.J. (2006). Critical molecular switches involved in BMP-2-induced osteogenic differentiation of mesenchymal cells. *Gene. 366*(1):51-7.

Ten Cate, A.R. (1996). The role of epithelium in the development, structure and function of the tissues of tooth support. *Oral Dis. 2*:55-62.

Terskikh,A.V., Bryant,P.J., and Schwartz,P.H. (2006). Mammalian stem cells. *Pediatr. Res. 59*, 13R-20R

Tomita M, Mori T, Maruyama K, Zahir T, Ward M, Umezawa A, Young MJ (2006). A comparison of neural differentiation and retinal transplantation with bone marrow-derived cells and retinal progenitor cells. *Stem Cells 24*(10),2270-8

Woodbury,D., Schwarz,E.J., Prockop,D.J., and Black,I.B. (2000). Adult rat and human bone marrow stromal cells differentiate into neurons. *J. Neurosci. Res. 61*, 364-370

Trounson,A. (2004). Stem cells, plasticity and cancer - uncomfortable bed fellows. *Development 131*, 2763-2768

Yoshimura,Y. (2006). Bioethical aspects of regenerative and reproductive medicine. *Hum. Cell 19*, 83-86.

Yoshizawa, T., Takizawa, F., Iizawa, F., Ishibashi, O., Kawashima, H., Matsuda, A., Endo, N., Kawashima, H. (2004). Homeobox protein MSX2 acts as a molecular defense mechanism for preventing ossification in ligament fibroblasts. *Mol Cell Biol. 24*(8):3460-72.

In: New Cell Differentiation Research Topics
Editor: Hitoshi Saitama, pp. 259-271

ISBN: 978-1-60021-937-5

Chapter IX

FOLLICLE CELL DIFFERENTIATION IN THE WALL LIZARD *PODARCIS SICULA*

Chiara Maria Motta**, Palma Simoniello, Stefania Tammaro and Silvana Filosa

Department of Biological Sciences, Evolutionary and Comparative Biology Unit, University of Naples Federico II, Naples, Italy.

ABSTRACT

In Podarcis, undifferentiated, pre-differentiated and differentiated follicle cells can be recognised by shape, size, position and function. Undifferentiated (stromal) and pre-differentiated (small) cells are stem cells destined to proliferate and differentiate while differentiated cells (pyriforms) are nurse cells programmed to sustain oocyte growth and then degenerate at the end of previtellogenesis.

The present work clarifies whether competence to undergo apoptosis in pyriforms is acquired at the time they originate from the small cells or later during aging. For this purpose the expression of Fas, Fas-L, caspase 3 and DNase I have been investigated by immunocytochemistry in small and pyriform cells. Results indicate that all these cell death factors are present exclusively in pyriforms. Their synthesis starts as soon as cells differentiate and that while aging, Fas and Fas-L are down and up regulated respectively. We conclude that pyriforms are committed for apoptosis since differentiation and that death occurs via mechanisms alternative to receptor or mitochondrial mediated pathways.

Keywords: apoptosis, pyriform cells, Fas and Fas-L, caspase 3, DNase I.

* Correspondence concerning this article should be addressed to: Chiara M. Motta, Dipartimento delle Scienze Biologiche, Via Mezzocannone, 8, 80134 Napoli, Italia. Tel: 39.81.2535175; Fax: 39.812535035; e-mail: chiaramaria.motta@unina.it.

INTRODUCTION

Evolution of the Follicular Epithelium during Previtellogenesis in Podarcis

Undifferentiated follicle cells in lizards are somatic cells dispersed in two germinal beds, together with clusters of oogonia and prefollicular oocytes (figure 1). They are small and elongated, have a dense cytoplasm and an irregular nucleus containing patches of condensed chromatin (Filosa and Taddei, 1976). Undifferentiated cells actively participate in the formation of the primordial follicle (figure 1) by encircling and separating the oocytes, proliferating and differentiating into small cuboidal cells that form a monolayered epithelium (Andreuccetti et al., 1990).

Small cells in primary follicles continue proliferation (figure 2) and in follicles approximately 100 μm in diameter, they further differentiate (figure 3) (Filosa et al., 1979). Cells enlarge and initially assume a bell-shape, with the base facing the oocyte (the intermediate stage) and, later, a flask shape (the pyriform stage). Cell cytoplasm becomes enriched in organelles that cluster between the nucleus and an intercellular bridge formed by secondary fusion with the oocyte membrane (Andreuccetti et al., 1978).

The appearance of the first intermediate cells is accompanied by changes in the oocyte nucleus. The nucleolus forms few large, transcriptionally inactive spherical bodies (figure 3) while the chromosomes stop the lampbrush stage and condense (Taddei and Andreuccetti, 1990; Tammaro et al., 1998). This oocyte inactivation has been correlated with the nurse function of intermediate and pyriform cells. These cells send organelles and RNAs to the oocyte via the intercellular bridge contributing to its growth (Taddei, 1972; Motta et al., 1995).

Small cell proliferation and differentiation continue in follicles ranging 150-1500 μm diameter (mid previtellogenesis) (Filosa et al., 1979). Several small cells incorporate bromodeoxyuridine (BrdU; figuer 4) and move toward the oocyte membrane to differentiate (figure 5). The epithelium thickens and show an outer layer of small cells, a middle layer of differentiated pyriforms and an inner layer of differentiating intermediate cells (figure 6). At this stage the oocyte nucleus still shows condensed chromosomes but transcriptionally active micronucleoli have formed (figure 7) (Tammaro et al., 1998).

When follicles reach 1500 μm in diameter (late previtellogenesis) small cells stop differentiating. Nuclei positive for BrdU disappear from the inner layers of the epithelium and the intermediate cells gradually decrease in number (figure 8). At the same time, pyriforms start regressing asynchronously (Motta et al., 1996). Dying cells show the typical hallmarks of apoptosis such cytoplasmic shrinkage (figure 9), chromatin condensation (figure 10), DNA fragmentation and membrane blebbings. Additionally and uniquely in pyriforms, cytoplasmic components escape degradation and are transferred into the oocyte via the intracellular bridge (Motta et al., 1996). By the beginning of vitellogenesis pyriforms have disappeared from the epithelium (figure 11) which gradually returns to a monolayered condition (figure 12) (Filosa, 1973).

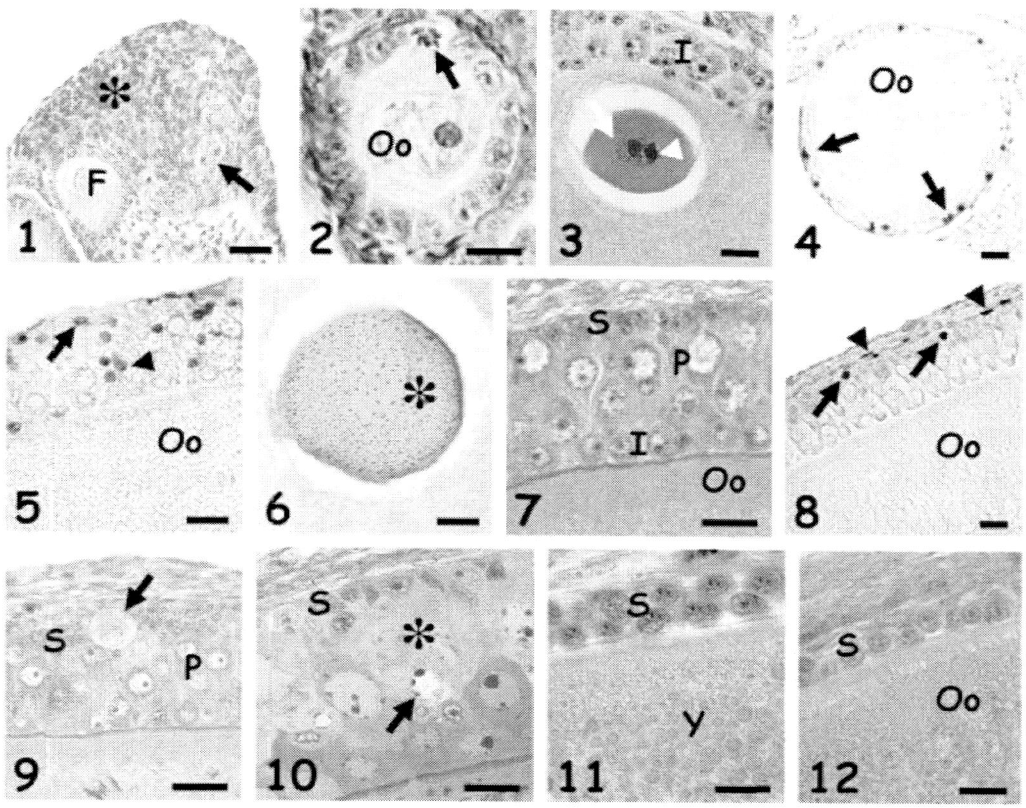

Figures 1-12. Changes in the organization of the follicular epithelium during previtellogenesis. (1) Germinal bed with prefollicular germ cells (arrow), undifferentiated follicle cells (*) and a primordial follicle F. (2) Primary follicle, 80μm diam. Growing oocyte Oo surrounded by a monolayered epithelium containing proliferating small S cells (arrow). (3) 150μm follicle. Growing oocyte with a fragmentating nucleolus (arrowhead) and condensed chromatin (arrow). The first intermediate I cells are recognizable. (4) BrdU incorporation. Early previtellogenic follicle with small cells showing labelled nuclei (arrows). (5) BrdU incorporation. Labelled nuclei of proliferating (arrow) and migrating (arrowhead) S cells in a mid previtellogenic follicle. (6) Fully differentiated epithelium. Note the S cells, the flask shaped pyriforms P and the intermediate, I cells close to the oocyte Oo. (7) Oocyte nucleus with micronucleoli (*). (8) BrdU incorporation. In late previtellogenic epithelia labelling is present on rare S cells (arrows) and on several thecal T cells (arrowheads). (9) Late previtellogenic epithelium showing a thick outer layer of S cells and occasional apoptotic P cells (arrow). (10) Apoptotic pyriform with altered cytoplasm (*) and chromatin in patches (arrow). (11) Bilayer of S cells in an early vitellogenic follicle. The oocyte contains yolk Y vesicles. (12) Monolayer of S cells in a vitellogenic follicle. Bars: 50μm (figure 4); 20μm (figures 1, 3, 6-9, 12); 10μm (figures 2, 5, 10-11);

The present work aims at determining when follicle cells become committed to apoptosis. For this purpose Fas, Fas-L, caspase 3 and DNAse I synthesis and/or activation have been studied by immunocytochemistry in small, intermediate and pyriform cells from mid and late previtellogenic follicles. Fas/Fas-L is the receptor/ligand complex triggering the apoptotic pathway I (Scaffidi et al., 1998). Caspase 3 and DNAse I are the key protease and endonuclease in pyriform apoptosis (unpublished results; Motta et al., 2002). For caspase 3 the presence of the inactive zymogen and of the active subfragment p17 have been investigated; for the nuclease, localization was determined since the enzyme is active only after nuclear translocation.

Further studies were performed to clarify whether follicular synthesis of the above mentioned factors may be altered by in vitro serum deprivation, a pro-apoptotic stress for Podarcis follicles (Tammaro et al., 2007). Secondly, these studies were extended to determine whether the presence of the anti apoptotic gonadotrophic factor, FSH, would protect against the effect of this induced pro-apoptotic stress.

Materials and Methods

Animals

Adult female lizards were captured close to Naples and housed in accordance with the Guideline for Animal Experimentation of the Italian Department of Health. Manipulations were carried out so to minimize the stress and the number of animals used.

Light Microscopy

Ovarian follicles were fixed in Bouin's solution for 6-12 hours, dehydrated in alcohol and embedded in wax. For morphological analyses sections were stained with Galgano trichrome (Beccari and Mazzi, 1972), (Beccari and Mazzi, 1972).

Epithelium Dissociation

Follicles manually deprived of the theca were suspended in 1 volume sterile medium. Mechanical dissociation of the follicle was carried out according to De Caro et al., 1998. Briefly, follicles were vortexed for 10 sec, the supernatant was recovered, centrifuged (3000 rpm, 4°C, 5 min) and pelletted small cells were resuspended in a small aliquot of culture medium. Follicles were washed, resuspended in fresh buffer and vortexed for another 20 seconds. This second supernatant, containing small and pyriform cells, was discarded. Follicles were recovered, resuspended in fresh buffer and vortexed for another 30 sec. This third supernatant was collected, centrifuged and pelletted pyriforms were resuspended in fresh medium.

In vitro Culture

Cell suspensions were diluted to a final concentration of 10^3/ml in HAM F-10 medium containing and 1% antibiotic solution (penicillin, 100 μg/ml, and streptomycin, 100 μg/ml, Sigma, St Louis, MO, USA) supplemented with 10% foetal bovine serum (BSA). Dissected follicles were pooled according to their size and randomly placed in Petri dishes containing 2 ml culture media prepared by omitting serum to expose cells to a pro apoptotic stress. All samples were maintained at 26°C for 60 hours under 5% CO_2 atmosphere. To test a possible recovery from serum deprivation, follicles were incubated in FSH (follicle stimulating hormone; 5 U/ml; Limatola and Filosa, 1989) supplemented media. Control experiments were conducted in the presence of media supplemented with 10% BSA. Results confirm that the culture method does not induce alterations and, therefore, they are not reported.

Bromodeoxyuridine Incorporation

To monitor cell proliferation, follicles in culture were treated with 50 μM BrdU (in DMSO, Sigma, St Louis, MO) for 24 hours. They were transferred into fresh medium, incubated for 36 additional hours, fixed and processed for wax embedding. Sections were denatured with 2 N HCl at 37°C for 30 min, washed in PBS 0.1 M pH 7.2, digested in protease XXIV for 10 sec at room temperature and rinsed in PBS. After washing in serum, sections were incubated with a mouse anti-BrdU antibody (Sigma) diluted 1:100 with PBS and 0.05% Tween20 overnight at 4°C. After washing in PBS, slides were incubated with a secondary POX-conjugated antibody (Dako, Glostrup, Denmark; 1:100, 2 hours) followed by incubation with a tertiary anti-PAP antibody (1:100; Sigma) and developed with Sigma Fast tablets.

Scanning Electron Microscopy (SEM)

Cells were fixed in Karnowsky (1965) solution, dehydrated in a graded sequence of ethanol, ethanol-Freon mixture to 100% Freon and critical point dried. After coating with carbon and gold (30 sec at a distance of 15 cm from the source, at 40kV and 2-3 mA) they were observed with a Cambridge Stereoscan 250 MK II microscope.

Immunocytochemistry

Sections were microwaved for 30 sec in citrate buffer pH 6.0, washed in 0.3% H_2O_2 in TBS, in 1% BSA in TBS and incubated overnight at 4° C with the primary antibodies diluted in 1% BSA in PBS. Primary antibodies were: anti Fas (1:200; Santa Cruz Biotechnology, Santa Cruz, CA); anti Fas-L (1:200; Santa Cruz Biotechnology); anti caspase 3 (1:100; Novocastra, Newcastle upon Tyne, UK); anti p17 (1:100; Novocastra), anti DNase I (1:50; Upstate Biotechnology, Billerica, MA). Antibody binding was revealed with a secondary POX-conjugated antibody (Dako, 1:100, 2 hours) followed by incubation with a tertiary anti-PAP antibody (1:100 in phosphate buffer, Sigma). POX was developed with Sigma Fast tablets. Negative controls were carried out by omitting the primary antibodies. Antibody cross reactivity with Podarcis tissues has been verified in preliminary experiments (Motta et al., 2002; Motta et al, unpublished results).

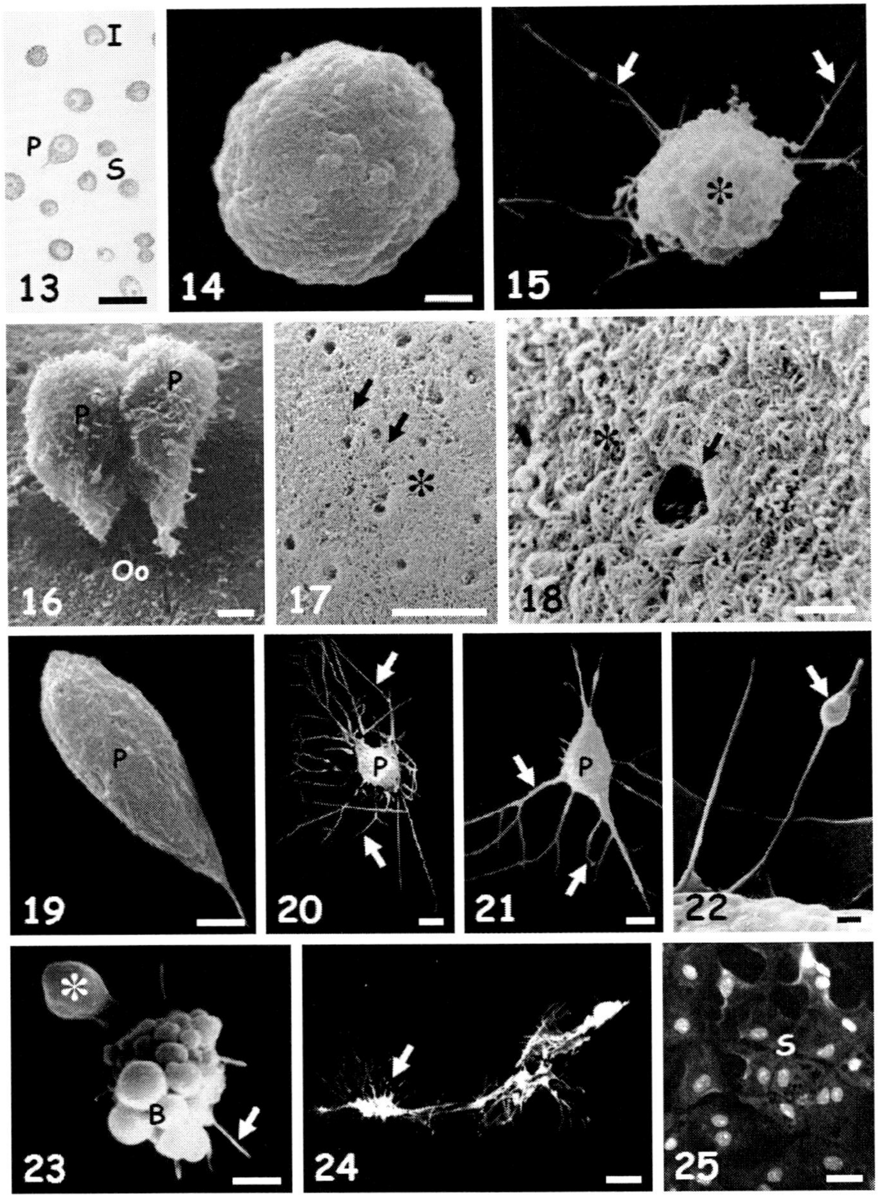

Figures 13-25. Cytological characterization of dissociated follicle cells in culture. (13) Suspension of dissociated follicle cells. Small S, intermediate I and pyriform P cells can be recognised. (14-15) S cells in suspension and after plating. The long filopodia (arrows) and the irregular surface (*) are evident. (16) Two P cells connected to the oocyte Oo. (17-18) Naked oocyte membrane showing long microvilli (*) and the interruptions (arrows) left by the intercellular bridges. (19) Flask shaped pyriform in suspension. (20) Plated mid previtellogenic P cell with several long filopodia (arrows). (21) plated late previtellogenic P cell with few filopodia (arrows). (22) tip of a filopodium showing a flat expansion (arrow). (23) Apoptotic pyriform with blebbings B and a detaching apoptotic body (*); short filopodia (arrow) are present. (24) Plated thecal cells showing an irregular contour and short lamellipodia (arrow). (25) Monolayer of S cells. Phase contrast microscope (figures 13; 25); scanning electron microscope (figures 14-24). Bars: 20μm (figures 13, 20-21, 23); 10μm (figures 16-17, 19, 24-25); 1μm (figures 14-15, 18, 22).

RESULTS

Cytological Characterization of small and pyriform Cells

Small cells dissociated from mid and late previtellogenic epithelia, under phase contrast (figure 13) or scanning (figure 14) microscopy appear round with approximate 10 μm diameters. In culture they rapidly adhere to the plate forming long filopodia (figure 15) and start dividing forming a monolayer (figure 25).

Dissociated pyriforms are flask shaped (figure 13) with long cytoplasmic protrusions (figure 16) connected to the oocyte plasma membrane (figures 17, 18). In culture they maintain shape for a short time (figure 19) then adhere to the substrate forming long filopodia (figures 20, 21) often ending with enlargements (figure 22). Filopodia are more developed in mid (figure 20) than in late (figure 21) previtellogenic pyriforms but never resemble the large lamellipodia formed by thecal cells (figure 24). Plated cells remain differentiated for a few days and then undergo apoptosis as demonstrated by the formation of membrane blebbings and apoptotic bodies (figure 23).

Molecular Characterization of small and pyriform Cells

Small cells never synthesise Fas (figures 26, 27), Fas-L (figures 29, 30), DNAse I (figures 32, 33) or caspase 3 (figures 35, 36). In addition, p17 (figures 38, 39), the active subfragment of caspase 3 is never found in these cells.

A completely different situation is observed for pyriforms. In mid previtellogenesis they contain Fas (figure 26), Fas-L (figure 29), DNAse I (figure 32) and caspase 3 (figure 35), dispersed in the cytoplasm. The cytoplasmic localisation of DNAse and the absence of p17 (figure 38) indicate that the two enzymes are inactive.

Pyriforms in late previtellogenesis show an apparent decrease in Fas (figure 27) and an increase in Fas-L (figure 30) synthesis. No changes are observed in DNase I (figure 33) or caspase 3 (figure 36) synthesis with respect to mid previtellogenic follicles. In these pyriforms caspase 3 is activated and p17 is present in the cytoplasm of all cells (figure 39). Pyriform nuclei, however, remain unstained and since they do not contain DNAse I (figure 33) they are not undergoing apoptosis.

Intermediate cells, the pyriforms undergoing differentiation, synthesise caspase 3 (figures 35, 36) but not Fas (figures 26, 27), Fas-L (figures 29, 30) or DNase I (figures 32, 33). The protease is dispersed in the cytoplasm in both mid (figure 35) and late (figure 36) previtellogenic follicles and is not activated since all cells do not contain p17 (figures 38, 39).

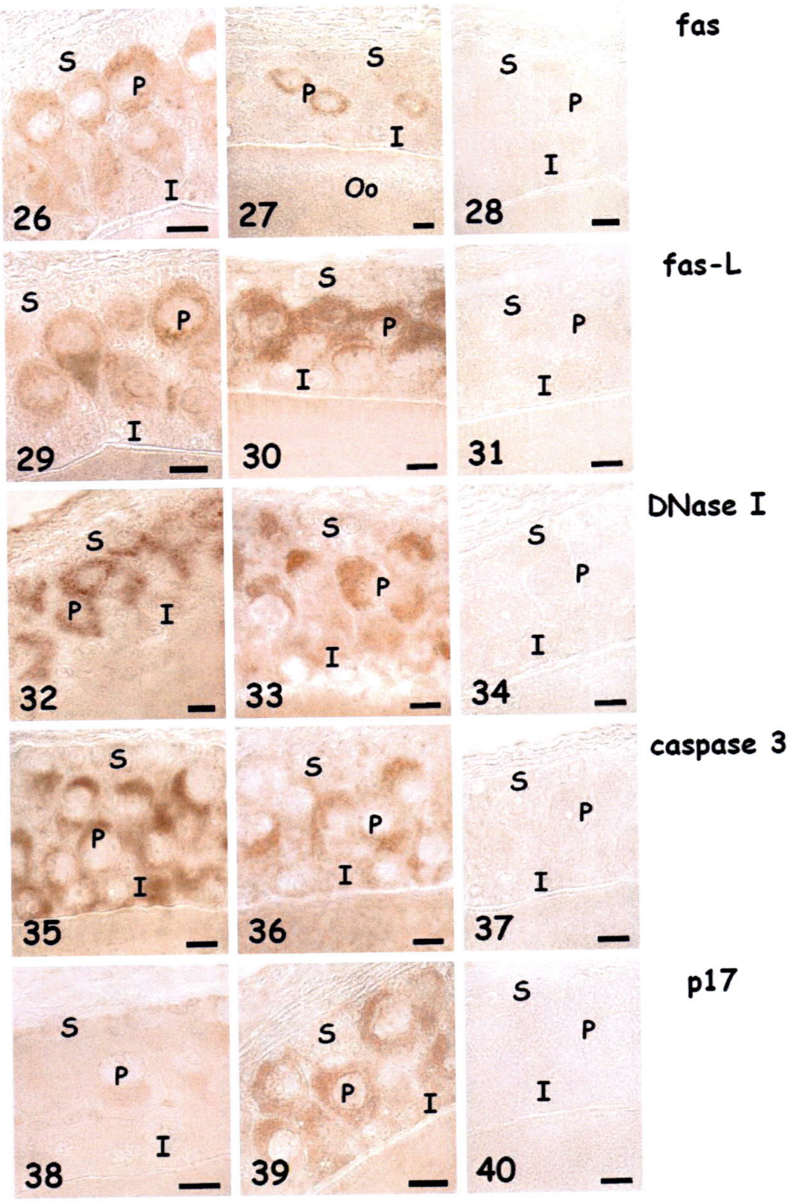

Figures 26-40. Expression of Fas, Fas-L, caspase 3 and DNase I in mid and late previtellogenic cells. Fas staining in mid (26) and late (27) previtellogenic follicles. The antigen is present in the cytoplasm of pyriforms P; small S and intermediate I cells are unstained as the negative control (28). Fas-L staining in mid (29) and late (30) previtellogenic follicles. The antigen is in the cytoplasm of P cells; S and I cells are unstained as the negative control (31). DNase I staining in mid (32) and late (33) previtellogenic follicles. The nuclease is in the cytoplasm of P cells; S and I cells are unstained as the negative control (34). Caspase 3 staining in mid (35) and late (36) previtellogenic follicles. In P cells the protease is concentrated close to the nucleus while in I cells it is uniformly dispersed in the cytoplasm. S cells are unstained as the negative control (37). p17 staining in mid (38) and late (39) previtellogenic follicles. The active subfragment of caspase 3 is present only the cytoplasm of P cells in late previtellogenic follicles. S and I cells are always unstained as the negative control (40). Bars: 10μm.

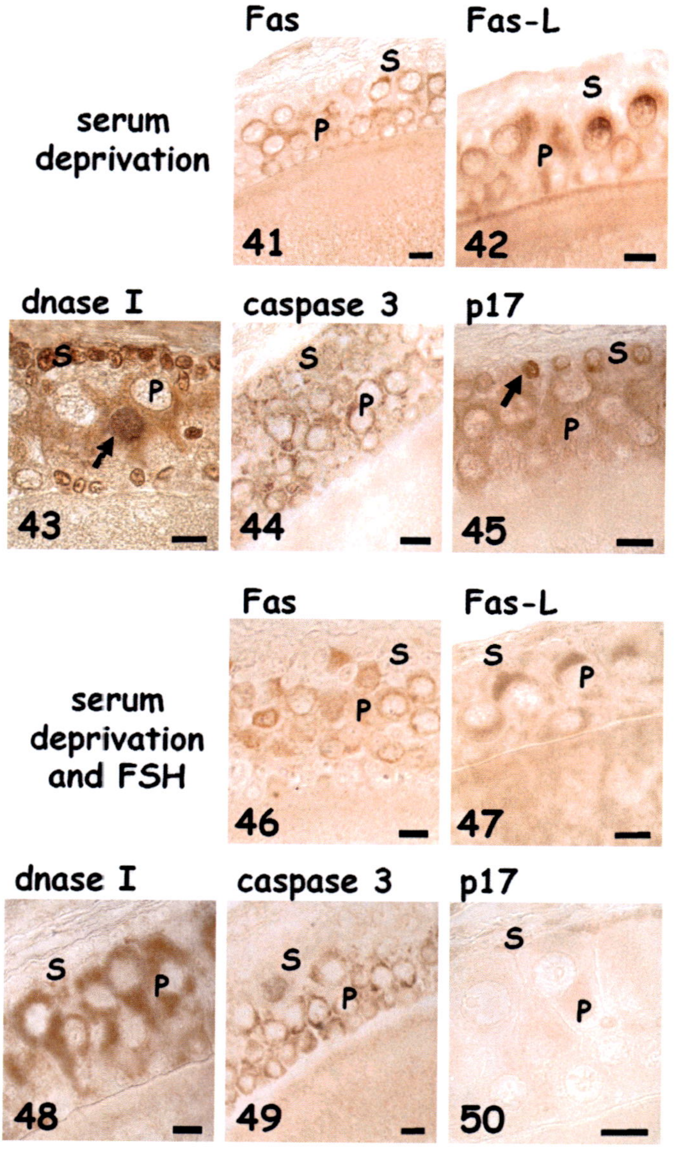

Figures 41-50. Effects of serum deprivation on Fas, Fas-L, caspase 3 and DNase I expression. Fas (41) and Fas-L (42) staining in mid previtellogenic follicles. Antigens are present in the cytoplasm of pyriform P cells; small S cells are negative. (43) DNase I staining in mid previtellogenic follicle. Nuclease is present in P cells cytoplasm, in all S cells nuclei and in occasional P cell nuclei (arrow). (44) Caspase 3 staining in mid previtellogenic follicles. Protease is present in the cytoplasm of all S and P cells. (45) p17 staining in mid previtellogenic follicles. The active subfragment of caspase 3 is present in P and S cells cytoplasm and in occasional nuclei of S cells (arrow). Effects of serum deprivation in presence of FSH on Fas, Fas-L, caspase 3 and DNase I expression. Fas (46), Fas-L (47), DNase I (48) and caspase 3 (49) staining in mid previtellogenic follicles. Antigens are present in P cells cytoplasm; S cells are negative. p17 (50) staining in mid previtellogenic follicle. The active protease is absent from all cells. Bars: 10μm.

Effects of Serum Deprivation of the Synthesis of Fas, Fas-L, Caspase 3 and DNase I in small and pyriform Cells

Serum deprivation does not change Fas (figure 41) and Fas-L (figure 42) synthesis; as in controls, the antigens are present exclusively in the cytoplasm of pyriforms while the small cells remain unstained.

Serum deprivation, by contrast, induces the synthesis and the activation of DNase I (figure 43) in small cells: nuclei, in fact, are intensely labelled. In pyriforms the enzyme remains cytoplasmic but in few cells in mid (figure 43) and late previtellogenic follicles in which nuclei are significantly labelled.

Serum deprivation induces caspase 3 synthesis in small cells that show intensely labelled cytoplasms (figure 44); in pyriforms no changes are evident. The active p17 appears in small cells and pyriforms cytoplasm and, also, in occasional small cell nuclei (figure 45) but only in mid previtellogenic follicles. In late previtellogenic follicles cells are labelled as in controls (data not shown).

When serum deprived media are supplemented with FSH the effects of serum deprivation on DNAse and caspase 3 synthesis/activation are completely inhibited. DNase I disappear from small cell nuclei (figure 48), caspase 3 from small cells cytoplasm (figure 49) and p17 from all small cells and pyriforms in mid previtellogenic follicles (figure 50). FSH treatment does not change Fas (figure 46) and Fas-L (figure 47) synthesis.

Conclusion

These results demonstrate that in Podarcis, follicle cells progress through two stages of development: a pre-differentiation, in which S cells continue proliferation and a terminal differentiation in which pyriforms become nurse cells, begin synthesizing cell death factors Fas, Fas-L, caspase 3 and DNase I (Nagata, 2000) thereby becoming committed for apoptosis. Both stages of differentiation occur during the mid previtellogenesis, period when epithelium reaches a maximum complexity (Filosa, 1973).

These observations indicate that during their life, pyriforms do not modify caspase 3 and DNase I expression but they do down and up regulate Fas and Fas-L respectively. Such evidence partly contrasts with literature reporting that both Fas and Fas-L increase during granulosa cell aging causing an enhanced susceptibility to apoptosis (Higami and Shimokawa, 2000). Fas-L decrease in pyriforms suggests a deficiency in this receptor mediated apoptotic pathway. This would parallel the deficiency of the mitochondrial apoptotic pathway already documented in the same cells (Tammaro et al., 2007). In pyriforms therefore, both conventional death pathways would be inhibited and regression of redundant pyriforms would occur via alternative pathways (paper in preparation). This hypothesis is sustained by the observation that the two enzymes can be translationally upregulated by serum deprivation in mid previtellogenic pyriforms in which Fas-L is not down regulated.

DNase I and caspase 3, key enzymes in pyriform apoptosis, are both synthesised in large advance - mid previtellogenesis - with respect to their use, in late previtellogenesis (Motta et al., 1996). An anticipated expression for DNase I has already been documented in several stratified epithelia showing a rapid turnover, for example, in the small intestine (Rauch et al., 1997). This stock of enzymes could be a specific adaptation that make cells particularly prone to elimination by apoptosis (Polzar et al., 1994). In Podarcis in particular, this facilitates the resorption of the over 10.000 pyriforms present in a fully differentiated, mid previtellogenic epithelium (Filosa et al., 1979).

The early synthesis of caspase 3 and DNase I in pyriforms is not accompanied by early activation of these enzymes. Caspase, in fact, is present as zymogen and DNase is exclusively cytoplasmic. This explains why in mid previtellogenesis there are no pyriforms in apoptosis (Motta et al., 1996). Apoptosis, by contrast, becomes evident in late previtellogenic pyriforms indicating that at this time the enzymes have been activated. Immunocytochemistry interestingly reveals that all pyriforms become positive for p17 prior to the appearance of apoptotic degeneration. This suggests an intriguing hypothesis: that the protease, before apoptosis, is involved in other functions and in particular that it could be responsible for cytoskeleton remodelling. During late previtellogenesis pyriforms empty their cytoplasm into the oocyte and caspase 3 could favour the intense trafficking of organelles by modifying the cytoskeleton (Acarin et al., 2007). In Podarcis, the high plasticity of the follicle cell cytoskeleton has already been shown (Maurizii et al., 2004) and is confirmed in this work, by the massive rearrangements in shape occurring in cells in culture after dissociation.

Being programmed for proliferation and differentiation (Filosa et al., 1979) it is not surprising that small cells under physiological conditions do not express caspase, DNase or the Fas/Fas-L. Similarly, it is not unexpected that caspase 3 and DNase I synthesis is activated following serum deprivation, a pro apoptotic stress and inhibited by FSH which up-regulates expression of intracellular anti-apoptotic proteins (Jiang et al., 2003). What is unexpected is that Fas/Fas-L are not upregulated by serum deprivation as already reported in granulosa cells (Hu et al., 2001). This complex is certainly involved in follicular cell apoptosis (Hakuno et al., 1996) and, therefore, its absence in small cells requires further investigation.

In conclusion, follicular epithelium in Podarcis is a highly dynamic structure. The components progress through an undifferentiated, a predifferentiated and a terminally differentiated stage, each characterised by specific cytological, functional and biochemical features. Pyriforms cells, in particular, the terminally differentiated cells, are characterised by the acquisition of the nurse function and the translational up-regulation of the cell death factors Fas, Fas-L, caspase 3 and DNase I. These, however, remain inactive until the nurse function is completed at the end of previtellogenesis. This shows that the epithelium is a highly coordinated structure in which apoptosis is regarded as a stage, fundamental for follicle survival and not simply the mean by which defective/redundant cells are eliminated by the epithelium.

ACKNOWLEDGMENTS

These studies were supported by a grant from the MIUR (PRIN project entitled: Molecular responses of embryonic, differentiated and tumoral cells exposed to cadmium intoxication. This paper is dedicated to Ted Howell.

References

Acarin, L; Villapol, S; Faiz, M; Rohn, TT; Castellano, B; Gonzalez, B. Caspase-3 activation in astrocytes following postnatal excitotoxic damage correlates with cytoskeletal remodeling but not with cell death or proliferation. *Glia*, 2007 55, 954-965. [Epub ahead of print]

Andreuccetti, P; Taddei, C; Filosa, S. Intercellular bridges between follicle cells and oocyte during the differentiation of follicular epithelium in *L. sicula*. *J Cell Sci*, 1978 33, 341-350.

Andreuccetti, P; Motta, CM; Filosa, S. Regulation of oocyte number during oocyte differentiation in the lizard *Podarcis sicula*. *Cell Differ Dev*, 1990 29, 129-41.

Beccari, N; Mazzi, V. Manuale di tecnica microscopica. Guida pratica alla ricerca istologica e istochimica. *Società Editrice Libraria*. 1972.

De Caro, M; Indolfi, P; Iodice, C; Spagnuolo, S; Tammaro, S; Motta, CM. How the ovarian follicle of P. sicula recycles the DNA of its nurse, regressing follicle cells. *Mol Reprod Dev*, 1998 51, 421-429.

Filosa, S. Biological and cytological aspects of the ovarian cycle in Lacerta sicula Raf. *Mon Zool It*, 1973 7, 151-165.

Filosa, S; Taddei, C. Intercellular bridges in lizard oogenesis. *Cell Diff*, 1976 5, 199-206.

Filosa, S; Taddei, C; Andreuccetti, P. The differentiation and the proliferation of the follicle cells during oocyte growth in P. sicula Raf. *J Embryol exp Morph*, 1979 54, 5-15.

Hakuno, N; Koji, T; Yano, T; Kobayashi, N; Tsutsumi, O; Taketani, Y; Nakane, PK. Fas/APO-l/CD95 system as a mediator of granulosa cell apoptosis in ovarian follicle atresia. *Endocrinology*, 1996 137, 1938-1948.

Hu, CL; Cowan, RG; Harman, RM; Porter, DA; Quir, SM. Apoptosis of bovine granulosa cells after serum withdrawal is mediated by Fas antigen (CD95) and Fas ligand. *Biol Reprod*, 2001 64, 518–526.

Higami, Y; Shimokawa, I. Apoptosis in the aging process. *Cell Tissue Res*, 2000 301, 125-132.

Jiang, JY; Cheung, CK; Wang, Y; Tsang, BK. Regulation of cell death and cell survival gene expression during ovarian follicular development and atresia. *Front Biosci*, 2003 8, d222-237.

Karnowsky, MJ. A formaldehyde-glutaraldehyde fixation of high osmolarity for use in electron microscope. *J Cell Biol*, 1965 27, 137-148.

Limatola, E; Filosa, S. Exogenous vitellogenesis and micropinocytosis in the lizard, Podarcis sicula, treated with follicle-stimulating hormone. *Gen Comp Endocrinol*, 1989 75, 165-176.

Maurizii, MG; Alibardi, L; Taddei, C. α-Tubulin and acetylated a-tubulin during ovarian follicle differentiation in the lizard Podarcis sicula Raf. *J Exp Zool*, 2004 301A, 532–541.

Motta, CM; Castriota Scanderberg, M; Filosa, S; Andreuccetti, P. Role of pyriform cells during the growth of oocytes in the lizard P. sicula. *J Exp Zool*, 1995 273, 247-256.

Motta, CM; Filosa, S; Andreuccetti, P. Regression of the epithelium in late previtellogenic follicle of P. sicula: a case of apoptosis. *J Exp Zool*, 1996 276, 233-241.

Motta, CM; Graziano, R; Tammaro, S; Filosa, S. Analisi comparativa della presenza/localizzazione di un enzima apoptotico, la DNasi I, nelle cellule follicolari di alcuni vertebrati ectotermi. In: Odierna G, Guarino FM, Editors. *I Vertebrati Ectotermi del Parco Regionale del Matese*, Napoli, Centro Stampa dell'Università di Napoli Federico II; 2002; 65-71.

Nagata, S. Apoptotic DNA fragmentation *Exp Cell Res*, 2000 256, 12-18.

Polzar, B; Zanotti, S; Stephan, H; Rauch, F; Peitsch, MC; Irmler, M; Tschopp, J; Mannherz, HG. Distribution of deoxyribonuclease I in rat tissues and its correlation to cellular turnover and apoptosis (programmed cell death). *Eur J Cell Biol*, 1994 64, 200-210.

Rauch, F; Polzar, B; Stephan, H; Zanotti, S; Paddenberg, R; Mannherz, HG. Androgen ablation leads to an upregulation and intranuclear accumulation of deoxyribonuclease I in rat prostate epithelial cells paralleling their apoptotic elimination. *Cell Biol*, 1997 137, 909-923.

Scaffidi, C; Fulda, S; Srinivasan, A; Friesen, C; Li, F; Tomaselli, KJ; Debatin, KM; Krammer, PH; Peter, ME. Two CD95 (APO-1/Fas) signaling pathways. *EMBO J*, 1998 17, 1675–1687.

Taddei, C. Significance of pyriform cells in ovarian follicle of L. sicula. *Exp Cell Res*, 1972 141, 562-566.

Taddei, C; Andreuccetti, P. Structural modifications of the nuclear components during lizard oogenesis in relation to the differentiation of the follicular epithelium. *Cell Differ Dev*, 1990 29, 205-215.

Tammaro, S; Andreuccetti, P; Filosa, S; Indolfi, P; Prisco, M; Motta, CM. Structural and functional modifications of the nucleolus during previtellogenic oocyte growth in the lizard Podarcis sicula. *Mol Reprod Dev*, 1998 51, 413-420.

Tammaro, S; Simoniello, P; Filosa, S; Motta, CM. Block of mitochondrial apoptotic pathways in lizard ovarian follicle cells as an adaptation to their nurse function. *Cell Tissue Res*, 2007 327, 625-635.

In: New Cell Differentiation Research Topics
Editor: Hitoshi Saitama, pp. 273-282

ISBN: 978-1-60021-937-5

Chapter X

Bone Marrow Stromal Cells Differentiated into Schwann Cells

Sai-yun Hou[1,2,*] ***and Jia-kai Zhu***[1]

[1] Department of Orthopedic and Microsurgery, the First Affiliated Hospital, SunYat-Sen University, GuangZhou, China

[2] Department of Orthopedic, Stanford University, Palo Alto, CA, USA

Abstract

Schwann cells (SCs) contribute to nerve regeneration by providing regeneration strands, many kinds of nerve growth factors and tight junctions to stabilize cell contact with injured nerve cells, as well as gap junctions to facilitate traffic of substances between the cells. It is recognized that Schwann cells play a crucial role in the process of peripheral nerve development and regeneration, even in the construction of tissue-engineered nerve grafts. However, the difficulty is to obtain sufficient numbers of autologous SCs in a short time. Use of allogeneic cells are involved in immunological rejection. Bone marrow stromal cells (MSCs) are nonhematopoietic multipotent stem cells that differentiate into connective tissue lineages including osteoblasts, chondrocytes, and adipocytes under special experimental conditions. In particular, MSCs showed potential to be differentiated to SCs in vitro and in vivo, which enhance peripheral nerve regeneration. In this charpter, we will summarize the various strategies that can be used to differentiate MSCs into Schwann cells in vitro and in vivo. Many different assays and techniques may be used to define and characterize MSCs which were differentiated along SCs lineage. Functional assays that may be used to further define the differentiated MSCs and determine if they can promote peripheral nerve regeneration are described.

Keywords: cell differentiation, bone marrow stromal cells, Schwann cells, nerve regeneration, implantation.

* Correspondence concerning this article should be addressed to: Dr. Saiyun Hou, Building 7-A132, 3801 Miranda Ave, Palo Alto, CA 94304, USA E-mail: housaiyun@hotmail.com

1. INTRODUCTION

Peripheral nerve defects are a commonly encountered clinical problem and often result in long-time functional deficits [1]. Treatment of lesions causing complete transection of peripheral nerve is primarily aimed at recreating continuity of the nerve to allow axonal regrowth into the nerve stump distal to the lesion. Currently autologous nerve grafts is still primary method to achieve the satisfied results [2].

But disadvantages with this method include harvesting the autografts from a second operative site, with consequent loss of end-organ innervation, prolonged surgical procedures, and shortage of autogenous nerve graft material for extensive nerve injuries [3]. Allografts, blood vessels, degenerated muscle fibres, vein conduits have also been employed as conduits for nerve regeneration, but all have very poor success rates [4,5]. Recently, many researchers focus on using tissue engineering bioartificial nerves as new alternatives. They have the advantages of unlimited availability and avoidance of donor-site morbidity. Since Schwann cells (SCs) are the primary structural and functional cells that play a crucial role in the process of peripheral nerve regeneration, they become the most important and commonly used seed cells supplemented in the construction of tissue-engineered nerve grafts. However, the culture of sufficient numbers of autologous SCs is time consuming, limiting the usefulness of this technique as delay compromises axonal regeneration [6]. Use of allogeneic SCs is involved in immunological rejection [7]. Therefore, considerable efforts have focused on inducing other cells to turn into SCs, so as to expand their sources for the need of nerve tissue engineering and cell therapy for the injury of PNS (peripheral nervous system) and CNS (Central nervous system).

An alternative to SCs transplantation is stem cell differentiation to achieve the same functional properties. Clinical application of embryonic stem cells raises ethical dilemmas [8]. Bone marrow stromal cells (MSCs) are well known as multipotential stem cells which, under specific inducing conditions, can differentiate into several cells types such as osteoblasts, adipocytes, chondrocytes, and even muscle cells [9,10].

Although quite different from neuronal lineage, the potential of MSCs to develop into neurons, astrocytes, oligodendrocytes, or microglia both in vivo [11-13] and in vitro [14-16] has been reported. MSCs were also differentiated in vitro to express markers of SCs and enhanced peripheral nerve regeneration [17-20].

Because MSCs are easy to isolate from bone marrow aspirates that can be obtained under local anesthesia in the clinic, and they can readily be expanded under culture conditions for transplantation. MSCs' differentistion into SCs, and ability to promote axonal regeneration are appealing for tissue engineering. So MSCs become the strongest candidates for seed cell of peripheral nerve tissue engineering and cell transplantation involving tissues of the nervous system. In this charpter, various strategies that can be used to differentiate MSCs into SCs in vitro and in vivo are summarized. At same time, many different assays and techniques may be used to define and characterize MSCs which were differentiated along SCs lineage. The effect on differentiation into SCs is discussed. We use different functional assays to test whether or not differentiated MSCs has the potential to express the markers of SCs and enhance peripheral nerve regeneration.

2. MSCs Differentiation into SC in vitro

In addition to stem cells for hematopoietic cells, bone marrow contains stem-like cells that are precursors of non-hematopoietic cells and can be cultivated in vitro [10]. They were initially referred to as plastic adherent cells or colony-forming-unit fibroblasts and recently as mesenchymal stem cells or marrow stromal cells. Marrow stromal cells are connective tissue elements that provide structural and functional support for hemopoiesis, but they are now recognized as progenitor cells capable of differentiating into mesenchymal tissues of bone, cartilage, fat, muscle and other connective tissue [10]. In addition to these tissues, transdifferentiation of MSCs into the neuronal lineage such as astrocytes and neurons has been reported [16,21,22]

Therefore, they are interesting targets for use in neural transplanation and regeneration. The strategies to induce and differentiate into SCs and functional assay for SC to promote nerve regeneration are presented.

Subconfluent cultures of MSCs were incubated for 24h with DMEM containing 1 mM beta-mercaptoethanol (BME), then replaced with new media consisting of DMEM, 10% FBS and 35 ng/ml all-trans-retinoic acid (RA) for 3 days, lastly transferred to DMEM containing 10% FBS, 5uM FSK, 10ng/ml recombinant human basicfibroblast growth factor (bFGF), 5ng/ml recombinant human platelet-derived growth factor-AA (PDGF) and 200 ng/ml recombinant human heregulin-beta 1(HRG). Such differentiated MSCs were morphologically different from the original undifferentiated MSCs and resembled primary cultured Schwann cells [20]. Long time induction, such as 2 weeks, was also executed [23]. The differentiated MSCs were assessed morphologically by immunohistochemistry of S100, glial fibrillary acidic (GFAP), P75 and O_4, all known as markers of Schwann cells [20,24-26]. The results were compared with those obtained from Schwann cells. Quantitative Western blot analysis showed that the levels of S100, GFAP protein were significantly elevated upon differentiation [20].

There are various times and passages for induction and differentiation in vitro. MSCs differentiated for 5 days achieved higher expression for S100 and GFAP than that of 2 and 8 days. It suggested the peak of S100 and GFAP expression was at 5 days [20]. However, undifferentiated MSCs cultures were negative for S100 and GFAP. Control SC cultures were immunoreactive for both markers. Jenny Caddick [23] reported the influence of cell passage on the differentiated MSCs ability to express SCs markers. A comparision of cells differentiated to either passage 2 or passage 5 showed no difference of expression for SC makers, suggesting that cell age, at least at this relatively early stage, does not affect their differentiation ability. Interestingly, there was occasional weak S100 expression in the undifferentiated MSCs expanded to passage 8, possibly due to the innate potential of a small cell proportion to differentiate spontaneously.

The ability of differentiation into SCs was evaluated by the ratio of S100 and GFAP positively stained cells. The differentiated MSCs (with SCs characteristics) at day 5 got higher ratio of positively stained cells than that of day 2 and day 8, up to 51-75% [20]. So the ability of differentiation in vitro is satisfied, but function of the differentiated MSCs (with SCs characteristics) still needs the further evidence.

What are the mechanisms in vitro that MSCs can be differentiated into SCs? It was demonstrated that sequential administration of various factors BME, RA, followed by a mixture of FSK, bFGF, PDGF and HRG effectively induces the differentiation of MSCs into SCs and promotes nerve regeneration. BME is a reducing agent, and previous reports showed that the addition of BME to cultured MSCs induced the formation of neurite-like processes [16,21]. RA is also an important regulatory molecule for controlling cell growth and differentiation in both the adult and the embryo. RA is a morphogen during development, and was reported to induce differentiation of embryonic stem (ES) and neural stem cells into nerve cells [27]. Retinoic acid RA also regulates the expressoion of various transcription factors that are crucial to an early neuronal determination such as MASH1 and NeuroD, and it furthermore plays a role in the acquision of the responsiveness to neurotrophins [28]. So BME and RA are presumed to serve as triggering factors, including changes in the morphological and transcriptional characteristics of MSCs. bFGF, PDGF and HRG are effective in differentiation and proliferation of glial and SCs [29]. It is well known that bFGF functions as a mitogen for mesenchymal cells including MSCs, and that MAP kinases are activated by PDGF and bFGF in MSCs [30]. HRG, a subtype of neuregulin, instructively influences the decision of cell fates, rather than permissively supporting the survival of lineage committed progenitors on neural crest cells, and was reported to induce neural crest cells to develop selectively into SCs [31]. FSK incresses the level of intracellular cyclic adenoses monophosphate (cAMP), which increases mitogenic responses at an early time, perhaps by stimulating the expression of growth factor receptors [32]. In the case of the SC lineage, an elevation of cAMP leads to the myelin-related differentiation. As cAMP elevation is reported to enhance the responsiveness of cells to trophic factors [33], FSK together with bFGF, PDGF, HRG could have a synergistic effect in enhancements of these factors to MSCs. These regents as used above seem to have cumulative effects for the differentiation of MSCs to express the SC phenotype, because an omission of any one of the factors BME, RA, FSK or HRG resulted in an incomplete differentiation of MSCs regarding morphology and immunoreactivity to SCs markers. However, further investigations on the mechanism of the MSCs differentiation are required.

3. MSCs Differentiation into SC in Vivo

The research on MSCs differentiation into SCs in vivo also has shown good results. So MSCs may likely be used as a candidate for cell transplantation in peripheral nerve repair. Commonly, the strategy for differentiation into SCs in vivo is to label MSCs with the label solutions and transplant into the injured site of nerve or nerve alternative. After several weeks, fluorescent double-staining for label marker and SCs markers such as S100, GFAP and P75 was sequentially performed, which suggested whether MSCs grafted can be differentiated into SCs in vivo. At the same time, functional assay was also done to evaluate the effectivity of nerve regeneration. We summarize the various methods that can be used to label MSCs in vitro prior to grafting. Brdu, bisbenzimide (Hoechst's 33342) and PKH67 were reported to be used in the experiment for differentiation in vivo [34-36].

Generally, Brdu and bisbenzimide are used to label nuclear. Though nuclear labels obviously will not provide any information about cell shape, excellent results have been obtained for these labels in grafting experiments, in which the cells do not remain in situ for more than 1 month. However, with the exception of Brdu and bisbenzimide, none of these labels can be used for chronic grafting experiments because the labels tend to leak out of the cells. If the grafts are expected to remain in vivo for longer than 6 week before analysis, Brdu, bisbenzimide or one of the reporter genes should be utilized. Otherwise, it becomes extremely difficult to unequivocally state that a label-expressing cell is of graft origin. The PKH-67-GL cell linker kit uses patented membrane labeling technology to stably incorporate a fluorescent dye with long aliphatic tials (PKH67) into lipid regions of the cell membrane. With reduced cell-cell tansfer, PKH67 has been found be useful both in vitro cell proliferation labeling Band in vivo cell tracking applications for a relatively short time [37,38].

What are the mechanisms or factors that promote nerve regeneration and reconnection with adequate targets with MSCs implantation? A reasonable hypothesis is that MSCs may adopt different cell fates under the effects of a local microenvironment [39]. MSCs when placed into an environment of injury express trophic factors and their receptors which may facilitate these cells differentiation into SCs lineage. These factors when expressed first in MSCs after implantation can rescue axonal diaback [40]. Stem cells are normally quiescent in tissues such as the brain, but are capable of constant renewal and replenishment in epithelia and of tissue repair following injury. The microenvironment is thought to control stem cell behaviour due to a combination of locally secreted factors, cell–cell interactions mediated by integral membrane proteins, and the extracellular matrix [41]. We cannot exclude that these factors may be responsible for some of the undifferentiated marrow stromal cells showing expression of S100 in vivo. Hinks et al [42] have suggested that stem cell migration to zones of injury is more likely due to local depletion of progenitors than due to increased availability of survival factors.

Compared with in vitro differentiation, the ratio of postive staining for SCs markers during in vivo differentiation reduced. Cuevas P et al [35] reported that MSCs pre-labeled with Brdu previously to be injected in the distal stump of transected sciatic nerve of the rats. Dual immunofluorescence labeling showed the Brdu reactive cells survive in the injected area of transected sciatic nerve at least 33 days after implantation, and almost 5% of Brdu cells express SC-like phenotype. However, MSCs-treated rats exhibit significant improvement on walking track test at day 18 and 33 compared to controls. Some research reported that at least a portion of transplanted MSCs express SCs-like traits after in vivo differentiation [43]. However, the functional assessment in vivo has achieved satisfied results. It seems that the ratio of differentiation in vivo is not consistent with the functional recovery of nerve graft. Therefore, further research needs to be performed to elucidate the mechanism of differentiation in vivo.

4. Functional Evaluation for MSCs Differentiated along a SCs Lineage

In this preliminary study we were unable to carry out a full molecular and functional characterization of differentiated MSCs, and care must be taken in interpreting the results of phenotypical markers as determinants of cell differentiation. It is still unclear how a stimulation process might direct a cell towards a given phenotype and morphological shape, and whether phenotypical and morphological expression really equates with functional ability. The interpretation of the results of various studies is dependent upon the assumption that marrow stromal cells have become glial-type cells following in vitro differentiation and transplantation, acquiring Schwann cell marker expression and morphology. Indeed their effects on improving regeneration is consistent with studies of Schwann cell physiology following injury and transplantation [44-46]. There are various methods of functional assay for MSCs differentiation into SC: 1. Electrophysiology 2. Walking track analysis 3. Weight of gastrocnimus 4. DRG functional assay.

MSCs were differentiated into SCs in vitro as described above, then suspended in DMEM containing 10% FBS at a concentration of $1\text{-}2\times10^7$ cells/ml, and deopped gently onto the ECM-gel-coated PLGA (Sigma) filaments in an Eppendorf tube. The culture was kept with daily changes of the medium until differentiated MSCs attached tightly with ECM-gel-coated filaments and formed cell-bridges with the shape of chains and sheets. Then they were put into the hollow PLGA tubes carefully by sterile forceps under an operating microscope. Twelve weeks after the operation. Signals to S100 and neurofilaments of nerve fiber on the distal parts of nerve grafts showed strong immunopostive. Differentiated MSCs grafts showed that myelinated axons had traversed the whole length of the tube in a good growth orientation [20]. The densely regenerating axons were observed throughout the entire tube. The newly forming myelin sheath displayed a typical multilamellar structure, and the myelinated nerve fiber axon was covered with an external basal lamina of SC, developing the normal physiological peripheral nerve structure. In contrast,acellular grafts exhibited poor nerve regeneration. Percentage of neural tissue, myelinated nerve fiber density and axon average diameter showed statistically significant differences between the differentiated BMSC-laden grafts and acellular grafts, but myelin sheath depth did not in each group. Also differentiated MSCs grafts achieved better electrophysiological effects. Nerve conduction velocities and peak amplitude in differentiated MSCs grafts were observed a little lower than that of autografts, with no statistically difference. Walking track analysis showed that function of animals receiving differentiated MSCs grafts recovered earlier than that of acellular grafts.

For DGR functional assay, Caddick et al. [23] described that neurons from DRG were co-cultured with SC, differentiated MSCs and undifferentiated MSCs. Then immunostaining for DGR neurons was performed. The functional capacity of cultured SCs, undifferentiated MSCs, and differentiated MSCs to promote neurite outgrowth was determined by quantification of their effects on co-cultured adult sensory neurons. Three separate parameters of neurite outgrowth were quantified: percentage of process-bearing neurons, length of longest neurite, and neurite density. The percentage of neurite-bearing cells induced by the differentiated MSCs was significantly greater than that observed with the undifferentiated MSCs. In addition, the length of the longest neurite developing from each cell increased significantly with the addition of both SCs or differentiated MSCs but not the undifferentiated MSCs. neurite density was also significantly increased by the addition of both SCs and differentiated MSCs.

5. Conclusion

SCs derive from the neural crest, then can differentiate into immature precursor cells and mature precursor cells. Though they are already fully differentiated and extending myelin sheaths, SCs rapidly change their structure and characteristics after injury to become undifferentiated, and then follow active mitosis and proliferation. Along with the well-known functions of SCs such as the production of trophic factors and cell adhesion molecules, SCs contribute to nerve regeneration by providing tight junctions to stabilize cell contact with injured nerve cells, as well as gap junctions to faciliate traffic of substances between the cells. So MSCs' differentiation into SCs is important for peripheral nerve regeneration and promoting the functional recovery. But the further evidence and mechanisms of differentiation are acquired. Long observation for nerve regeneration is required. The influence of long distance defect of nerve on nerve regeneration with MSCs differentiation into SCs has not been reported. Therefore, continued evidence for the ability of marrow stromal cells to trans-differentiate will highlight their potential use for therapeutic strategies in a wide range of injuries and diseases. MSCs are certainly one of the most attractive cells for an application of neural transplantation and nerve tissue engineer because they exhibit several important and potential advantageous features in peripheral nerve regeneration. In addition, they are relatively easy to isolate from small aspirates of bone marrow that can be obtained under local anesthesia, and they can be expanded easily in vitro. They could provide a vast source of autologous cells to prevent immunological rejection and need of a second surgical step. Therefore, differentiated MSCs are candidates for cell transplantation and tissue engineer involving tissue of nervous system.

References

[1] Madison RD, Archibald SJ, Krarup C. Peripheral nerve injury. In: Cohen IK, Diegelman F, Lindbald WJ, eds. *Wound Healing:Biochemical and Clinical Aspects*, Philadelphia: W.B. Saunders, 1992: pp. 450–487

[2] Millesi H. Progress in peripheralnerve reconstruction. *World J Surg* 1990; 14: 733–747

[3] Trumble TE, Shon FG. The physiology of nerve transplantation. *Hand Clin* 2000;16:105–122.

[4] Pu LL, Syed SA, Reid M, Patwa H, Goldstein JM, Forman DL, Thomson JG. Effects of nerve growth factor on nerve regeneration through a vein graft across a gap. *Plast Reconstr Surg* 1999; 104: 1379–1385.

[5] Genden EM, Mackinnon SE, Yu S, Flye MW. Induction of donor-specific tolerance to rat nerve allografts with portal venous donor alloantigen and anti-ICAM-1/LFA-1 monoclonal antibodies. *Surgery* 1998; 124:448–456.

[6] Sulaiman OAR, Gordon T. Effects of short- and long-term Schwann cell denervation on peripheral nerve regeneration, myelination and size. *Glia* 2000; 32:234–246.

[7] Guenard V, Kleitman N, Morrissey TK, Bunge RP, Aebischer P. Syngenic Schwann cells derived from adult nerves seeded in semipermeable guidance channels enhance peripheral nerve regeneration. *J Neurosci* 1992, 12:3310–3320.

[8] Gilbert DM. The future of human embryonic stem cell research: Addressing ethical conflict with responsible scientific research. *Med Sci Monit* 2004;10:RA99-103

[9] Pereira RF, Halford KW, Hara MD, O'Hara MD, Leeper DB, Sokolov BP, Pollard MD, Bagasra O, Prockop DJ. Cultured adherent cells from marrow can serve as long-lasting precursor cells for bone, cartilage, and lung in irradiated mice. *Proc Natl Acad Sci USA* 1995;92:4857-4861.

[10] Prockop DJ. Marrow stromal cells as stem cells for nonhematopoietic tissues. *Science* 1997;276:71-74.

[11] Kopen GC, Prockop DJ, Phinney DG. Marrow stromal cells migrate throughout forebrain and cerebellum, and they differentiate into astrocytes after injection into neonatal mouse brains. *Proc Natl Acad Sci* 1999;96:10711-10716.

[12] Mezey E, Chandross KJ, Harta G, Maki RA, Mckercher SR. Turning blood into brain: cells bearing neuronal antigens generated in vivo from bone marrow. *Science* 2000;290:1779-1782.

[13] Nakano K, Migita M, Mochizuki H, Shimada T. Differentiation of transplanted bone marrow cells in the adult mouse brain. *Transplantation* 2001;71:1735-1740.

[14] Sanchez-Ramos, J., Song, S., Cardozo-Pelaez, F., Hazzi, C., Stedeford, T., Willing, A., Freeman, T. B., Saporta, S., Janssen, W., Patel, N., Cooper, D. R., Sanberg, P. R. Adult bone marrow stromal cells differentiate into neural cells in vitro. *Exp. Neurol* 2000;164:247–256.

[15] Kim, B. J., Seo, J. H., Bubien, J. K., Oh, Y. S. Differentiation of adult bone marrow stem cells into neuroprogenitor cells in vitro. *Neuroreport* 2002; 13(8):1003–1004.

[16] Woodbury D, Schwarz EJ, Prockop DJ, Black IB. Adult rat and human marrow stromal cells differentiate into neurons. *J Neurosci Res* 2000;61:364_370.

[17] Dezawa M, Takahashi I, Esaki M, Takano M, Sawada H. Sciatic nerve regeneration in rats induced by transplantation of in vitro differentiated bone-marrow stromal cells. *Eur J Neurosci* 2001;14:1771-1776.

[18] Tohill M, Mantovani C, Wiberg M, Terenghi G. Rat bone marrow mesenchymal stem cells express glial markers and stimulate nerve regeneration. *Neurosci Lett* 2004;362:200–203.

[19] Sasaki,M., Honmou, O., Akiyama, Y., Uede, T., Hashi, K., Kocsis, J. D. Transplantation of an acutely isolated bone marrow fraction repairs demyelinated adult rat spinal cord axons. *Glia* 2001;35:26–34.

[20] Hou SY, Zhang HY, Quan DP, Liu XL, Zhu JK. Tissue-engineered peripheral nerve grafting by differentiated bone marrow stromal cells. *Neuroscience* 2006;140,110-110

[21] Deng W, Obrocka M, Fischer I, Prockop DJ. In vitro differentiation of human marrow stromal cells into early progenitors ofneural cells by conditions that increase intracellular cyclic AMP. *Biochem Biophys Res Commun* 2001;282:148-152.

[22] Sanchez-Ramos J, Song S, Cardozo-Pelaez F, Hazzi C, Stedeford T, illing A, Freeman TB, Saporta S, Janssen W, Patel N, Cooper DR, Sanberg PR. Adult bone marrow stromal cells differentiate into neural cells in vitro. *Exp Neurol* 2000;164:247-256.

[23] Caddick J, Kingham PJ, Gardiner NJ, et al. Phenotypic and functional characteristics of mesenchymal stem cells differentiated along a Schwann cell lineage. *Glia* 54:840-49.

[24] Shah NM, Groves AK, Anderson DJ. Alternative neural crest cell fates are instructively promoted by TGFbeta superfamily members. *Cell* 85,331-43.

[25] Jessen KR, Mirsky R. Developmental regulation in the Schwann cell lineage. *Adv Exp Med Biol* 1999:468,3-12.

[26] Morrison SJ, White PM, Zock C, et al. Prospective identification, isolation by flow cytometry, and in vivo self-renewal of multipotent mammalian neural crest stem cells. *Cell* 1999:96,737-49.

[27] Soprano DR, Teets BW, Soprano KJ. Role of retinoic acid in the differentiation of embryonal carcinoma and embryonic stem cells. *Vitam Horm*. 2007;75:69-95

[28] Ikeda R, Kurikawa MS, Chiba S, et al. Transplantation of neural cells derived from retinoic acid-treated cynomolgus monkey embryonic stem cells successfully improved motor function of hemiplegic mice with experimental brain injury. *Neurobiol Dis* 2005 Oct;20(1):38-48.

[29] Yan H, Bunge MB, Wood PM. Mitogenic response of adult rat olfactory ensheathing glia to four growth factors. *Glia* 2001 Mar 15;33(4):334-42

[30] Benavente CA, Sierralta WD, Conget PA, et al. Subcellular distribution and mitogenic effect of basic fibroblast growth factor in mesenchymal uncommitted stem cells. *Growth Factors*. 2003 Jun;21(2):87-94

[31] Sakurai H, Bush KT, Nigam SK. Heregulin induces glial cell line-derived neurotrophic growth factor-independent, non-branching growth and differentiation of ureteric bud epithelia. *J Biol Chem*. 2005 Dec 23;280(51):42181-7. Epub 2005 Sep 23.

[32] Kim HA, Ratner N, Roberts TM, et al. Schwann cell proliferative responses to cAMP and Nf1 are mediated by cyclin D1. *J. Neurosci* 2001; 21,1110-1116.

[33] Lisak RP, Bealmear B, Benjamins JA, et al. Interferon-gamma, tumor necrosis factor-alpha, and transforming growth factor-beta inhibit cyclic AMP-induced Schwann cell differentiation. *Glia* 2001; 36(3):354-63

[34] Chen X, Wang X, Chen G, et al. Study of in vivo differentiation of rat bone marrow stromal cells into Schwann cell-like cells. *Microsurgery* 2006;26:111-115.

[35] Cuevas P, Carceller F, Dujovny M, et al. Peripheral nerve regeneration by bone marrow stromal cells. *Neurol Res* 2002;24:634-638

[36] Zhang P, He XJ, Liu KY, et al. Bone marrow stromal cells differentiated into functional Schwann cells in injured rats sciatic nerve. *Artificial cells, Blood Substitutes, and Biotechnology* 2004;32: 509-518.

[37] Boutonnat J., Barbier M., Muirhead K., et al. Response of chemosensitive and chemoresistant leukemic cell lines to drug therapy: simultaneous assessment of proliferation, apoptosis, and necrosis. *Cytometry* 2000;15;42(1):50–60.

[38] Askenasy N., Daniel L., Farkasa C. Antigen barriers or available space do not restrict in situ adhesion of hemopoietic cells to bone marrow stroma. *Stem Cells* 2002;20:80–85.
[39] Tohilla M, Mantovanib C, Wiberg M, et al. Rat bone marrow mesenchymal stem cells express glial markers and stimulate nerve regeneration. *Neuroscience Letters* 2004; 362:200–203
[40] Ansselin AD, Fink T, Davey DF. Peripheral nerve regeneration through nerve guides seeded with adult Schwann cells. *Neuropathol Applied Neurol* 1997; 23: 387–398
[41] Jan YN, Jan LY. Asymmetric cell division, *Nature* 1998;392:775–778.
[42] Hinks GL, Chari DM, O'Leary MT, et al. Depletion of endogenous oligodendrocyte progenitors rather than increased availability of survival factors is a likely explanation for enhanced survival of transplanted oligodendrocyte progenitors in X-irradiated compared to normal CNS, *Neuropathol. Appl. Neurobiol.* 2003;27:59–67.
[43] Chen X, Wang X, Chen G, et al. Study of in vivo differentiation of rat bone marrow stromal cells into Schwann cell-like cells. *Microsurgery* 2006;26:111-115.
[44] Guenard V, Kleitman N, Morrissey TK, et al. Syngeneic Schwann cells derived from adult nerves seeded in semipermeable guidance channels enhance peripheral nerve regeneration. *J. Neurosci* 1992;12:3310–3320.
[45] Hall S, The effect of inhibiting Schwann cell mitosis on the reinnervation of acellular autografts in the peripheral nervous system of the mouse. *Neuropathol. Appl. Neurobiol.* 1986;12:401–414.
[46] Mosahebi A, Woodward B, Wiberg M, et al. Retroviral labeling of Schwann cells: In vitro characterization and in vivo transplantation to improve peripheral nerve regeneration, *Glia* 2001;34:8–17.

INDEX

A

B

D

E

F

G

H

I

J

K

L

P

Q

R

S

T

U

V

W

X

Y

Z